High Efficient Buildings in Mediterranean Area: Challenges and Perspectives

High Efficient Buildings in Mediterranean Area: Challenges and Perspectives

Editors

Natale Arcuri
Roberto Bruno
Piero Bevilacqua

MDPI • Basel • Beijing • Wuhan • Barcelona • Belgrade • Manchester • Tokyo • Cluj • Tianjin

Editors
Natale Arcuri
University of Calabria
Italy

Roberto Bruno
Università della Calabria
Italy

Piero Bevilacqua
Università della Calabria
Italy

Editorial Office
MDPI
St. Alban-Anlage 66
4052 Basel, Switzerland

This is a reprint of articles from the Special Issue published online in the open access journal *Energies* (ISSN 1996-1073) (available at: https://www.mdpi.com/journal/energies/special_issues/ High_Efficient_Buildings_in_Mediterranean_Area_Challenges_and_Perspectives).

For citation purposes, cite each article independently as indicated on the article page online and as indicated below:

LastName, A.A.; LastName, B.B.; LastName, C.C. Article Title. *Journal Name* **Year**, *Volume Number*, Page Range.

ISBN 978-3-03943-955-3 (Hbk)
ISBN 978-3-03943-956-0 (PDF)

Cover image courtesy of Piero Bevilacqua, Roberto Bruno and Natale Arcur.

Contents

About the Editors

Natale Arcuri is an Associate Professor at the Department of Mechanical, Energy and Management Engineering of the University of Calabria, where he was also a nominated delegate for international relationships with Latin American countries. He has been a member of the Energy Commission of the same university since 2014. He was responsible for different national research projects in partnership with industrial companies of national relevance, ranging from cognitive dynamic systems for buildings to innovative solutions for the building envelope and smart cities for the energy governance of the territory. He holds a degree in Industrial Engineering (1989) from the University of Calabria and is the author of more than 80 papers in international journals and conferences. His interests include energy system optimization for building applications, energy flow management in smart grids, cognitive buildings, solar energy modeling, building energy demand reduction, thermal characterization of materials, and energy regulations.

Roberto Bruno is a Researcher in Environmental Technical Physics at the University of Calabria, Department of Mechanical, Energy and Management Engineering. He has been a speaker at national and international congresses concerning the sector of Environmental Technical Physics and the author/co-author of over 100 scientific papers, of which 34 are published in Scopus indexed journals. He has been an effective member of the national association AICARR ("Italian Air Conditioning, Heating and Cooling") since 2005, a full member of the Italian Thermotechnical Association (ATI) since 07/15/2006, and a member of IBPSA ("International Buildings Performance Simulation Association"). His interests include thermal analysis of building components; passive systems; net- and near-zero energy buildings; double skin systems; innovative technologies for building air conditioning; cogenerative and trigenerative photovoltaic systems; innovative photovoltaic generators; models for the intelligent management of building–plant systems with cognitive algorithms; green roofs; solar-assisted heat pumps; energy recovery from the LNG regasification process; and bio-fuel production from microalgae.

Piero Bevilacqua is a Postdoc Researcher at the University of Calabria, Department of Mechanical, Energy and Management Engineering, since 2014. He has co-authored several journal papers and two book chapters and has participated in numerous national research projects. He received his PhD in Mechanical Engineering in 2014. He has conducted research at the University of Belgrade (2014) and at the GREA Center, University of Lleida (2013). He holds a master's degree in Energy Engineering (2011). His interests include passive systems for the building envelope, green roofs, photovoltaic systems, thermal comfort of indoor spaces, NZEB in the Mediterranean area, innovative air-conditioning plants, integrated thermal storage systems, solar cooling, thermal properties of building materials, and experimental analysis.

Preface to "High Efficient Buildings in Mediterranean Area: Challenges and Perspectives"

At present, worldwide energy legislation is pushing toward high targets and imposing strict regulations in order to mitigate environmental pressure and climate change issues. The building sector represents one of the major players in this field, often requiring a considerable amount of energy for services related to annual air-conditioning and indoor thermal comfort. The design of highly efficient low-energy buildings is often a challenging task, especially in the Mediterranean area, where the balanced energy requirement for heating and cooling does not usually allow for a high level of envelope insulation, which would lead to summer overheating. Several systems and technologies have been recently proposed in order to optimize building envelopes and air-conditioning plants and to opportunely control and manage energy requirements in order to attain valuable design procedures according to the building typology and the climatic context. This book will be of interest to architects and building designers who aim to reach high standards of efficiency through the use of innovative solutions, as well as to researchers and students who are interested in strengthening aspects related to building energy simulations.

Natale Arcuri, Roberto Bruno, Piero Bevilacqua
Editors

Article

On the Energy Performance of an Innovative Green Roof in the Mediterranean Climate

Luca Evangelisti, Claudia Guattari, Gianluca Grazieschi, Marta Roncone * and Francesco Asdrubali

Department of Engineering, Roma TRE University, Via Vito Volterra 62, 00146 Rome, Italy; luca.evangelisti@uniroma3.it (L.E.); claudia.guattari@uniroma3.it (C.G.); gianluca.grazieschi@uniroma3.it (G.G.); francesco.asdrubali@uniroma3.it (F.A.)
* Correspondence: marta.roncone@uniroma3.it

Received: 19 July 2020; Accepted: 25 September 2020; Published: 3 October 2020

Abstract: Green roofs have a thermal insulating effect known since ancient times. In the building sector, green roofs represent a sustainable passive solution to obtain energy savings, both during winter and summer. Moreover, they are a natural barrier against noise pollution, reducing sound reflections, and they contribute to clean air and biodiversity in urban areas. In this research, a roof-lawn system was studied through a long experimental campaign. Heat-flow meters, air and surface temperature sensors were used in two buildings characterized by different surrounding conditions, geometries and orientations. In both case studies, the thermal behaviors of the roof-lawn system were compared with the conventional roofs. In addition, a dynamic simulation model was created in order to quantify the effect of this green system on the heating and cooling energy demands. The roof-lawn showed a high thermal inertia, with no overheating during summer, and a high insulating capacity, involving energy savings during winter, and consequently better indoor thermal conditions.

Keywords: green roof; measurements; thermal behavior; monitoring; dynamic model

1. Introduction

Greenhouse gas emissions (GHG), mainly produced by anthropogenic activities, can be considered to be largely responsible for the global average temperature increase, with a growth of about +1 °C compared to the pre-industrial era [1]. Therefore, many countries could become inhospitable because of climate change [2].

The effects can be clearly observed in densely populated cities, where the "urban heat island" (UHI) phenomenon has been rising during the last decades [3].

It is well-known that UHIs are correlated with urban overheating phenomena, which are commonly contrasted by applying cool or retroreflective materials or expanding urban green areas, or, alternatively, by resorting to green roofs [4–6]. It is essential to study UHIs and suggest interventions for the mitigation of these phenomena, thereby reducing their strength [7,8]. Achieving this goal is essential for reducing the growing energy consumption of buildings energy, in particular during summer [9–12]. Moreover, the high temperatures in cities during the hottest months can involve significant and negative effects on daily life [13]. It is therefore clear that UHI phenomenon and its countermeasures are interesting issues for the scientific community [14–17]. One of the most significant solutions for counteract UHIs is represented by green roofs. They are a passive solution characterized by multiple advantages. Green roofs have a thermal insulation function that has been known since ancient times. It should be noted that these systems involve higher initial costs, which, however, can be amortized quickly [18]. Green roofs can be also considered a natural shield against noise pollution [19,20]. Furthermore, a green roof generates oxygen, reducing, at the same time, CO_2 and pollutants, thus representing a natural countermeasure against atmospheric pollution. Therefore, green roofs represent a viable and effective solution for reducing the effects of UHI.

In further detail, some scientific works showed the capability of green roofs to mitigate the UHI phenomenon [4–6,21,22], also improving the thermal comfort of a building over a whole year [7]. In fact, this solution allows the achievement of particularly high-performance dynamic thermal characteristics [8,9], with higher thermal insulation during winter and improved inertial behavior during summer. The evapotranspiration effects allow for the external surface temperature reduction during the warmer months, with consequent advantages in terms of energy saving [10–12]. Moreover, green roofs can upgrade the management of rainwater, decreasing the volume of run-off and causing an attenuation of the peak to consequently mitigate the generation of rainwater run-off [14–17]. Further works underlined their sound insulation capacity compared to a traditional roof [13,19,20] and also their importance for safeguarding the aesthetics of areas and wildlife [23] with the aim of avoiding habitat loss in urban areas [24,25].

However, today, the high costs of construction, maintenance and roof dispersion problems are the main challenges associated with the application of green roofs [18]. Although the cost of a green roof is higher than the cost of a conventional roof, the use of this green solution is becoming a common practice in many developing countries [26]. On the other hand, European countries have been a model in implementing these mitigation techniques in the building design and construction strategy.

Starting from the above, this study focused on the experimental investigation of the thermal behavior of an innovative roof-lawn system installed on the roofs of two buildings placed in central Italy. Long-lasting measurement campaigns were carried, out and simulations were performed in order to better understand the thermal behavior and the effects of this passive green solution on the energy performance of buildings.

2. Materials and Methods

2.1. The Roof-Lawn System: Two Cases of Study

The experimental investigation consisted of analyzing two buildings where the roof-lawn system was installed: Case 1 is represented by a single-story building, located about 70 km from Rome and characterized by a tilted roof (Figure 1a); Case 2 is represented by a small building placed in a central neighborhood of Rome (Figure 1b) and characterized by a horizontal roof. The roof-lawn system is distinguished by the species of the Zoysia genus, which has a slow growth. Zoysia is able to tolerate wide variations in temperature, sunlight and water, and it is generally used for lawns in temperate climates. The irrigation system only reintegrates the evapotranspiration losses. The required maintenance is very low, because the lawn can reach a maximum growth of about 0.25 m, with a distinctive wave effect.

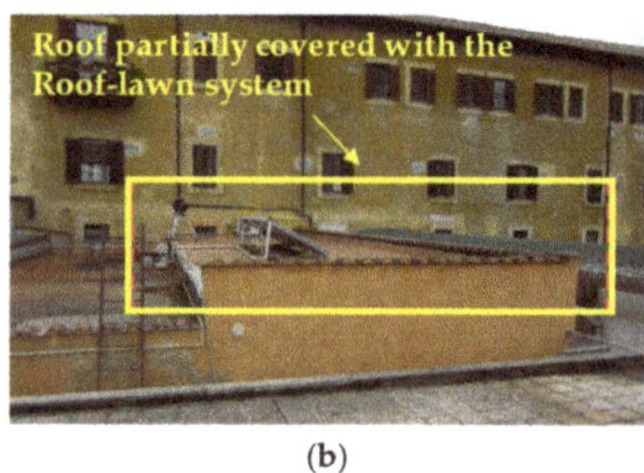

(a)   (b)

Figure 1. The case studies: (**a**) Case 1; (**b**) Case 2.

The roof-lawn system is composed of the following layers: on the higher part of the roof, a waterproofing sheath avoids water infiltrations; over that, the green roof is built on a draining mat and an inorganic substrate. In Case 1, a reinforced concrete layer (about 0.08 m thick) characterizes the structural part of the roof. On the other hand, in Case 2, the structural part of the roof is characterized by hollow tiles, with a thickness of about 0.10 m.

In order to make a comparison, the roof-lawn system was installed covering only a part of the roofs. In Case 1, the traditional roof is characterized by a reinforced concrete slab, covered with tiles. In Case 2, the original roof is finished on the outer side with pavement.

The indoor environments under the roofs have the same orientation and occupation rate.

2.2. Experimental Setup

In order to assess the stationary and the dynamic thermal performance of the green roof, heat-flow meters, air and surface temperature sensors were employed, as shown in Figure 2. The schematic representation of the installed sensors is descriptive for both Case 1 (sloped roof) and Case 2 (horizontal roof).

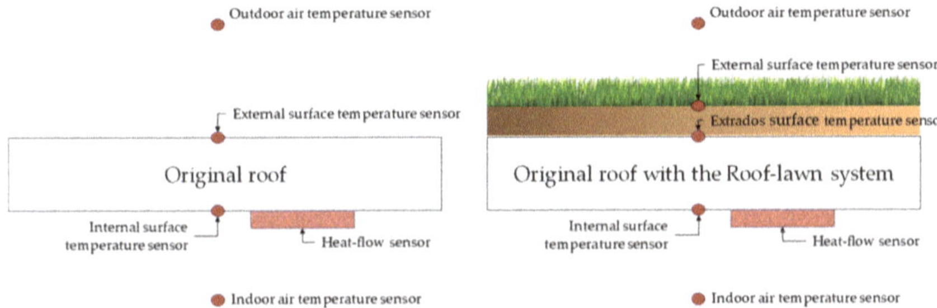

Figure 2. Schematic view of the installed sensors for both Case 1 and Case 2.

It is worthy to mention that aiming at ensuring the best thermal contact, the external surface temperature sensor was fixed into the higher layer of the green roof, under a thin layer of soil (represented by the brown layer in Figure 2).

Data-loggers were used for connecting all sensors and logging heat fluxes, air and surface temperatures with a time step of 10 min, for 24 h per day. The measurement survey began in October 2018 for a period of one year, finishing in September 2019 for Case 1. On the other hand, for Case 2, the experimental survey started in May 2019 and finished in May 2020.

2.3. Methodology

With the aim of evaluating the efficiency of the green roofs, two parts of the roof were analyzed and compared. The stratigraphy of the roofs was identified, but the thermophysical properties of each layer were unknown. The roof-lawn system is characterized by five layers, of which the upper part is non-homogeneous, consisting of grass and soil.

Starting from this, all the acquired data from Case 1 and Case 2 were employed for the calculation of the thermal transmittances (also known as U-value or just U) of the green roofs and the original ones. It is known that heat flux densities and indoor/outdoor air temperatures are employed to compute the U-value according to the following equation:

$$q = U(T_i - T_e) \tag{1}$$

where q is the heat flux density, and T_i and T_e are the indoor and outdoor air temperatures, respectively. In compliance with ISO 9869-1 [27], the stationary U-value of building components (that is, the U-value of the roofs) can be obtained by means of the average progressive method with the following equation:

$$U = \frac{\sum_{j=1}^{N} q_j}{\sum_{j=1}^{N} \left(T_{ij} - T_{ej} \right)} \tag{2}$$

where N is the whole recorded samples.

Furthermore, the thermal conductance (C) of the roofs can be calculated employing surface temperatures instead of air temperatures according to the following formula:

$$C = \frac{\sum_{j=1}^{N} q_j}{\sum_{j=1}^{N} \left(T_{sij} - T_{sej} \right)} \tag{3}$$

where T_{si} and T_{se} are the internal and external surface temperatures, respectively.

Evidence about the dynamic thermal behavior of the roofs (phase shift (PS) and decrement factor (DF)) was evaluated considering internal and external surface temperatures.

The PS is the time difference between the highest internal surface temperature and the highest external surface temperature of the roof [28]:

$$PS = t_{T_{si}^{MAX}} - t_{T_{se}^{MAX}} \tag{4}$$

The DF can be calculated as [28]:

$$DF = \frac{T_{si}^{MAX} - T_{si}^{MIN}}{T_{se}^{MAX} - T_{se}^{MIN}} \tag{5}$$

where T_{si}^{MAX} and T_{si}^{MIN} are the highest and the lowest internal surface temperature registered during a day, respectively, and, in turn, T_{se}^{MAX} and T_{se}^{MIN} are the highest and the lowest external surface temperature recorded during a day, respectively.

Data obtained from the green and the original roofs were used for simulating the heating and cooling energy needs of a building through the energy simulation tool DesignBuilder (Version 6.1.5.002, DesignBuilder Software Ltd., Stroud, UK) [29]. Since the solar reflectance of green roofs ranges from 0.3 to 0.5 in function of the type of plant [30], a reflectivity of 0.2 (typical value for grass) was used in the model. The building model is characterized by walls with a simple stratigraphy (0.22 m of concrete and 0.04 m of extruded polystyrene, plastered on both sides), distinguished by a thermal transmittance of 0.600 W/(m^{2}K). The windows (total area of 18 m^2) have a thermal transmittance of 5.61 W/(m^{2}K). A solar absorptance of 0.6 was set for walls.

An infiltration rate of 0.3 1/h was used. The set-point temperatures for heating and cooling were set to 20 °C and 26 °C, respectively. A typical schedule of an office building was selected to define the heating and cooling functioning hours. The heating generation is guaranteed by a gas boiler (mean efficiency of 0.9), while a split system (mean energy efficiency ratio: 2.21) performs the cooling requirements.

Case 1 was selected as reference to run the simulation; a similar geometry was created in the 3D modeler of DesignBuilder, and the equivalent thermophysical properties found experimentally in a previous study [10] were employed to define the roof characteristics (see Table 1). According to the data reported, the green roof has a U-value of 1.361 W/(m^{2}K), a thermal resistance of 0.735 W/(m^{2}K), a heat capacity of 840 J/(kgK) and a density of 1100 kg/m^3. These data do not include the vegetation layer. The external layer of the roof was simulated using the green roof module of DesignBuilder. Table 2 lists the thermophysical properties considered for the simulation. In particular, the typical value of the leaf area index (LAI) of a grass surface of 2.51 [31,32] was set. The LAI depends on the plants' density and on their height, [33] however, adopting an average value, it is possible to consider it acceptable for the Zoysia grass coverage of the roof. The high maximum moisture content of the terrain adopted is typical of a very porous terrain like that adopted in the case study. Moreover, the choice of a relatively high initial moisture content (0.50) is coherent with the regular irrigation schedule of the green roof, even during the days before the measurement campaign. Within the simulation, we considered a continuous irrigation during the day that is always able to balance the evapotranspiration and the grass water absorption losses.

Table 1. Case 1: Thermophysical properties of the roof employed in the simulation.

Case 1	Thermal Transmittance (W/m^{2}K)	Heat Capacity (J/kgK)	Density (kg/m^3)
Original roof	3.021	840	1000
Green roof	1.361	840	1100

Table 2. Thermophysical properties of the vegetation layer. LAI: leaf infiltration rate.

Property	Value
Height of the grass	20 cm
LAI	2.51 [31–33]
Leaf reflectivity	0.22 [34,35]
Leaf emissivity	0.95 [34–36]
Minimum stomatal resistance	180 s/m [34–36]
Maximum volumetric moisture content of the soil layer	0.7 [34,35]
Minimum volumetric moisture content of the soil layer	0.010 [34,35]
Initial volumetric moisture content of the soil layer	0.5 [34,35]

3. Results and Discussion

3.1. Case 1—Experimental Investigation

The experimental results are reported in this section. One-year monitoring is characterized by a very large amount of data. It is worthy to observe that, over a whole year, recording heat fluxes and temperatures every 10 min does not allow a clear graphical representation of the sinusoidal trends for readers. Thus, in order to provide a clear view of the yearly behaviors of the investigated green roofs, the results are here presented in terms of average data and standard deviations (SDs). It is well-known that SD is a measure of the dispersion of a set of values. A low SD indicates that the values tend to be close to the mean of the set. Conversely, a high SD indicates that the values are spread over a wider range. Thus, using average data and SDs, it is possible to make a simplified direct comparison between the thermal behavior of the green roof and the original one, thus observing the fluctuations of the single quantities on a monthly basis. Starting from September 2018 to August 2019, it is possible to analyze the performance of the green roof considering diverse weather conditions.

Figures 3–5 show the comparison between the green roof and the original one in Case 1.

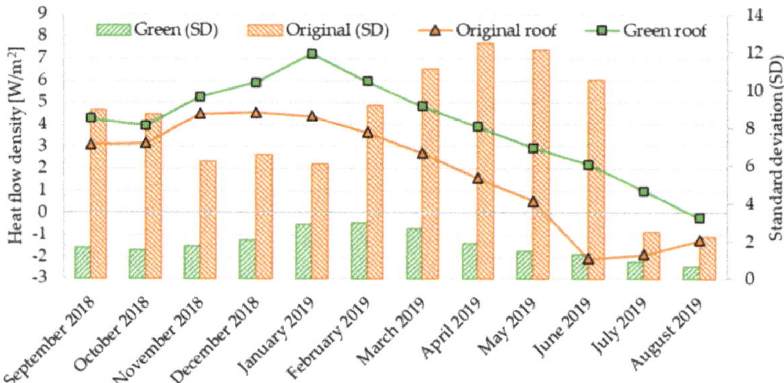

Figure 3. Comparison between the green roof and the original one in Case 1: average heat flow densities and their standard deviations (SDs). The convention used is positive flux in the outgoing direction and negative flux in the inward direction.

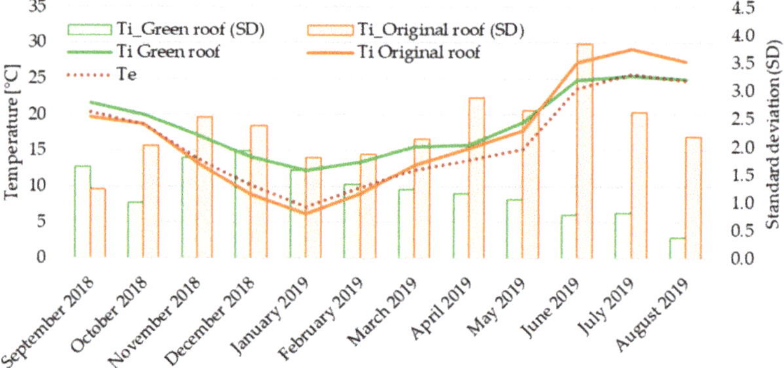

Figure 4. Comparison between the green roof and the original one in Case 1: average outdoor air temperatures and average indoor air temperatures with their SDs.

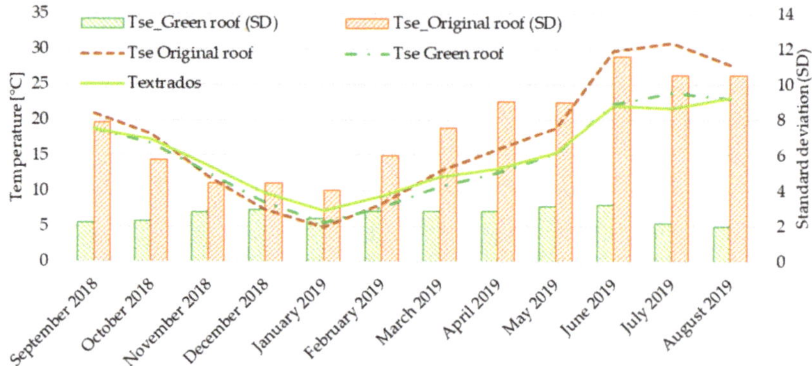

Figure 5. Comparison between the green roof and the original one in Case 1: extrados' temperature and average external surface temperatures with their SDs.

Figure 3 shows the average values of the heat flow densities and their SDs. The average monthly heat flow densities are represented in the graph by continuous lines on the main axis: the green curve is relative to the green roof (indicated in the legend with "Green roof"), while the orange one refers to the original roof (indicated with "Original roof").

In the secondary axis, the monthly SD of the heat flow are represented: the green columns are relative to the green roof (indicated by "Green (SD)"), while the orange ones are relative to the original roof ("Original (SD)").

The analysis of the heat flow densities reveals low average values for the original roof, but it is worthy to note that it is characterized by SD values always higher than those related to the green roof. Therefore, when compared with the original one, a stable thermal performance of the green roof can be observed. Taking into consideration December 2018 (representative of the winter season), the external air temperatures ranged between 3.83 °C and 14.16 °C. Throughout this month, the highest heat flux value was about 14 W/m^2, much lower than that recorded for the original roof (with a heat flux higher than 18 W/m^2). The low thermal inertia of the original roof revealed a strong heat fluxes oscillation, also with negative values. The greater inertial effects of the roof-lawn system can be observed even more clearly during the middle season (from March 2019 to April 2019). Taking into consideration April 2019, the external air temperatures ranged between 5.53 °C and 26.23 °C. For the green roof, a maximum heat flux of about 10.7 W/m^2 was recorded. On the other hand, a maximum heat flow value of about 19.1 W/m^2 was registered for the original roof.

During the summer season, considering June 2019 as a representative month, the external air temperatures varied between 16.0 °C and 33.4 °C. In this month, for the green roof, the highest heat flux was about 6 W/m^2. Conversely, for the original roof, the highest heat flow was about 15 W/m^2. In order to provide an overview, Table 3 summarizes the average, the minimum and the maximum heat fluxes registered during each month.

Table 3. Average, minimum and maximum heat fluxes registered during each month. The convention used is positive flux in the outgoing direction and negative flux in the inward direction.

Case 1	Green Roof—Heat Flux (W/m^2)			Original Roof—Heat Flux (W/m^2)		
	MIN	AVG	MAX	MIN	AVG	MAX
September 2018	1.38	1.69	9.05	−21.02	3.11	8.95
October 2018	1.26	3.97	10.21	−18.06	3.16	18.54
November 2018	1.94	5.27	12.01	−11.88	4.52	19.01
December 2018	2.01	5.91	14.30	−10.42	4.56	19.61
January 2019	2.57	7.23	14.25	−10.15	4.39	14.80
February 2019	1.37	5.97	14.59	−23.61	3.64	19.52
March 2019	0.20	4.84	13.79	−25.38	2.71	20.13
April 2019	1.48	3.92	10.66	−32.56	1.59	19.07
May 2019	1.01	2.95	8.47	−29.17	0.54	17.81
June 2019	−0.61	2.20	5.59	−33.79	−2.06	15.04
July 2019	−0.19	0.99	5.36	−7.63	−1.87	2.38
August 2019	−3.69	−0.22	1.16	−6.47	−1.24	2.57

The average outdoor air temperatures, the average indoor air temperatures and their SD are reported in Figure 4.

The monthly average indoor air temperatures are represented in the graph by continuous lines on the main axis: the green curve is relative to the green roof (indicated in the legend with "T_i Green roof") while the orange one refers to the original roof (indicated with "T_i Original roof"). On the main axis, the red dashed curve of the outside air temperature (indicated with "T_e") is also represented.

The secondary axis shows the monthly SDs of the monthly indoor air temperature: the green columns refer to the green roof (indicated by "T_i_Green (SD)"), while the orange ones refer to the original roof ("T_i_Original (SD)").

The constant thermal behavior of the green roof can be deduced also by observing the average internal air temperatures (Figure 4). From September 2018 to May 2019, they are always characterized by higher values when compared with the average indoor air temperatures of the original roof. On the contrary, during June, July and August 2019 the indoor air temperatures are lower than those registered for the original roof, with subsequent advantages in terms of indoor thermal comfort. Furthermore, in this case, SDs calculated for the original roof are greater than those computed for the roof equipped with the roof-lawn system. The behavior of the green roof is related both to the additional thermal resistance caused by the additional layer represented by the roof-lawn system and to the increased thermal inertia. Indeed, considering December 2018, the green roof caused an average indoor air temperature of 14.14 °C. Conversely, for the original roof, indoor air temperature showed an average value of 8.95 °C. During the middle season (April and May 2019) the average indoor air temperatures are similar, but it is possible to observe better performance with the roof-lawn system; during April 2019, a minimum internal air temperature of 13.91 °C caused by the green roof, and a minimum value of 9.52 °C caused by the original roof were obtained. Moreover, for the indoor air temperatures, aiming at providing an outline, Table 4 recaps the average, the minimum and the maximum values registered during each month.

Table 4. Average, minimum and maximum indoor air temperatures registered during each month.

Case 1	Green Roof—Air Temperature (°C)			Original Roof—Air Temperature (°C)		
	MIN	AVG	MAX	MIN	AVG	MAX
September 2018	19.6	21.7	24.2	17.7	19.7	22.5
October 2018	17.3	20.0	21.5	13.0	18.8	23.0
November 2018	13.7	17.2	20.6	7.2	13.4	18.3
December 2018	10.2	14.1	17.2	4.1	9.0	13.7
January 2019	10.0	12.4	15.6	3.9	6.2	9.9
February 2019	9.9	13.5	16.3	4.3	9.0	132.0
March 2019	12.8	15.6	19.2	7.1	13.0	18.1
April 2019	13.9	15.8	19.0	9.5	15.4	24.6
May 2019	16.2	19.0	22.2	10.9	17.9	24.4
June 2019	23.1	24.8	26.7	15.9	27.3	35.3
July 2019	23.7	25.5	27.5	23.4	29.2	34.8
August 2019	24.2	25.0	25.7	24.1	27.4	31.6

Finally, the average external surface temperatures and their SDs are shown in Figure 5, along with the extrados' average temperatures measured under the roof-lawn system. The monthly average temperatures of the external surface are represented in the graph by dashed curves on the main axis: the green curve is relative to the green roof (indicated in the legend with "T_{se} Green roof"), while the brown one refers to the original roof (indicated with "T_{se} Original roof"). On the main axis, the green continuous curve of the extrados' temperature is also shown (indicated by "$T_{extrados}$").

The secondary axis shows the monthly SDs of the external surface temperatures: the green columns are related to the green roof (denoted by "T_{se}_Green (SD)"), while the orange ones are related to the original roof ("T_{se}_Original (SD)").

The outer side of the investigated roofs is completely different in terms of materials and heat transfer mechanisms. The effect of the thermophysical features of the employed materials played an essential role. The tiles on the original roof absorbed solar radiation because of their high absorptance coefficient. For the green roof, this did not happen because of the evapotranspiration phenomena of the greenery. Moving from the colder months to the warmer ones, it is possible to observe much higher average external surface temperatures in the case of the original roof. Analyzing the summarizing data listed in Table 5, the original roof reached about 55 °C, while the external temperatures of the roof-lawn system reached values which do not exceed about 30 °C.

Table 5. Average, minimum and maximum external surface temperatures registered during each month.

Case 1	Green Roof—Surface Temp. (°C)			Original Roof—Surface Temp. (°C)		
	MIN	AVG	MAX	MIN	AVG	MAX
September 2018	12.0	18.8	24.0	8.2	20.9	42.2
October 2018	5.3	16.7	21.4	7.1	18.1	35.5
November 2018	5.7	12.5	18.4	1.5	12.0	23.4
December 2018	2.3	8.4	13.9	-2.4	7.4	17.8
January 2019	2.3	5.6	11.5	-2.5	4.8	15.8
February 2019	-0.5	7.7	14.1	-2.9	8.1	24.4
March 2019	0.7	10.7	18.6	-0.9	12.8	35.1
April 2019	6.4	12.6	21.4	2.5	15.9	42.7
May 2019	7.4	15.3	23.6	3.8	18.9	43.7
June 2019	13.3	22.3	28.8	8.9	29.7	52.9
July 2019	18.4	23.9	30.3	16.0	30.9	55.0
August 2019	20.4	23.1	27.1	16.5	27.8	49.5

The recorded data during winter allowed the calculation of the U-value of the roofs. For the green roof, a U-value of 1.361 W/(m^{2}K) was found and, for the original one, a value of 3.021 W/(m^{2}K) was obtained. Thus, making a comparison between the roofs, a percentage difference in terms of thermal transmittance of about −55% can be pointed out. Moreover, applying Equation (3), the thermal conductance of the roofs was computed, finding 1.168 W/m^{2}K for the green roof and 2.811 W/m^{2}K for the original one, with a resulting percentage difference of −58.45%. The measurement of the green roof extrados' temperature and the external surface temperature allowed a preliminary calculation of the roof-lawn thermal conductance, which was equal to 4.487 W/m^{2}K. All these results are summarized in Table 6.

Table 6. Case 1: C-values and U-values of the original and green roofs.

Case 1	Thermal Conductance (W/m^{2}K)	Thermal Transmittance (W/m^{2}K)
Original roof	2.811	3.021
Green roof	1.168	1.361
Roof-lawn	4.487	-

3.2. Case 2—Experimental Investigation

The experimental results of Case 2 are reported in this section. Again, the results of a one-year monitoring are presented here in terms of average data and SDs. Therefore, it is possible to make a direct comparison between the green roof and the original roof over time, also observing the fluctuations of the individual quantities on a monthly basis. From May 2019 to May 2020, it is possible to comprehend the thermal behavior of the green roof under different weather conditions.

It is necessary to specify that this monitoring has some missing data caused by the malfunction of the installed sensors.

Figures 6–8 show the comparison between the green roof and the original one.

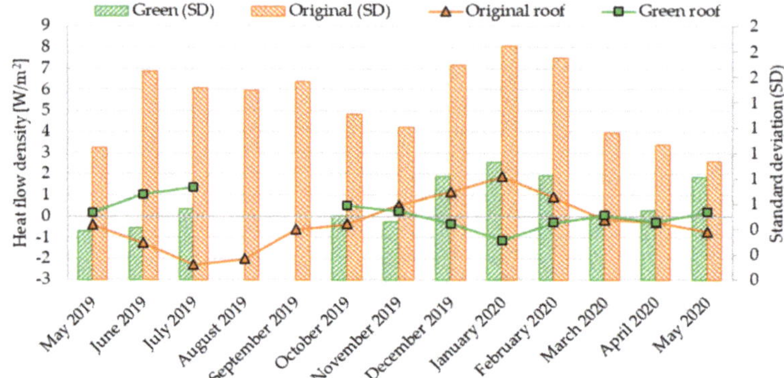

Figure 6. Comparison between the green roof and the original one in Case 2: average heat flow densities and their SDs. The convention used is positive flux in the outgoing direction and negative flux in the inward direction. Instrument failure caused missing data.

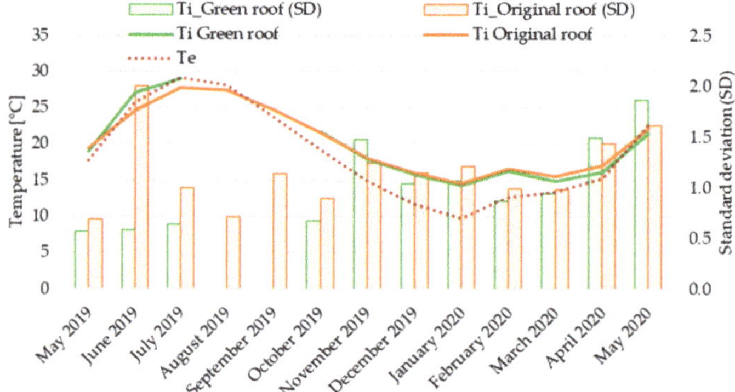

Figure 7. Comparison between the green roof and the original one in Case 2: average outdoor air temperatures and average indoor air temperatures with their SDs. Instrument failure caused missing data.

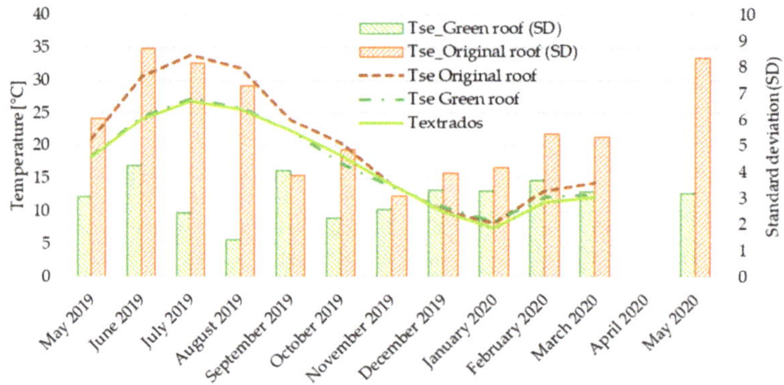

Figure 8. Comparison between the green roof and the original one in Case 2: extrados' temperature and average external surface temperatures with their SDs. Instrument failure caused missing data.

Figure 6 shows the average values of the heat flow densities and their SDs.

The average monthly heat flow densities are represented in the graph by continuous lines on the main axis: the green curve is relative to the green roof (indicated in the legend with "Green roof"), while the orange one refers to the original roof (indicated with "Original roof").

In the secondary axis, the monthly SDs of the heat flow are represented: the green columns are relative to the green roof (indicated by "Green (SD)"), while the orange columns are relative to the original roof ("Original (SD)").

The analysis of the heat flow densities reveals low average values for the original roof, but it is important to note that it is characterized by SD values always higher than those related to the green roof.

The greater inertial effects of the roof-lawn system can be observed even more clearly during the middle season. Taking into consideration May 2019, the external air temperatures ranged between 12.58 °C and 25.77 °C. For the green roof, a maximum heat flux of about 1.33 W/m^2 was recorded. On the other hand, a maximum heat flow value of about 2.88 W/m^2 was registered for the original roof. Taking into consideration May 2020, the external air temperatures ranged between 13.82 °C and 34.53 °C. For the green roof, a maximum heat flux of about 2.66 W/m^2 was recorded. On the other hand, a maximum heat flow value of about 3.06 W/m^2 was registered for the original roof.

During summer, considering June 2019 as a representative month, the external air temperatures ranged from 14.06 °C to 37.99 °C. In this month, for the green roof, the highest heat flux was about 2.38 W/m^2. Conversely, for the original roof, the highest heat flow was about 4.0 W/m^2.

In order to provide an overview, Table 7 summarizes the average, the minimum and the maximum heat fluxes registered during each month.

Table 7. Average, minimum and maximum heat fluxes registered during each month. The convention used is positive flux in the outgoing direction and negative flux in the inward direction. Instrument failure caused missing data.

Case 2	Green Roof—Heat Flux (W/m^2)			Original Roof—Heat Flux (W/m^2)		
	MIN	AVG	MAX	MIN	AVG	MAX
May 2019	−0.84	0.18	1.33	−1.96	−0.40	2.88
June 2019	0.13	1.04	2.38	−4.13	−1.25	4.00
July 2019	0.24	1.39	3.09	−5.18	−2.28	3.36
August 2019	−	−	−	−4.88	−2.00	3.81
September 2019	−	−	−	−3.65	−0.63	6.06
October 2019	−2.17	0.52	1.99	−3.91	−0.36	4.11
November 2019	−1.84	0.26	1.65	−1.58	0.53	4.00
December 2019	−3.04	−0.33	1.39	−1.77	1.17	5.20
January 2020	−3,82	−1.12	1.27	−1.21	1.89	6.10
February 2020	−2.89	−0.29	1.87	−3.65	0.92	5.27
March 2020	−1.46	0.05	1.65	−3.61	−0.16	4.04
April 2020	−2.17	−0.26	0.79	−2.97	−0.26	4.56
May 2020	−2.29	0.21	2.66	−2.15	−0.71	3.06

Comparing the green and the original roofs, a more stable thermal performance of the green roof can be observed. Taking into consideration December 2019 (representative of the winter season), the external air temperatures ranged between 2.42 °C and 19.23 °C. Throughout this month, the highest heat flux was about 1.39 W/m^2, a much lower value than that recorded for the original roof (with a heat flux higher than 5.20 W/m^2). The low thermal inertia of the original roof allowed us to observe a strong heat flows oscillation, also showing negative values.

The average outdoor air temperatures, the average indoor air temperatures and their SDs are reported in Figure 7. The monthly average indoor air temperatures are represented in the graph by continuous lines on the main axis: the green curve is relative to the green roof (indicated in the legend with "T_i Green roof"), while the orange one refers to the original roof (indicated with "T_i Original roof"). On the main axis, the red dashed curve of the outside air temperature (indicated with "T_e") is also represented.

The secondary axis shows the monthly SDs of the monthly indoor air temperature: the green columns refer to the green roof (indicated by "T_i_Green (SD)"), while the orange ones refer to the original roof ("T_i_Original (SD)").

The more stable thermal behavior of the green roof can be deduced also observing the average indoor air temperatures (Figure 7). From May 2019 to May 2020, indoor air temperatures' SDs calculated for the original roof are greater than those computed for the roof equipped with the roof-lawn system, except for November 2019 and for April and May 2020. In this case, the behavior of the green roof is also related both to the additional thermal resistance caused by the additional layer represented by the roof-lawn system and to the increased thermal inertia. Indeed, considering December 2019, the green roof caused an average indoor air temperature of 16.08 °C. On the contrary, for the original roof, an average indoor air temperature of 15.82 °C can be observed. During the middle season (April and May 2020) the average indoor air temperatures are similar: during April 2020, a minimum internal air temperature of 12.81 °C caused by the green roof, and a minimum value of 13.51 °C caused by the original roof were obtained. Additionally, for the indoor air temperatures, aiming at providing an outline, Table 8 recaps the average, the minimum and the maximum values registered during each month.

Table 8. Average, minimum and maximum indoor air temperatures registered during each month. Instrument failure caused missing data.

Case 2	Green Roof—Air Temperature (°C)			Original Roof—Air Temperature (°C)		
	MIN	AVG	MAX	MIN	AVG	MAX
May 2019	18.1	19.1	20.1	18.0	19.5	21.0
June 2019	26.0	27.2	28.2	18.9	24.7	27.7
July 2019	27.6	29.0	30.4	25.7	27.8	30.1
August 2019	-	-	-	26.3	27.5	30.2
September 2019	-	-	-	22.5	24.8	27.8
October 2019	20.0	21.7	23.2	19.8	21.7	24.5
November 2019	15.7	18.0	21.4	16.0	18.1	21.3
December 2019	13.1	15.8	18.4	13.1	16.1	18.6
January 2020	12.0	14.4	17.0	12.1	14.6	17.7
February 2020	13.7	16.3	18.0	13.9	16.6	18.6
March 2020	13.0	15.0	17.7	13.5	15.6	18.8
April 2020	12.8	16.2	18.4	13.5	17.1	19.3
May 2020	18.1	21.6	25.0	19.0	22.2	25.8

Finally, the average external surface temperatures and their SDs are shown in Figure 8 along with the extrados' average temperatures measured under the roof-lawn system. The monthly average temperatures of the external surface are represented in the graph by dashed curves on the main axis: the green curve is relative to the green roof (indicated in the legend with "T_{se} Green roof"), while the brown one refers to the original roof (indicated with "T_{se} Original roof"). On the main axis, the green continuous curve of the extrados' temperature is also shown (indicated by "$T_{extrados}$").

The secondary axis shows the monthly SDs of the external surface temperatures: the green columns are related to the green roof (denoted by "T_{se}_Green (SD)"), while the orange ones are related to the original roof ("T_{se}_Original (SD)").

The outer side of the investigated roofs is completely different in terms of materials and heat transfer mechanisms. The effect of the thermophysical features of the employed materials played an essential role. Indeed, while in the original roof cover the solar radiation was absorbed by the tiles due to their high absorption coefficient, this phenomenon does not occur in the green roof due to the evapotranspiration phenomena of the greenery. Moving from the colder months to the warmer ones, it is possible to observe much higher average external surface temperatures in the case of the original roof. Analyzing the summarizing data listed in Table 9, the original roof reached about 54 °C, while the external temperatures of the roof-lawn system reached values which do not exceed about 40 °C.

Table 9. Average, minimum and maximum external surface temperatures registered during each month. Instrument failure caused missing data.

Case 2	Green Roof—Surface Temp. (°C)			Original Roof—Surface Temp. (°C)		
	MIN	AVG	MAX	MIN	AVG	MAX
May 2019	13.3	18.3	30.1	13.2	20.8	45.7
June 2019	14.4	24.4	37.3	13.8	30.5	52.8
July 2019	22.3	27.2	34.0	21.7	33.9	54.3
August 2019	22.9	25.9	31.3	22.9	32.0	53.9
September 2019	15.1	22.4	40.0	19.6	23.9	45.6
October 2019	12.1	17.4	24.5	13.5	20.4	35.4
November 2019	8.2	13.8	20.5	8.4	14.2	26.2
December 2019	2.4	10.7	19.0	0.2	16.1	20.6
January 2020	2.1	8.6	17.5	0.5	10.1	22.5
February 2020	2.8	12.2	21.4	1.5	13.2	31.2
March 2020	7.2	12.7	28.3	6.2	14.4	33.2
April 2020	-	-	-	-	-	-
May 2020	12.9	19.5	31.0	12.0	25.9	48.8

The recorded data during winter allowed for the calculation of the thermal transmittance of the roofs. Applying Equation (2), a U-value of 0.190 W/(m^{2}K) was obtained for the green roof, and a value of 0.362 W/(m^{2}K) was found for the original one. Therefore, making a comparison between the green and the original roof, a percentage difference in terms of thermal transmittance of about -47.5% can be highlighted. Moreover, applying Equation (3), the thermal conductance of the roofs was computed, finding a value of 0.162 W/m^{2}K for the green roof and a value of 0.277 W/m^{2}K for the original one, with a resulting percentage difference of -41.4%. The measurement of the green roof extrados' temperature and the external surface temperature allowed a preliminary calculation of the roof-lawn thermal conductance, which was equal to 0.423 W/m^{2}K. All these results are summarized in Table 10.

Table 10. Case 2: C-values and U-values of the original and green roofs.

Case 2	Thermal Conductance (W/m^{2}K)	Thermal Transmittance (W/m^{2}K)
Original roof	0.277	0.362
Green roof	0.162	0.190
Roof-lawn	0.423	-

3.3. Building Energy Simulation

A dynamic energy simulation was performed in DesignBuilder in order to simulate the performance of the green roof. The resulting effects deriving from the installation of the green roof are shown in Figures 9–11. In particular, Figure 9 shows the monthly thermal energy requirement of the building, considering as positive the amount of energy that needs to be added in the internal spaces and negative the energy that has to be removed to guarantee comfortable conditions in summer. Comparing the monthly thermal heating and cooling requirements (see Figure 9), it is possible to observe a relevant reduction of peak values in summer and winter design months: in particular, we observed a percentage difference of -31% in January (-339 kWh) and of -48% in July ($+224$ kWh). Instead, on an annual basis, the installation of the green roof is able to decrease the thermal energy requirement by 31% during the heating season (from October to April) and by 50% during the cooling period (form June to September). It should be noted that the building is characterized by a good summer performance, even if the original roof has poor thermal behavior. This can be linked to the use of high reflective solar blinds, to avoid summer overheating in internal spaces, and of natural ventilation systems.

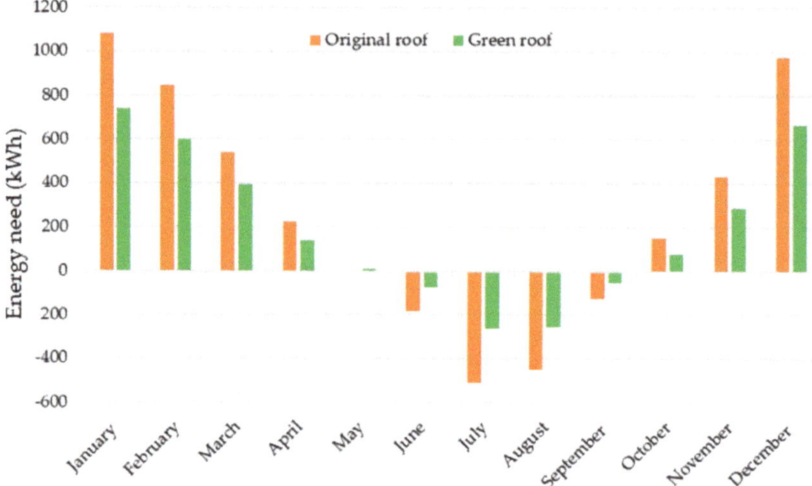

Figure 9. Monthly thermal energy needs of the building with the original and green roof.

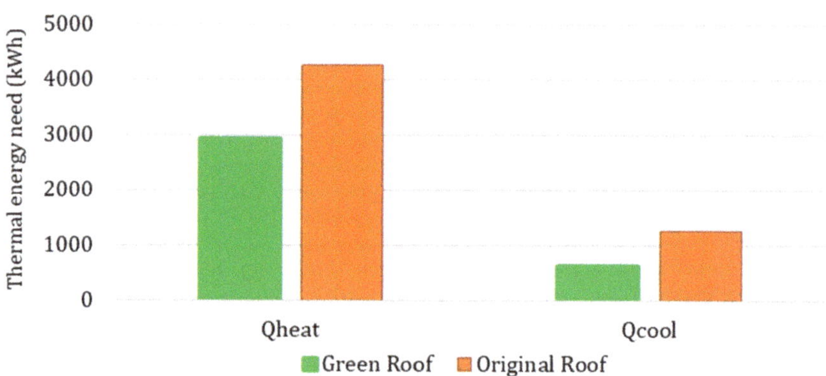

Figure 10. Annual thermal energy requirement for heating (Qheat) and cooling (Qcool) of the building with the original and green roof.

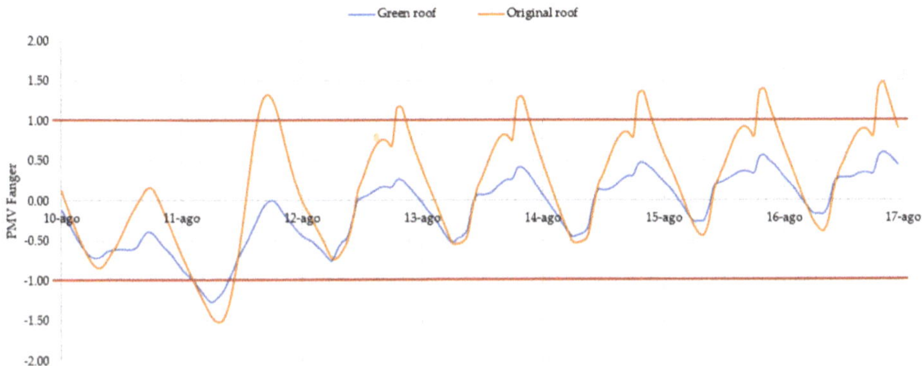

Figure 11. Comfort evaluation (using Fanger's Predicted Mean Vote (PMV) index) during a typical week in summer.

In addition, since the best performance of the green roof was found in summer, an evaluation of comfort conditions was performed using the indicator predicted mean vote (PMV Fanger) [37].

The PMV Fanger is equal to zero when thermal neutrality is reached, while the comfort zone is defined as the interval between the recommended PMV limits (e.g., $-0.5 < \text{PMV} < + 0.5$). The index is calculated from Fanger's equations that are characterized by six input parameters: air temperature, mean radiant temperature, relative humidity, air speed, metabolic rate, and clothing insulation.

A typical summer week was chosen (from 10 August till to 17 August), and the hourly PMV Fanger index is plotted in Figure 11. As can be noted, the installation of the green roof is able to sensibly improve the comfort conditions during the working days, keeping the index in the range between -1 and $+1$. Since the air speed, metabolic rate and the clothing insulation are considered constant in the simulation, and the air temperature is controlled by the cooling system, the installation of the green roof results to improve the distribution of surface temperatures and the values of relative humidity in the building rooms during the analyzed period.

Analyzing the results deriving from the simulation model, it is possible to affirm that the installation of a green roof is able to reduce energy consumptions in both heating and cooling seasons due to its properties of insulation and thermal inertia. Furthermore, a sensible improvement of comfort conditions in the building's internal spaces was found during the summer period.

4. Conclusions

This study aimed to investigate the thermal behavior of an innovative roof-lawn system, tested on two buildings placed in central Italy. Long-lasting measurement campaigns were carried out, and simulations were performed in order to understand the influence of this passive technique on the energy performance of buildings.

In both case studies, the green roof showed more stable thermal behavior, characterized by lower heat fluxes and temperature fluctuations. This is related to the additional massive layer represented by the roof lawn system, which influenced the steady state and the dynamic thermal performance of the roof.

For Case 1, the green roof showed a thermal transmittance of 1.361 W/(m^{2}K), and the original one a value of 3.021 W/(m^{2}K). Thus, comparing the green and the original roof, a percentage difference of about -55% was found. In terms of thermal conductance of the roofs, a value of 1.168 W/m^{2}K for the green roof and a value of 2.811 W/m^{2}K for the original one was identified, with a resulting percentage difference of about -58%. The measurement of the green roof extrados' temperature and the external surface temperature allowed a preliminary calculation of the roof-lawn thermal conductance, which was equal to 4.487 W/m^{2}K.

For Case 2, the green roof showed a thermal transmittance of 0.190 W/(m^{2}K) and the original one a value of 0.362 W/(m^{2}K). In this case, a percentage difference in terms of U-value of about -47.5% was found. Moreover, the thermal conductance of the roofs was computed, finding a value of 0.162 W/m^{2}K for the green roof and a value of 0.277 W/m^{2}K for the original one, with a resulting percentage difference of about -41%. In this case, the calculation of the roof-lawn thermal conductance was equal to 0.423 W/m^{2}K.

The effects of the roof-lawn system in terms of annual energy needs were simulated by means of a dynamic code, achieving significant reduction in terms of heating and cooling energy demands. Simulating the roof-lawn system and making a comparison with the original roof, primary energy reductions of 30% for heating and 51% for cooling were obtained, respectively. Benefits were found considering indoor thermal comfort.

Since ancient times, green roofs have always been considered thermal insulators, representing a sustainable answer to the energy saving problem. Future developments of this study will regard the optimization of the green roof performance, focusing on the structural part of the roof, identifying a roof composition able to work better with the roof-lawn, and taking advantage of the thermophysical properties of the system. For this purpose, it would be interesting to carry out the prolonged monitoring of several case studies, considering different geographical areas and the orientations of other roofs, as well as the study of buildings with different areas, uses and occupancy rates.

A broader, transversal and complex study can lead to a more reliable estimation of both energy savings and comfort increase. Finally, a financial analysis could be included, investigating installation, operation and maintenance costs.

Author Contributions: Conceptualization, F.A.; methodology, F.A., L.E. and C.G.; software, G.G.; formal analysis, L.E. and M.R.; investigation, L.E., C.G. and M.R.; resources, F.A.; data curation, L.E. and M.R.; writing—original draft preparation, L.E, M.R. and G.G.; writing—review and editing, F.A. and C.G.; supervision, F.A. All authors have read and agreed to the published version of the manuscript.

Funding: This research received no external funding.

Conflicts of Interest: The authors declare no conflict of interest.

Nomenclature

C	Thermal conductance (W/(m^{2}K))
h_{int}	Internal heat transfer coefficient (W/(m^{2}K))
q	Heat flux density (W/m^2)
T	Temperature at the boundary of the geometry (K, °C)
T_e	Outdoor air temperature (K, °C)
T_{env}	Temperature outside the simulated domain (K, °C)
T_i	Indoor air temperature (K, °C)
T_{se}	External surface temperature (K, °C)
T_{si}	Internal surface temperature (K, °C)
U	Thermal transmittance (W/(m^{2}K))
T	Time (h, min)
MAX	Maximum value
MIN	Minimum value
AVG	Average value
PS	Phase Shift (h)
DF	Decrement Factor (-)
PMV Fanger	Fanger's Predicted Mean Vote

References

1. IPCC. Global Warming of 1.5 °C. Available online: https://www.ipcc.ch/sr15 (accessed on 14 May 2020).
2. United Nations. UNSD Environmental Indicators. Available online: https://unstats.un.org/unsd/envstats/qindicators.cshtml (accessed on 14 May 2020).

3. Mohajerani, A.; Bakaric, J.; Jeffrey-Bailey, T. The urban heat island effect, its causes, and mitigation, with reference to the thermal properties of asphalt concrete. *J. Environ. Manag.* **2017**, *197*, 522–538. [CrossRef]

4. Bevilacqua, P.; Mazzeo, D.; Bruno, R.; Arcuri, N. Surface temperature analysis of an extensive green roof for the mitigation of urban heat island in southern mediterranean climate. *Energy Build.* **2017**, *150*, 318–327. [CrossRef]

5. Wong, N.H.; Chen, Y.; Ong, C.L.; Sia, A. Investigation of thermal benefits of rooftop garden in the tropical environment. *Build. Environ.* **2003**, *38*, 261–270. [CrossRef]

6. He, Y.; Yu, H.; Ozaki, A.; Dong, N.; Zheng, S. Long-term thermal performance evaluation of green roof system based on two new indexes: A case study in Shanghai area. *Build. Environ.* **2017**, *120*, 13–28. [CrossRef]

7. Bevilacqua, P.; Bruno, R.; Arcuri, N. Green roofs in a Mediterranean climate: Energy performances based on in-situ experimental data. *Renew. Energy* **2020**, *152*, 1414–1430. [CrossRef]

8. Bevilacqua, P.; Mazzeo, D.; Bruno, R.; Arcuri, N. Experimental investigation of the thermal performances of an extensive green roof in the Mediterranean area. *Energy Build.* **2016**, *122*, 63–79. [CrossRef]

9. Bevilacqua, P.; Mazzeo, D.; Arcuri, N. Thermal inertia assessment of an experimental extensive green roof in summer conditions. *Build. Environ.* **2018**, *131*, 264–276. [CrossRef]

10. Guattari, C.; Evangelisti, L.; Asdrubali, F.; De Lieto Vollaro, R. Experimental Evaluation and Numerical Simulation of the Thermal Performance of a Green Roof. *Appl. Sci.* **2020**, *10*, 1767. [CrossRef]

11. Asdrubali, F.; Evangelisti, L.; Guattari, C. Green roof for Zero Energy Buildings: A pilot project. *IOP Conf. Ser. Mat. Sci. Eng.* **2019**, *609*, 072011. [CrossRef]

12. Asdrubali, F.; Evangelisti, L.; Guattari, C.; Marzi, A.; Roncone, M. Monitoraggio e simulazione dinamica di un edificio pilota dotato di tetto verde. *AiCARR J.* **2019**, *59*, 40–44. [CrossRef]

13. Peng, L.L.H.; Jim, C.Y. Green-roof effects on neighborhood microclimate and human thermal sensation. *Energies* **2013**, *6*, 598–618. [CrossRef]

14. Piro, P.; Carbone, M.; De Simone, M.; Maiolo, M.; Bevilacqua, P.; Arcuri, N. Energy and Hydraulic Performance of a Vegetated Roof in Sub-Mediterranean Climate. *Sustainability* **2018**, *10*, 3473. [CrossRef]

15. Chen, X.-P.; Huang, P.; Zhou, Z.-X.; Gao, C. A review of green roof performance towards management of roof runoff. *Chin. J. Appl. Ecol.* **2015**, *26*, 2581–2590.

16. Nagase, A.; Dunnett, N. Amount of water runoff from different vegetation types on extensive green roofs: Effects of plant species, diversity and plant structure. *Landsc. Urban Plan.* **2012**, *104*, 356–363. [CrossRef]

17. Berndtsson, J.C.; Emilsson, T.; Bengtsson, L. The influence of extensive vegetated roofs on runoff water quality. *Sci. Total Environ.* **2006**, *355*, 48–63. [CrossRef]

18. Shafique, M.; Kima, R.; Rafiq, M. Green roof benefits, opportunities and challenges—A review. *Renew. Sustain. Energy Rev.* **2018**, *90*, 757–773. [CrossRef]

19. Connelly, M.; Hodgson, M. Experimental investigation of the sound transmission of vegetated roofs. *Appl. Acoust.* **2013**, *74*, 1136–1143. [CrossRef]

20. Yang, H.S.; Kang, J.; Choi, M.S. Acoustic effects of green roof systems on a low-profiled structure at street level. *Build. Environ.* **2012**, *50*, 44–55. [CrossRef]

21. Morakinyo, T.E.; Dahanayake, K.K.C.; Ng, E.; Chow, C.L. Temperature and cooling demand reduction by green-roof types in different climates and urban densities: A co-simulation parametric study. *Energy Build.* **2017**, *145*, 226–237. [CrossRef]

22. Shafique, M.; Reeho, K. Application of green blue roof to mitigate heat island phenomena and resilient to climate change in urban areas: A case study from Seoul, Korea. *J. Water Land Dev.* **2017**, *33*, 165–170. [CrossRef]

23. Niu, H.; Clark, C.; Zhou, J.; Adriaens, P. Scaling of economic benefits from green roof implementation in Washington, DC. *Environ. Sci. Technol.* **2010**, *44*, 2–8. [CrossRef] [PubMed]

24. Francis, R.A.; Lorimer, J. Urban reconciliation ecology: The potential of living roofs and walls. *J. Environ. Manag.* **2011**, *92*, 1429–1437. [CrossRef] [PubMed]

25. MacIvor, J.S.; Lundholm, J. Insect species composition and diversity on intensive green roofs and adjacent level-ground habitats. *Urban Ecosyst.* **2011**, *14*, 225–241. [CrossRef]

26. Teotónio, I.; Cabral, M.; Cruz, C.O.; Silva, C.M. Decision support system for green roofs investments in residential buildings. *J. Clean. Prod.* **2020**, *249*, 119365. [CrossRef]

27. International Organization for Standardization (ISO). *Thermal Insulation: Building Elements—In-Situ Measurement of Thermal Resistance and Thermal Transmittance. Part 1: Heat Flow Meter Method*; ISO 9869-1:2014; ISO: Geneva, Switzerland, 2014; Available online: https://www.iso.org/standard/59697.html (accessed on 12 June 2020).

28. Kontoleon, K.J.; Bikas, D.K. The effect of south wall's outdoor absorption coefficient on time lag, decrement factor and temperature variations. *Energy Build.* **2007**, *39*, 1011–1018. [CrossRef]

29. DesignBuilder Software, Version 6.1.5.004. Available online: http://designbuilderitalia.it (accessed on 28 June 2020).

30. Cool Roofs, Cool Cities, Cool Planet-Heat Island Group. Available online: https://heatisland.lbl.gov/resources/presentations (accessed on 21 June 2020).

31. Ramirez-Garcia, J.; Almendros, P.; Quemada, M. Ground cover and leaf area index relationship in a grass, legume and crucifer crop. *Plant Soil Environ.* **2012**, *58*, 385–390. [CrossRef]

32. Global Leaf Area Index Data from Field Measurements, 1932-2000. Available online: https://daac.ornl.gov/VEGETATION/LAI_support_images.html#table (accessed on 26 May 2020).

33. He, Y.L.; Su, D.R.; Liu, Z.X.; Liu, Y.S. Relationship between leaf area index and plant density of *Zoysia japonica* Steud. under different trimming heights. *Acta Agrestia Sin.* **2009**, *17*, 527–531.

34. Gagliano, A.; Nocera, F.; Detommaso, M.; Evola, G. Thermal Behavior of an Extensive Green Roof: Numerical Simulations and Experimental Investigations. *Int. J. Heat Mass Tran.* **2016**, *34*, 226–234.

35. Mehrinejad, E.; Abdollah, K.; Daemeib, B.; Malekjahana, F.A. Simulation study of the eco green roof in order to reduce heat transfer in four different climatic zones. *Results Eng.* **2019**, *2*, 100010.

36. Poddar, S.; Park, D.; Chang, S. Energy performance analysis of a dormitory building based on different orientations and seasonal variations of leaf area index. *Energy Effic.* **2017**, *10*, 887–903. [CrossRef]

37. Fanger, P.O.; Toftum, J. Extension of the PMV model to non-air-conditioned buildings in warm climates. *Energy Build.* **2002**, *34*, 533–536. [CrossRef]

Article

Assessing the Energy Performance of Prefabricated Buildings Considering Different Wall Configurations and the Use of PCMs in Greece

Stella Tsoka *, Theodoros Theodosiou, Konstantia Papadopoulou and Katerina Tsikaloudaki

Department of Civil Engineering, Aristotle University of Thessaloniki, P.O. BOX 429, 54124 Thessaloniki, Greece; tgt@civil.auth.gr (T.T.); pkonstanti@civil.auth.gr (K.P.); katgt@civil.auth.gr (K.T.)
* Correspondence: stsoka@civil.auth.gr

Received: 23 August 2020; Accepted: 22 September 2020; Published: 24 September 2020

Abstract: Despite the multiple advantages of prefabricated compared to conventional buildings, such as significant reductions in cost and time, improved quality and accuracy in manufacture, easy dismantling and reuse of components, reduction in environmental degradation, increase of productivity gains, etc., they still share a small part of the European building stock, mainly in the Mediterranean. This paper attempts to highlight the potential of prefabricated buildings to achieve advanced levels of performance, particularly as regards their thermal and energy behavior. More specifically, in this paper the energy needs of a single-family building constructed with prefabricated elements is analyzed, considering different climate contexts. The prefabricated elements comprising the building envelope were developed in order to address specific requirements with respect to their structural, hygrothermal, energy, fire, acoustical, and environmental performance, within the research project SUPRIM (sustainable preconstructed innovative module). The new multifunctional building element, also incorporating phase change materials for increased latent thermal heat storage, has been proven to be beneficial in all the examined climate zones. The results of the relevant studies will highlight the contribution of the new prefabricated element to the sustainability of the overall construction, as well as its advantages when compared with conventional constructions.

Keywords: prefabricated buildings; SUPRIM; EnergyPlus; building energy performance; phase change materials

1. Introduction

In the Mediterranean countries, residential buildings represent almost 80% of the total building stock, with the onsite process, and reinforced concrete and brick masonry, being the dominant construction solutions [1]. Although industrialization and off-site, precast construction is a growing sector in Central Europe and a reality in many countries all over the world, such as Singapore, China, and the United States [2–4], this is not the case for the countries of the Mediterranean area, such as Greece, where the share of prefabricated buildings accounts for less than 2% [5]. It is true that for many years, the precast construction method, involving the manufacturing of the building modules off site and their transportation and assembly on site, has been regarded as of inferior quality or was dedicated to temporary constructions. However, today, things have changed; prefabricated construction methods compete with conventional ones in every aspect of performance [6]; they also present considerable reductions in cost, installation time, and noise; moreover, they are also considered as a cost effective and environmentally friendly solution that does not have to compromise the architectural design and the building shape [7,8].

To date, several studies have investigated the performance of prefabricated buildings with respect to energy consumption and waste during the construction phase [9,10], the construction quality and safety [11,12], and of course with regard to their energy and environmental performance [13–18]. However, only a small number of relevant studies, assessing the energy performance of prefabricated constructions, have been conducted in the Mediterranean area [16,19], indicating a gap in the existing literature and the need for supplementary scientific analysis. Especially for the warm Mediterranean conditions, the establishment of strategies that would further improve the thermal performance of the prefabricated buildings, generally characterized by less thermal inertia, compared to conventional heavyweight ones, is of crucial importance. In this regard, the use of latent heat storage components in the building walls, such as phase change materials (PCMs), could considerably reduce the heat transmission and control the peak cooling loads in summer, while their incorporation could also be very useful for indoor temperature regulation during the winter period [20]. Up to the present time, many scientific studies have evaluated the role of PCM in building applications in the Mediterranean area with respect to the indoor thermal comfort conditions, the heating and cooling energy savings, and the reduction of temperature fluctuations of the envelope surfaces ([21–27]). Despite this, most of them mainly focus on conventional or lightweight constructions rather than on prefabricated buildings, while the emphasis is mainly given to the summer period. Moreover, most of the existing relevant studies, assess the PCMs' application either as a combination with cement plaster for exterior/interior facades, or in the form of mats, consisting of rectangular pouches that are filled with the PCMs and installed as a single layer behind wall boards ([21–27]). On the other hand, and to the authors knowledge, the use of the phase change materials as a composite concrete layer has been far less evaluated.

Based on the above remarks, the aim of the present paper is to provide further knowledge on the energy performance and the indoor thermal conditions of prefabricated structures in the Mediterranean context, not only in summer, but also in the winter period. More precisely, the study attempts to evaluate the energy performance of a one-story, prefabricated family building compared to a conventional construction, under different climatic conditions in the Mediterranean area. For this purpose, dynamic energy performance simulations with the EnergyPlus tool were conducted. The energy performance of the building, involving the annual heating and cooling needs, and the indoor thermal conditions were examined for different wall configurations (i.e., conventional and prefabricated building elements), whereas the effectiveness of the PCMs as a latent thermal heat storage strategy was also evaluated as an alternative supporting the prefabricated construction.

In this study, the investigated prefabricated building incorporated the new prefabricated building element that was developed within the context of the research project SU.PR.I.M. (SUstainable PReconstructed Innovative Module) [28], which is co-funded by the European Union and national sources. The new building module was developed in order to satisfy specific needs as regards its mechanical strength, hygrothermal performance, energy behavior, response to sound, reaction to fire, and environmental impact. The new prefabricated building element is made of steel hollow elements, positioned vertically at a distance between 0.70 m and 1.00 m, and two concrete panels, positioned parallel to each other, on either side of the steel elements. The gap between the steel elements is filled with expanded or extruded polystyrene. Each layer (concrete panel, thermal insulation/steel hollow element) is 0.05 m thick. The whole wall element is insulated with external thermal insulation, usually extruded polystyrene, covered with organic mortar. It should be mentioned that the type and the thickness of the external thermal insulation materials may vary, in order to satisfy different needs, i.e., with regard to the climate, the requirements for acoustic insulation, or fire performance, etc.

In order to further enhance the thermal performance of the examined wall element, especially as regards its thermal mass and its dynamic behavior, the integration of PCMs into the concrete mixture was studied. Actually, as discussed in Section 2 of the paper, it was decided to add PCMs only on the inner concrete panel, in order to contribute to the control of the interior surface temperature and, as a result, to the improvement of the indoor comfort conditions and the building's energy performance. The study considered the addition of two types of powder PCMs in the concrete mixture with melting temperatures equal to 24 °C and 28 °C. The selection of the melting points was based on the recommendations of Soares et al. [29] and Ascione et al. [30] regarding the optimum phase change temperature for PCM in the Mediterranean area, targeting higher energy efficiency both in the winter and summer periods. Regarding the proper concentration of the PCMs, values ranging between 5% and 25% [31–33] have been often reported for building envelope applications, with the concentration of 20% presenting considerable energy savings and an improved indoor thermal environment. Additionally, experimental tests showed that the mechanical strength of the concrete panels is significantly reduced when the concentration of PCMs in the concrete mixture is higher than 20%. Based on the above, the proportion of the PCMs in the concrete mixture was considered equal to 20% in this study. The developed building module (with and without the addition of PCMs) was considered as the main wall element of the investigated prefabricated buildings.

The paper is organized as follows: In Section 2, an overview of previous studies, assessing the role of PCMs on the improvement of the buildings' energy performance is given, whereas the detailed description the various examined scenarios and the simulations' input parameters are presented in Section 3. Section 4 provides the presentation and justification of the simulation output and the discussion of the performance of the prefabricated building, compared to the conventional one, while in Section 5, the main conclusions are summarized.

2. Phase Change Materials Applications in Buildings of the Mediterranean Area

The incorporation of phase change materials into the components of the building envelope constitutes a solution that has gained increasing scientific interest, to increase the thermal heat storage and avoid overheating and increased cooling loads in summer, but also to improve the buildings' winter energy performance. Especially for lightweight buildings, PCM applications are considered as an effective solution to increase the buildings' thermal inertia, and to eliminate excessive indoor "Tair" (air temperature) fluctuations, which often compromise thermal comfort conditions [34,35]. Up to the present time, an increasing amount of literature, evaluating the improvement of the energy performance of Mediterranean buildings using latent thermal heat storage techniques has been published. Focusing on building wall applications, the respective studies generally evaluate the effect of PCMs' applications at various locations (i.e., internally, externally, and within the wall section), with the optimum position strongly depending on the combination of various parameters, including the weather conditions of the examined area, the considered melting points, the main goal of the PCMs' application (i.e., emphasis on winter or summer energy demand and indoor thermal comfort) etc. [36].

Indicatively, Saafi et al. [24] evaluated the role of PCMs integrated into conventional building envelopes under the Tunisian warm climate. Several cases of PCM integration in the building shell (inside, outside brick wall, combined with EPS insulation, roof), various peak melting temperatures between 20 °C and 30 °C, as well as different wall orientations were examined. The analysis suggested that the PCM layer, when applied on the outside face of a brick wall, provides better energy performance efficiency compared to its application on the inside, with cooling energy savings rising to 13.4% for a south orientation. In a brick wall, better efficacy of PCMs is observed in the absence of insulation, with energy reduction around 12.21%. In the same context, Panayiotou et al. [37] studied the integration of PCMs with a melting temperature of 29 °C on the envelope of a typical dwelling in Cyprus, though dynamic numerical simulations. The authors examined various PCM placements in the building shell. The obtained simulation results suggested the outside face of the external building wall as the optimum placement of the PCM layer in a typical Mediterranean building located in the warm region of Cyprus. The achieved annual energy savings due to PCMs (no insulation) ranged between 21.7% and 28.6%, with the higher efficiency being noticed in the summer rather than in the winter period. On the other hand, when PCMs were combined with thermal insulation the maximum energy savings per year reached 66.2%, while the PCMs presented higher performance in summer rather than in winter. Other studies have mainly focused on the effect of PCMs when applied at the inner face of the building walls. For example Ozdenefe et al. [38] examined, via numerical simulation means, the effect of PCMs when they are integrated into the inner side of the wallboards in a typical building in

Cyprus. The analysis focused on the cooling season, while four scenarios were considered in terms of construction styles (perforated clay brick walls with reinforced concrete slabs, and cellular concrete block walls with cellular concrete slabs) and layer thicknesses. Simulation results indicated that the PCMs were most effective in the scenario of thinner walls and lightweight cellular concrete slabs; in this case, indoor air temperatures were reduced by up to 1.7 °C, and the cooling energy needs decreased by 14.0%.

Soares et al. [29] conducted multi-dimensional research on the effect of PCMs integrated in the inner side of a lightweight, steel-framed residential building on an annual and monthly base, while different case study cities in Spain, Portugal, Italy, and France were examined. Parameters that were taken into consideration were different melting temperatures ranging from 18 °C to 28 °C, as well as seven different countries (climate locations) from an energy saving point of view. The results indicated that the optimum melting point of the PCMs for the warmer Mediterranean climates ranged between 22 °C and 26 °C, while their application resulted in a peak energy efficiency gain of about 62% for Portugal's climate. Similar results have also been mentioned by Ascione et al. [30] who evaluated the effect of a PCM wallboard, positioned on the inner face of the outside building walls of an office building, with respect to indoor thermal comfort and the building's energy needs. Different melting points, ranging from 26 °C to 29 °C were considered, while dynamic energy performance simulations were conducted for Ankara, Naples, Athens, Marseille, and Seville. The obtained simulation results indicated higher summer energy savings and more favorable indoor thermal conditions for the higher melting points, with the cooling energy savings reaching 7.2% in Ankara, 4.1% in Marseille, and almost 3% in Seville and Naples. Moreover, according to the authors suggestions, when PCMs are implemented in buildings of the Mediterranean area, the optimal PCM melting point in the winter period ranges between 18 °C and 22 °C, while in summer, suitable values vary between 25 °C and 30 °C.

To continue, Guarino et al. [25] numerically assessed the effectiveness of PCMs integrated into the inner face of the exterior walls of a lightweight structure in Palermo, both during heating and cooling season. However, as the authors mention, the application of PCMs to the inner side of the exterior building walls would lead to a heat release in the indoor thermal environment during the nighttime (i.e., discharge period), increasing the overheating risk. The acquired simulation results thus highlighted the need to combine PCMs with natural ventilation in summer, so as to increase the heat discharge, effectively remove heat release, and reduce cooling energy needs; in winter, PCMs again proved an efficient solution towards the decrease of the heating energy demand, with their effect being, however, less prominent. Similarly, Costanzo et al. concluded in their research [26] that the implementation of a natural ventilation strategy at nighttime is necessary to enable a full discharge of the PCM on a daily basis in a lightweight buildings under Mediterranean conditions. More specifically, they examined the use of PCMs in a typical lightweight office building shell during the cooling period. Parameters, such as different melting peak temperatures (23 °C, 25 °C, 27 °C), thickness values, and night ventilation rates (2 ACH, 4 ACH) were considered for the cooling season. Peak cooling loads, energy needs, as well as operative temperatures were examined for all parameters stated above.

Based on the scientific evidence mentioned above, it can be generally said that:

- the applications of PCMs in the Mediterranean context mainly concerned conventional and lightweight constructions, realized onsite, rather than prefabricated buildings, while the emphasis was mainly on the cooling period.
- most of the existing studies assessed the PCMs combined with the plaster layer, either on the inner or the external part of the wall, whereas applications on composite concrete panels were rarer.
- in hot regions such as the Mediterranean area, and when the emphasis is placed on the summer period, the most favorable position of the PCM layer is the exterior face of the external building walls, as this reduces the external heat gains of the thermal zone. On the other hand, during winter the optimal PCM location is the inner part of the external building wall as this facilitates the storage and release of heat back to the thermal zone.

In this context, the present paper attempts to examine the role of PCMs in enhancing the performance of prefabricated buildings, located in the Mediterranean climate of Greece. Two building types, a prefabricated and a conventional one, made of reinforced concrete and brick masonry, are analyzed, both with respect to their annual energy performance and indoor thermal conditions, while the effectiveness of the PCMs as a latent thermal heat storage strategy in the prefabricated building is also considered. Moreover, given that the application aims at the improvement of the building's energy performance, both in winter and in summer, the PCM is applied at the internal face of the exterior wall, while summer night ventilation will be also considered to facilitate both the heat removal and the solidification process.

3. Materials and Methods

3.1. Case Study Presentation and Simulation Scenarios

The study concerns the energy performance of a small, single-family building, covering an area of 47.3 m², as shown in Figure 1. The plan is rectangular, expanded along the south-north axis, with openings only on the south and north walls. The roof is inclined and is covered with clay tiles. Below the inclined roof of the building, there is a horizontal slab, made of reinforced concrete, above which a thick layer of 10.0 cm thermal insulation (XPS) is positioned. The floor of the house, in contact with the ground, is constructed with reinforced concrete and is insulated with a 10.0 cm thick XPS layer. The windows comprise a PVC frame with a double, low-e glazing. The U-value of the transparent elements is equal to 2.0 W/(m² K).

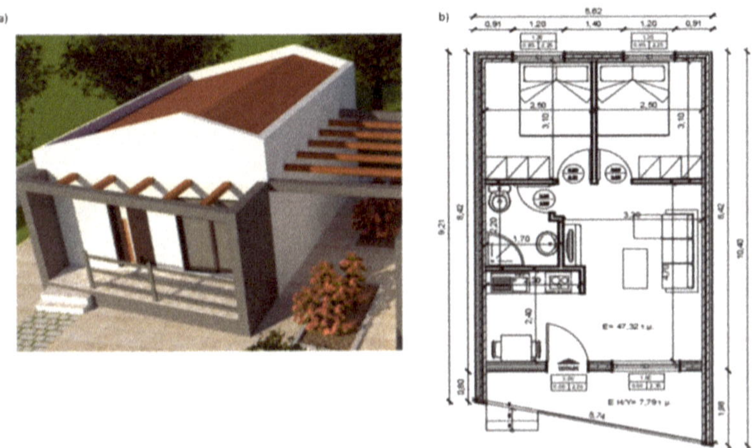

Figure 1. (**a**) External 3D view and (**b**) plot of the examined building.

To comparatively assess the performance of the prefabricated construction with the new module developed through the SUPRIM project, the building is examined for three different wall types:

- Considering the conventional wall construction made of reinforced concrete and hollow clay bricks (ETICS).
- Incorporating the new prefabricated module developed through the SUPRIM project, without the use of phase change materials (SUPRIM).
- Incorporating the new prefabricated module developed through the SUPRIM project, with the additional use of phase change materials (SUPRIM-PCM).

For the analysis of the energy performance, different levels of XPS thermal insulation were studied for all three wall types. More specifically, for the three types of the vertical elements' construction, the thickness of the XPS thermal insulation, positioned on the external surface of the vertical elements, was set equal to 5.0, 10.0, 15.0, and 20.0 cm. For the sake of clarity, it should be noted that the results concerning the examined configurations henceforward mentioned in the text, are named after their type and the insulation thickness. For example, "ETICS 5 cm" corresponds to a conventional wall construction of reinforced concrete and bricks, having 5 cm external thermal insulation.

The thermophysical parameters of the building envelope materials of the investigated vertical elements are given in Table 1. With regards to the 3rd configuration scenario of the building walls (i.e., SUPRIM-PCM), a mix of two PCMs has been considered, with melting points at 24 °C and 28 °C and phase change enthalpies at 140 kJ/kg and 185 kJ/kg, respectively. The PCMs are directly incorporated into the concrete mixture (including aggregates) at a concentration of 20% to further modify the energy storage capacity of the inner concrete panel. In order to define the thermal properties of the concrete panel containing PCMs accurately in the Energy Plus simulation tool, the results of a relevant experimental campaign, conducted in the context of the SUPRIM research program, were used. However, given that the focus of the present paper is mainly given to the energy performance analysis, only the basic experimental output is provided here.

Table 1. Geometrical and thermophysical parameters of the building envelope materials of the examined vertical walls (starting from the outside towards the inside face).

Examined Wall Scenario	Thickness x (m)	Specific Heat Capacity Cp (J/KgK)	Thermal Conductivity λ (W/mK)
1) ETICS			
Organic plaster	0.01	1100	0.87
Insulation layer XPS	0.05–0.20	1450	0.034
Bricks	0.19	1000	0.58
Internal plaster	0.02	1100	0.87
2) SUPRIM			
Organic plaster	0.01	1100	0.87
Insulation layer XPS	0.05–0.20	1450	0.034
Concrete panel	0.05	1000	2.1
Insulation layer EPS	0.05	1450	0.036
Concrete panel	0.05	1000	2.1
Internal plaster	0.02	1100	0.87
3) SUPRIM-PCM			
Organic plaster	0.01	1100	0.87
Insulation layer XPS	0.05–0.20	1450	0.034
Concrete panel	0.05	1000	2.1
Insulation layer EPS	0.05	1450	0.036
Concrete panel with PCM	0.05	-	1.95
Internal plaster	0.02	1100	0.87
Organic plaster	0.01	1100	0.87

More precisely, the thermal conductivity of the concrete samples (dimensions $10 \times 10 \times 10$ cm) with and without the incorporation of PCMs were measured with a KD2 Pro handheld device, using a single needle TR-1 sensor (Figure 2). The experimental campaign was conducted according to the Standards IEEE 442, "guide for soil thermal resistivity measurements" and ASTM 5334, "standard test method for determination of thermal conductivity of soil and soft rock by thermal needle probe procedure" [39,40]. The experimental output suggested a thermal conductivity of 1.95 W/mK for the concrete sample including the PCMs, whereas the respective value for the reference sample without PCMs was 2.1 W/mK. The latter values were considered for the energy performance simulations, while it was also assumed that the PCM is uniformly distributed throughout the thickness of the concrete panel.

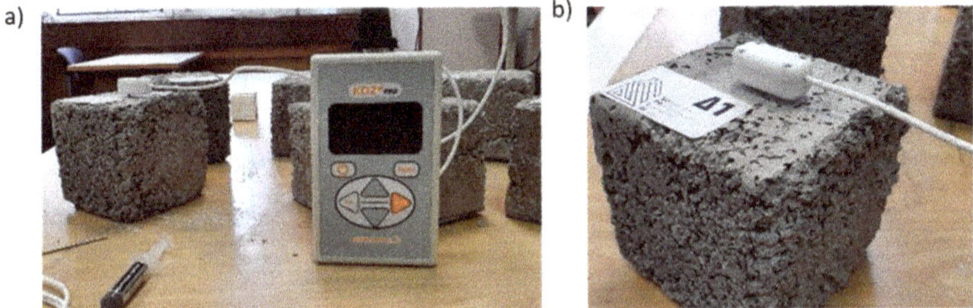

Figure 2. (**a**) experimental device and samples for the thermal conductivity measurements and (**b**) TR-1 sensor installation into the material.

Moreover, to evaluate the buildings' energy performance under different climatic conditions, it was assumed that the three examined building types were located in four different cities in Greece, with respect to the four climatic zones defined in the country: Heraklion (Lat. 35°19′ N; Long. 25°08′ E) for zone A (the warmest), Athens (Lat. 37°58′ N; Long. 23°42′ E) for zone B, Thessaloniki (Lat. 3740°38′ N; Long. 22°56′ E) for zone C, and Grevena (Lat. 40°05′ N; Long. 21°25′ E) for zone D (the coldest). The location of the four cities and their average monthly outdoor air temperatures are presented in Figure 3 and Table 2 respectively.

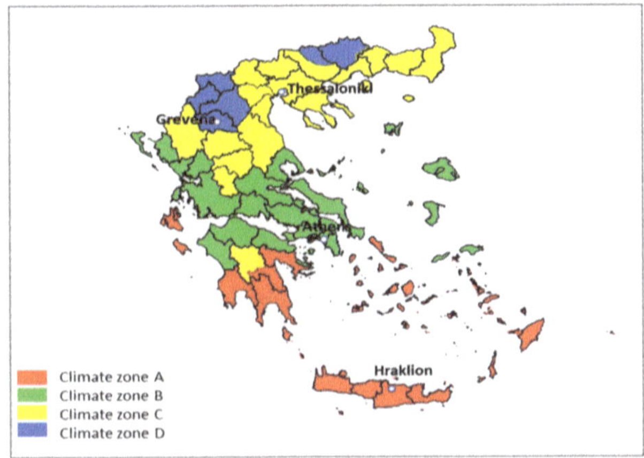

Figure 3. Climatic zones of Greece according to the Greek Regulation on the Energy Performance of Buildings, and locations of the case study cities [41].

Table 2. Monthly average air temperature (Tair) of the 4 investigated cities, corresponding to the 4 climate zones.

City/Zone	Jan	Feb	Mar	Apr	May	June	July	Aug	Sept	Oct	Nov	Dec
Heraklion Clim. Zone A	13.3	12.39	13.92	15.45	18.83	22.68	26.09	26.2	23.71	21.07	17.26	14.81
Athens Clim. Zone B	9.85	9.78	12.7	15.52	20.61	25.3	28.52	28.4	23.37	19.66	14.98	11.44
Thessaloniki Clim. Zone C	5.39	6.47	10.14	13.67	19.37	23.89	26.79	26.48	21.14	16.93	11.28	6.98
Grevena Clim. Zone D	3.94	5.33	9.65	13.38	19.25	23.42	26.97	26.71	20.39	15.73	9.61	5.31

Finally, aiming at a comparative assessment of the energy performance and the indoor thermal conditions of the building incorporating the SUPRIM module, an analysis was conducted: (a) for controlled indoor temperature, assuming the use of an HVAC (Heating, Ventilation, Air Conditioning) system, and (b) for free floating conditions, where the indoor temperature is free running and the PCMs are expected to improve the indoor thermal conditions.

All considered simulation scenarios are presented in the diagram of Figure 4.

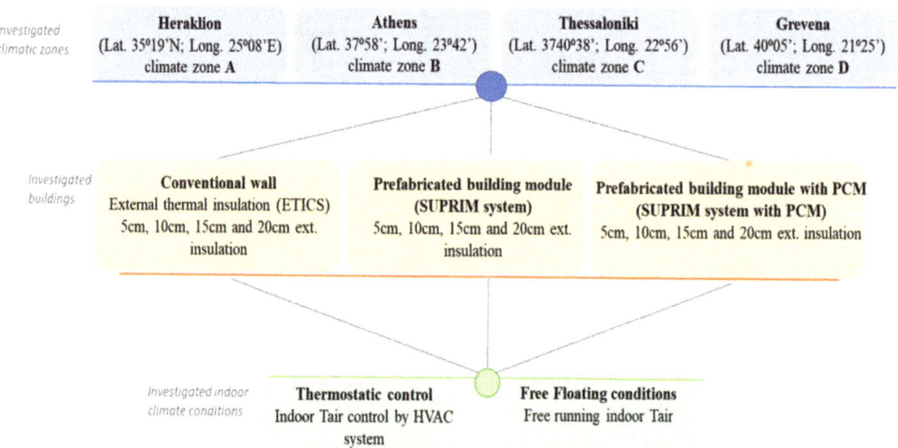

Figure 4. Examined simulation scenarios involving climatic conditions, building typologies, and control of indoor climate.

3.2. Dynamic Energy Performance Simulations and Modelling Set-Up

In this study, dynamic energy performance simulations were conducted with the EnergyPlus tool, a software developed by the US Department of Energy and continuously used and validated by many previous scientific studies worldwide [42–45]. Given its increased simulation capabilities, the model has been successfully used in several research fields, including the estimation of energy needs of residential [46–48] or tertiary buildings [49–51], indoor thermal comfort evaluation [52,53], and daylight analysis [54,55], but also in energy performance assessment of buildings with PCMs in their envelope components [30,38,56,57].

In EnergyPlus, the calculation of the thermal loads of buildings is by default based on the ASHRAE (American Society of Heating, Refrigerating and Air-Conditioning Engineers) heat balance method, accounting for the heat fluxes on outdoor and indoor surfaces, and also heat conduction through the building elements [58]. The principal advantage of the conduction transfer function (CTF) method lies in the reduction of the transient heat transfer equations into simple linear equations, having constant coefficients, which can be finally easily solved for both inside and outside face temperatures and heat fluxes [38]. The CTF method can be thus efficiently applied for simulating buildings having constant material properties (i.e., ETICS and SUPRIM scenarios in this study), but this is not the case for the PCMs, the heat storage capacity of which is not constant and varies as a function of temperature.

To overcome this barrier, and accurately simulate materials with variable thermal properties, EnergyPlus has incorporated the conduction finite difference solution algorithm (CondFD) [24], a model complementing the CTF method for cases where materials with variable thermal properties are evaluated. The CondFD method has been tested and validated by previous studies [42,59], while a detailed description of the governing equations is provided in [26,38,56]. When a phase change material is thus modeled in EnergyPlus, the user needs to provide a set of temperature/enthalpy values of the PCM, by using the dedicated module "material property: phase change".

In this study, the accurate determination of the thermal properties of the PCM in the SUPRIM-PCM scenario was crucial for the correct evaluation of the building's energy performance. However, given that the PCM was considered to be directly incorporated in the inside concrete panel, the enthalpy–temperature relationship of the mixture, rather than of the pure PCM, had to be known and introduced into the EnergyPlus model. In this context, the temperature-enthalpy values of the inside concrete panel of the prefabricated module were defined through the results of a measurement campaign, conducted in the context of the SUPRIM research program.

More precisely, samples of concrete incorporating microencapsulated phase change materials (manufacturer Encapsys) with melting points at 24 °C and 28 °C, and at a 20% concentration, were evaluated through differential scanning calorimetry (DSC), a widely applied method to test the phase change performance of a sample PCM [60]; the examined samples were submitted to a controlled temperature increase of 5 °C/min and the corresponding heat fluxes were recorded, providing relevant information on latent heat and specific enthalpy, with regards to the material's phase change. The heat flow (mW) at each temperature provided by the DSC was then converted to energy (J/g) and the estimated values were introduced into the EnergyPlus model [61]. It is also important to mention that, given that 2 PCMs with different melting points were considered, the phase transition did not occur at a specific temperature but a melting range was considered, starting from 21 °C to 28.5 °C. The PCMs' selection was made so as to find a suitable compromise between summer and winter period for the Mediterranean area, where the optimal melting points of the incorporated PCMs should be within the range of 24–30 °C and 18–22 °C respectively [30].

To continue, other important boundary parameters for the simulations involved:

- The infiltration rate. It was defined in accordance with the Hellenic Regulation on the Energy Performance of Buildings [41]; the respective value was set to 0.5 Air Change per Hour (ACH) and it remained constant for all the examined scenarios.
- The operational schedules concern the occupancy profile, the lighting and equipment usage, the ventilation rates, as well as the heating and cooling setpoints. They were all defined in accordance with the Hellenic Regulation on the Energy Performance of Buildings [13], as shown in Table 3.

Table 3. Boundary conditions and operational schedules, considered for the simulations.

Parameter	Unit	Value	Schedule Type
Occupancy	Persons/100 m^2	5	7/7; full occupancy: 00:00–07:00, 17:00–00:00; 50% occupancy: 07:00–17:00
Air change/ventilation	m^3/s/person	0.042	According to the usage profile.
Night ventilation	ACH	15	During the cooling period; from 00:00 till 08:00 a.m.; only if the indoor air temperature is higher than the outdoor air temperature by 1.0 °C.
Lighting	W/m^2	6.4	12/12; 0 W/m^2: 00:00–08:00; 0.3 W/m^2: 08:00–17:00; 0.75 W/m^2 17:00–00:00
Heating setpoint	°C	20	Heating period (according to the Climate zone)
Cooling setpoint	°C	26	Cooling period (according to the climate zone)
Heat gains from occupants	W/person	80	Follows the usage profile
Heat gains from equipment	W/m^2	4.0	Follows the usage profile

Finally, as previously mentioned, the energy performance of the one-story house was calculated for three different types of wall, for the four climatic zones, and considering two scenarios of the indoor microclimate involving:

1. Controlled indoor temperature, where the indoor air temperature is controlled by an HVAC system. In this case, the heating setpoint is set to 20 °C, while the cooling setpoint is set to 26 °C (from 9:00 a.m. to 24:00 during the cooling season). Heating and cooling seasons are defined for each case study city, according to the recommendations of the Technical Guides of the Greek Building Energy Performance Regulation [41]. Moreover, since the aim of the study is the analysis of the performance of a building without modeling a full HVAC system, the "ideal loads air system" was used in the EnergyPlus simulations

2. Free-floating temperature, where the indoor thermal conditions are not controlled by an HVAC system and the temperature is free running. In this way, the effect of the PCMs towards the improvement of the indoor thermal conditions both in winter and summer could be evaluated. In this case, a night ventilation with a constant air change per hour (ACH) of 15 h^{-1} was assumed during the cooling period to enhance the PCM discharge, provided that the outdoor temperature was lower than the indoor Tair by at least 1.0 °C

4. Results and Discussion

4.1. Thermal Performance under Controlled Indoor Temperature Conditions

The annual heating and cooling energy needs of the examined buildings are shown in Figure 5, with respect to the thermal insulation thickness, the wall configuration, and the climate zone. In parallel, Figure 6 presents the corresponding heating and cooling energy savings for the prefabricated building (with and without PCMs) compared to the conventional ETICS solution.

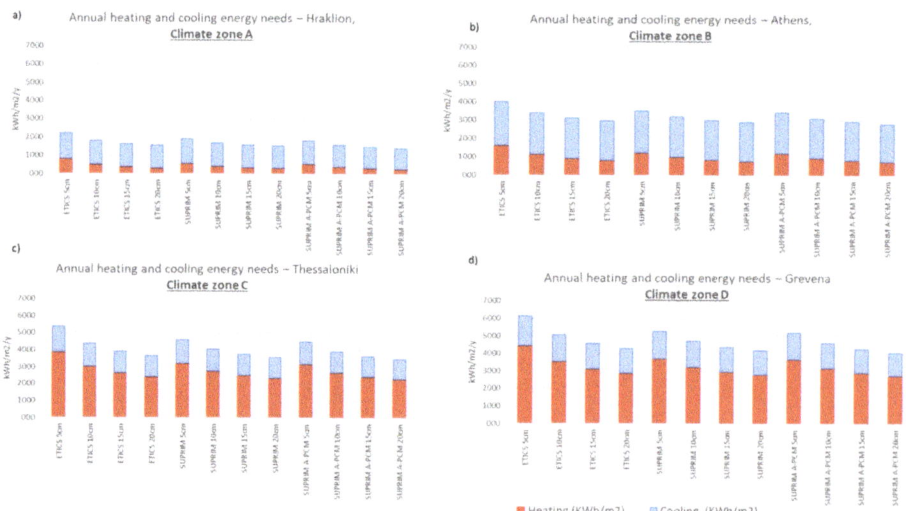

Figure 5. The heating and cooling needs calculated for the three examined construction types (conventional with exterior insulation, sustainable preconstructed innovative module (SUPRIM) and SUPRIM-PCM) for climate zone A, B, C, and D.

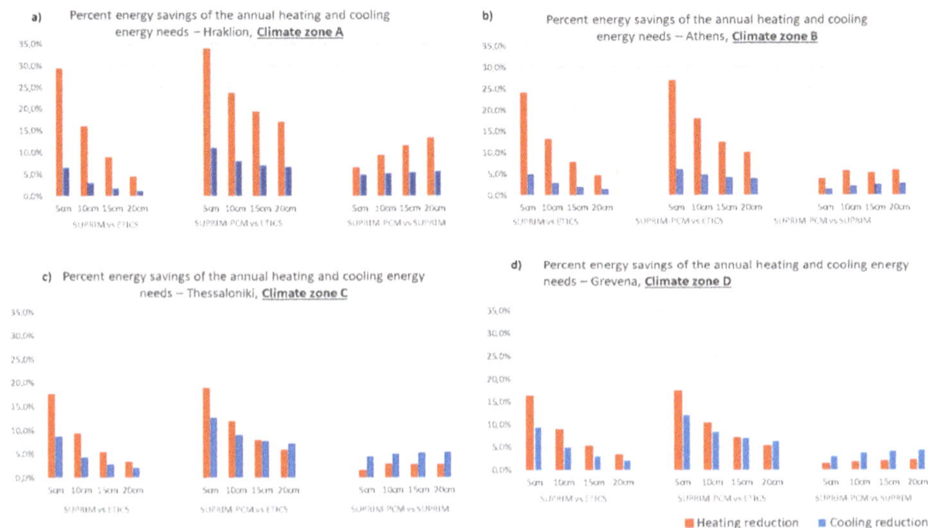

Figure 6. Percentage energy saving of the annual heating and cooling energy needs estimated for the various examined scenarios and for climate zone A, B, C, and D.

First, it can be generally seen that the energy needs of the prefabricated building are always lower when the innovative module (i.e., SUPRIM) is incorporated, compared to the respective values for the conventional ETICS construction. More precisely, in the warmest climate zone (i.e., Zone A), the energy required for covering the total heating and cooling needs is higher by 2–15% for the ETICS scenario compared to SUPRIM, with the difference varying as a function of the thermal insulation thickness; it is interesting to notice that as the thickness of the thermal insulation increases, the difference in energy needs between the two wall configurations becomes lower. Furthermore, it can be seen that in Zone A, the cooling needs are significantly higher than the heating ones. In fact, the cooling needs can be even four times higher than the heating ones. The gap becomes larger as the thermal insulation thickness increases. This can be attributed to the fact that the maximization of the thermal resistance of the walls leads to the minimization of heat losses during winter, but it does not cause an equivalent decrease of the cooling needs, as the solar heat gains, its major component, stem mainly from the transparent building elements. For the same reason, the maximum differences between the cooling needs for the two construction types (i.e., SUPRIM and ETICS) hardly exceed 6%, while for the heating loads the maximum difference reaches 29% (see Figure 6a). Again, the difference on the heating needs between the two examined construction types is not the same for every thickness of thermal insulation. As the thermal insulation thickness increases, its impact on heating load reduction weakens.

For Climate Zone A, it can also be seen that the differences on the total energy needs are rather moderate when the PCMs are integrated in the inner concrete panel. In fact, when comparing SUPRIM with SUPRIM-PCM, the observed reductions range between 5 and 6% and 6 and 13% in cooling and heating loads, respectively. Especially for the cooling needs, the monthly analysis of the derived results showed that the PCMs perform better in June, when a reduction of up to 16% is observed for all insulation thicknesses, while their performance falls in the following months and does not exceed 5%. In more detail, the city of Heraklion is characterized by warm summers, with an average daily maximum temperature in June and July close to 32 °C and average daily minimum close to 21 °C. Thus, in spite of the frequent activation of the phase change, the warm climatic conditions do not permit a full heat discharge of the materials, a phenomenon that has also been observed by Ascione et al. in relevant research in the Mediterranean area [30]. Additionally, as can be seen in

Figure 5a, the PCMs can have a more significant impact on the formation of the cooling needs for lower insulation levels. This can be attributed to the heat discharge and recharge phases that are more intense when the thermal insulation is not increased.

Similar trends are observed when the examined buildings are located in the second warmest zone of Greece, Zone B. Again, the annual energy needs are lower for the building that integrates the SUPRIM element compared to the ETICS, but the differences range over lower levels, i.e., between 2% and 12%. Higher differences are observed for the heating loads (4.7% to 24%, with regard to the thermal insulation thickness), while the difference in cooling needs are lower (1.3% to 4.7%), although the cooling needs are almost doubled (Figure 6b). As in the case of Heraklion, SUPRIM-PCM presents marginally lower cooling energy needs compared to the SUPRIM case, ranging between 1.3 and 3%, depending on the insulation thickness. On a monthly basis, the derived results showed that the SUPRIM-PCM again performed better in June, when a reduction of up to 6% and 7% for insulation thicknesses of 10 cm and 15 cm, respectively, was observed compared to the SUPRIM building. The performance, in the case of SUPRIM-PCM, falls considerably in August, probably because the PCM does not perform a complete discharge. Again, the PCMs present a more significant impact on the formation of the cooling needs for the lower insulation levels, and similar magnitudes of energy savings have also been reported in a previous study in Athens [30].

To continue, the energy need profile changes drastically when the buildings are located in a colder location, such as Thessaloniki in zone C (Figure 5c). The amount of energy needs increases significantly, and the heating needs prevail, as they are almost doubled with respect to cooling. Again, the SUPRIM building always performed better than the conventional ETICS one, as its energy needs are always lower. The difference on total energy needs ranges between 3% and 15%, with the heating loads being reduced by 3% to 17.5% with respect to the thickness of thermal insulation (Figure 6c). As with the previous climatic zones, it can be seen that the maximization of thermal resistance of the walls (as the insulation thickness increases) would contribute to lower heat losses during winter, but the decrease of the cooling needs are of lower importance, again due to the prevailing role of the solar heat gains; the maximum differences between the cooling needs for the two construction types (i.e., SUPRIM and ETICS) is close to 6%. Similarly, in zone D, the coldest of Greece, the energy needs reach their highest levels, with the cooling loads accounting for only one third of the heating ones (Figure 5d). The performance of the SUPRIM wall elements is better in the colder climate of zone D compared to the ETICS construction, as the decrease on the energy needs ranges from 3% to 14%. Again, the heating reduction is more substantial, reaching 16%. Interestingly, the incorporation of the PCMs seems to provide higher cooling energy savings in the two colder zones, ranging between 4.0 and 5.3% and 3.2 and 4.3% for climate zone C and D, respectively. This is mainly attributable to the milder summer conditions in the two colder zones, enabling the full discharge of the PCMs.

4.2. Thermal Performance under Free Floating Conditions

As the next step, the thermal performance of the examined buildings under free floating conditions was evaluated. Then, the assessment of the annual performance of the three examined buildings types was performed (i.e., ETICS, SUPRIM and SUPRIM-PCM). In this case, the lower limit of the thermal comfort (i.e., in the winter period) was 18 °C, while the upper acceptable limit was set to 28 °C (i.e., in the summer period). Figure 7 depicts the percentage period of the total occupancy time that the indoor T_{air} was within the comfort range (i.e., 18 °C $\leq T_{zone} \leq$ 28 °C) for the four examined climatic zones, and accounting for 10 cm external thermal insulation for all three building configurations.

First, it can be seen that the prefabricated building, SUPRIM, performed marginally better than the ETICS one, regarding the percent time that the T_{air} of the zone was within the comfort range. However, in all climatic zones, the incorporation of PCMS can further improve the thermal response of the prefabricated building under free floating conditions. Especially in the two warmer climatic zones (i.e., A and B), the addition of PCMs allows for a longer duration of comfort conditions of around 3%, corresponding to 250 h and 150 h, respectively. However, the PCMs' positive effect was of lower importance in the two colder zones (i.e., C and D). Given the low winter temperatures and the absence of a heating system, the resulting indoor Tair values very often fell below 15 °C, even, which cannot be compensated by the PCMs incorporation.

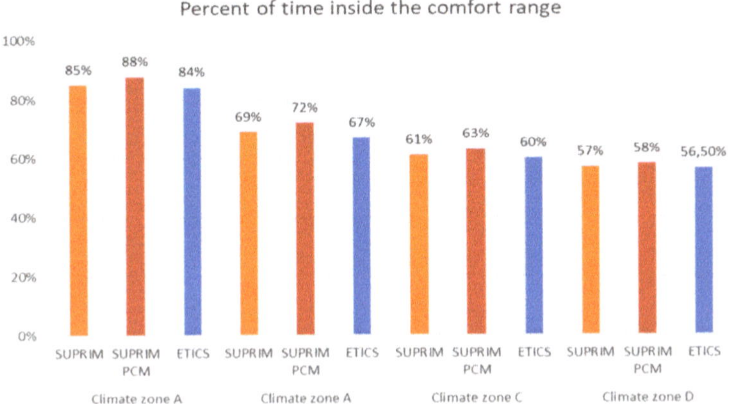

Figure 7. Annual percentage of the total occupancy time inside the comfort conditions for the three examined wall types, assuming 10 cm external thermal insulation.

Given the above mentioned results it should be emphasized that, in spite of the positive effect of PCMs, there is still an important part of the total occupancy period when the indoor Tair value is out of the comfort range, a fact that has to be taken into account when PCMs are considered in the design process of the buildings and their HVAC systems. Still, the beneficial effect of the PCMs could have been considerably higher if different melting points had been examined, depending on the climatic conditions. In this study, a mixture of two PCMs was evaluated (i.e., melting points at 24 °C and 28 °C), for all the climatic zones; while a future analysis of multiple melting points would provide the optimal solution for each climatic zone, also enhancing the performance of the SUPRIM building.

To continue, as previously mentioned, the performance of the phase change materials did not present similar characteristics during the summer months, with higher performance rates being reported for all climate zones in June. For the following months (i.e., July and August) the daytime and nighttime temperatures were higher, suggesting unfavorable conditions for a full discharge of the PCMs. In this context, and in order to better evaluate the impact of PCMs on the indoor thermal environment under free running conditions, the indoor Tair and Tsurf (surface temperature) profiles during typical weeks in June were analyzed. More precisely, Figures 8–11 depict the profiles of the indoor air temperature and the surface temperature on the internal side of the southern walls of the three examined building types and with reference to each climatic zone.

For every climatic zone, the provided surface and air temperature profiles corresponded to typical weeks during June, when no extreme heat wave phenomena and extreme temperatures are observed. In all cases, free floating conditions are considered, while the scenario of the 10 cm external insulation is here analyzed for all building configurations. It can be generally seen that during the examined days, in all climatic zones, the air temperature profiles of the SUPRIM-PCM building were always lower compared to the SUPRIM and the conventional ETICS building. More precisely, in Heraklion, the PCMs incorporation resulted in lower indoor Tair values by 0.2–0.8 °C compared to the prefabricated building without PCMs, with the peak daily indoor Tair values not exceeding 27.5 °C (Figure 8a).

Hraklion, Climate Zone A

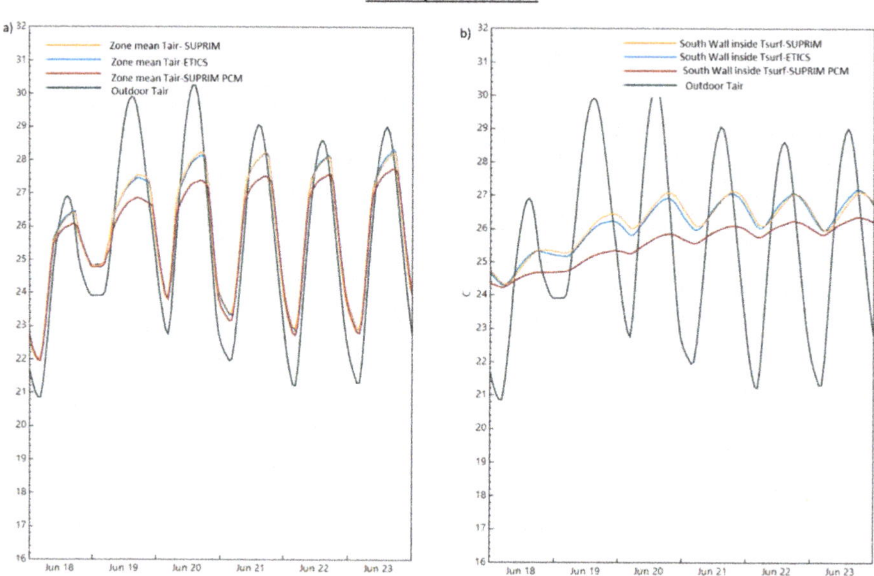

Figure 8. Climate zone A: (**a**) outdoor Tair and indoor Tair and (**b**) outdoor Tair and Tsurf evolution of the south facing wall during a typical week in June for the three investigated wall configurations and for 10 cm external insulation.

Athens, Climate Zone B

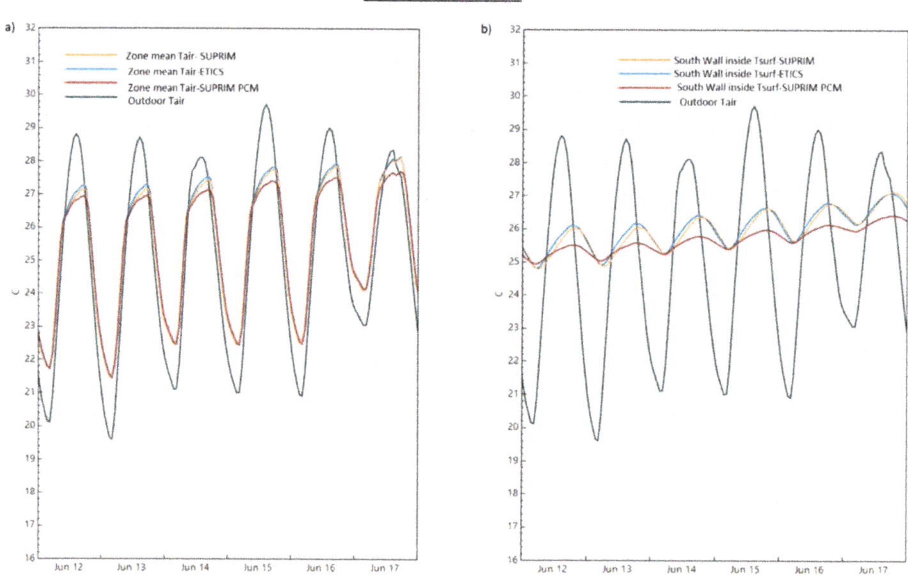

Figure 9. Climate zone B: (**a**) outdoor Tair and indoor Tair and (**b**) outdoor Tair and Tsurf evolution of the south facing wall during a typical week in June for the three investigated wall configurations and for 10 cm external insulation.

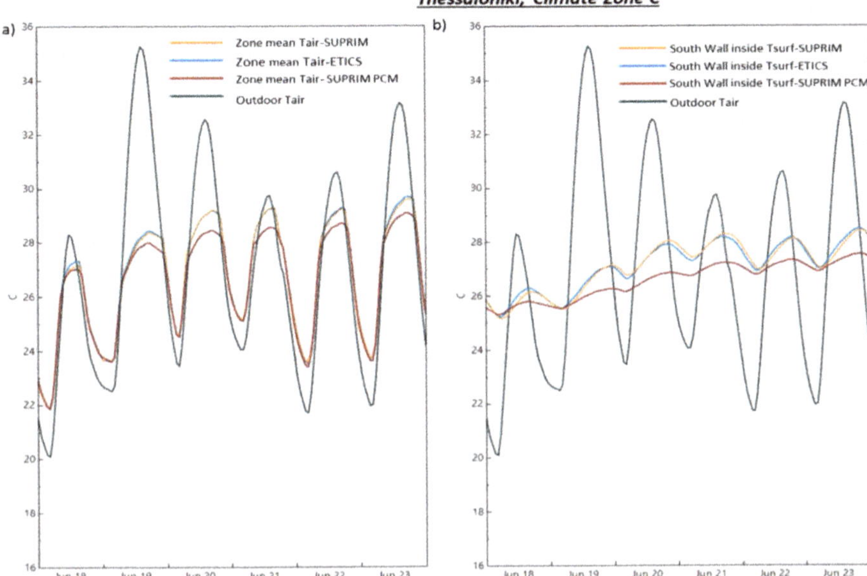

Figure 10. Climate zone C: (**a**) outdoor Tair and indoor Tair and (**b**) outdoor Tair and Tsurf evolution of the south facing wall during a typical week in June for the three investigated wall configurations and for 10 cm external insulation.

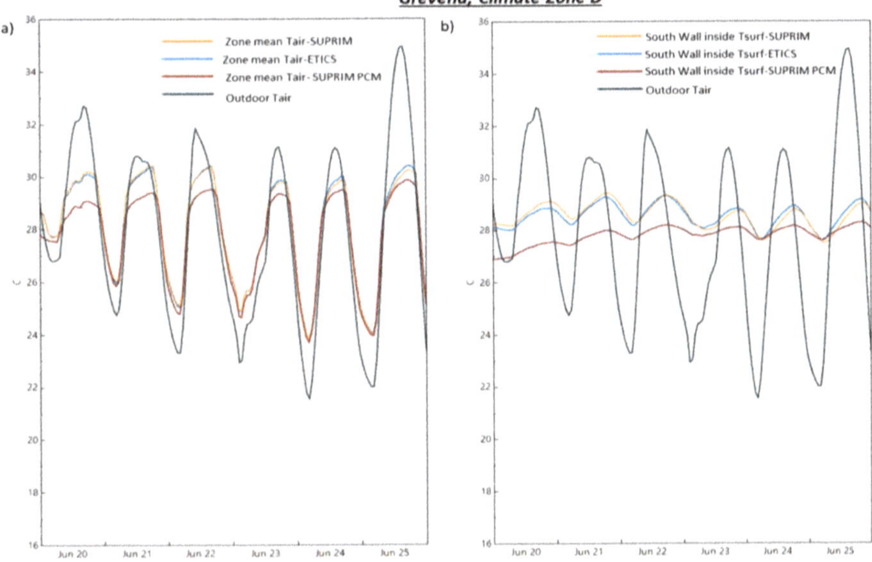

Figure 11. Climate zone D: (**a**) outdoor Tair and indoor Tair and (**b**) outdoor Tair and Tsurf evolution of the south facing wall during a typical week in June for the three investigated wall configurations and for 10 cm external insulation.

In Athens, the respective reduction due to PCMs ranged between 0.4 and 0.7 °C, and the indoor Tair fluctuated between 21.2 °C and 27.6 °C, whereas in the SUPRIM building, the zone Tair range fluctuated between 21.1 °C and 28.8 °C, with peak differences reported in the early afternoon (Figure 9a). In both cases, the lower indoor air temperatures were mainly attributed to the lower surface temperatures of the inner face of the building walls; indicatively, in climate zone A, during daytime and given the selected melting points of the PCM mixture (i.e., 24 °C and 28 °C), the phase change was activated and the south wall inside face temperature was lower by almost 1.0 °C compared to the no PCM building, thus leading to reduced convective heat transfer into the building.

Similar tendencies were also noticed in the colder climatic zones of Greece, with the beneficial effect of PCMs being slightly lower compared to the findings in Heraklion and Athens. More precisely, in the case of Thessaloniki, in climate zone C, the indoor Tair in the SUPRIM-PCM building was lower by 0.12–0.51 °C compared to the prefabricated building without PCMs during the examined week in June, with the peak daily indoor Tair values ranging between 27.2 °C and 28.9 °C (Figure 10a). A similar tendency was also noticed in Grevena, zone D, where the respective reduction due to PCMs, ranged between 0.12 and 0.8 °C. As with the two warmer zones, the PCM activation during the daytime, when the Tsurf exceeded 21 °C (a melting range is considered for the PCM mixture, starting from 21 °C to 28.5 °C), contributed to lower surface temperatures (Figures 10b and 11b) compared to the two other examined building configurations, and thus, to lower sensible heat transfer into the zone.

5. Conclusions

This paper evaluated the energy performance of a prefabricated building, incorporating the innovative SUPRIM building module in comparison to a conventional heavyweight construction under Mediterranean climatic conditions. The incorporation of PCMS, as a way to improve both the energy needs and the indoor thermal conditions, was also considered. A numerical simulation analysis with the EnergyPlus simulation model was conducted for a precast small family house, considering different thermal insulation thicknesses and various climatic zones in Greece.

The obtained simulation results suggested that the building constructed with the new prefabricated building elements had a better energy performance than the conventional one, made of reinforced concrete and bricks. This behavior was observed regardless of the thickness of the external thermal insulation, which was common for both constructions. More specifically, the improved energy performance concerned mainly the heating loads, while the cooling loads were not significantly reduced. Although the heating loads can be further reduced, through additional measures on the building envelope, such as windows with lower U-values, or thicker insulation on the roof, it is questionable whether the cooling loads can be similarly decreased.

As this work showed, the decrease of cooling loads cannot be achieved by further increasing the thermal resistance of the building envelope. Other, non-conventional measures should be introduced, which can be efficient in excessive heat management through the warmer periods of the year. Apart from solar shading, which is nevertheless beneficial, the heat capacity of the structure plays an important role. To this aim and to further increase the heat storage capacity of the examined prefabricated building, the incorporation of phase change materials into the concrete mix of the inside concrete panel was evaluated. The numerical results presented in this study indicated the potential of the PCMs incorporation to reduce the energy needs, both for heating and cooling periods on the one hand, and to improve the indoor comfort conditions on the other hand. However, their efficiency was found to be higher in warmer than in colder climatic zones. In light of the above, it is important to mention that the proper selection of the melting point of the PCMs, depending on the climatic conditions, would be necessary for the optimization of their performance, indicating a new field for future research. In other words, even if Greek cities are generally characterized by Mediterranean climatic conditions, important deviations still occur among them during the year, from the northern to the southern part of the country. More specifically, selecting the PCM melting temperature for each city, in line with the evolution of the ambient air temperature could further improve the energy performance and the indoor thermal comfort conditions of the examined buildings.

In parallel, to obtain a global perspective of the PCMs efficiency on the buildings' overall environmental and energy performance, a cost effectiveness analysis should be performed in the future, evaluating the pay-back periods, as well as the durability of PCMs after years of charging and discharging.

Finally, even if the conducted analysis was inevitably not exhaustive, the obtained results are expected to be a significant contribution on the respective scientific field. Considering that the share of prefabricated buildings and the relevant scientific studies are today very low in the Mediterranean area, despite their advantages, the present paper provides additional information on their performance, highlighting their enhanced energy behavior, especially when combined with latent heat storage systems such as PCMs.

Author Contributions: S.T., T.T., K.P. and K.T. conceived and design the study. S.T. and T.T. performed the numerical simulations and the results analysis. S.T. and K.P. wrote the paper. K.T. and T.T. reviewed and edited the manuscript. All authors have read and agreed to the published version of the manuscript.

Funding: This research has been co-financed by the European Regional Development Fund of the European Union and Greek national funds through the Operational Program Competitiveness, Entrepreneurship and Innovation, under the call RESEARCH-CREATE-INNOVATE (project code:T1EDK-03042).

Conflicts of Interest: The authors declare no conflict of interest.

References

1. EU Building Stock Observatory. Available online: https://ec.europa.eu/energy/topics/energy-efficiency/energy-efficient-buildings/eu-bso_en (accessed on 29 July 2020).
2. Mao, C.; Xie, F.; Hou, L.; Wu, P.; Wang, J.; Wang, X. Cost analysis for sustainable off-site construction based on a multiple-case study in China. *Habitat Int.* **2016**, *57*, 215–222. [CrossRef]
3. Steinhardt, D.A.; Manley, K. Exploring the beliefs of Australian prefabricated house builders. *Constr. Econ. Build.* **2016**, *16*, 27–41. [CrossRef]
4. Generalova, E.M.; Generalov, V.P.; Kuznetsova, A.A. Modular buildings in modern construction. *Proc. Eng.* **2016**, *153*, 167–172. [CrossRef]
5. Apaydin, F. Effectiveness of prefabricated house industry's marketing activities and Turkish consumers' buying intentions towards prefabricated houses. *Asian Soc. Sci.* **2011**, *7*, 267. [CrossRef]
6. Tam, V.W.; Tam, C.M.; Zeng, S.; Ng, W.C. Towards adoption of prefabrication in construction. *Build. Environ.* **2007**, *42*, 3642–3654. [CrossRef]
7. Jaillon, L.; Poon, C.S. The evolution of prefabricated residential building systems in Hong Kong: A review of the public and the private sector. *Autom. Constr.* **2009**, *18*, 239–248. [CrossRef]
8. Pan, W.; Gibb, A.G.; Dainty, A.R. Strategies for integrating the use of off-site production technologies in house building. *J. Constr. Eng. Manag.* **2012**, *138*, 1331–1340. [CrossRef]
9. Lu, N. The current use of offsite construction techniques in the United States construction industry. In Proceedings of the Construction Research Congress 2009: Building a Sustainable Future, Seattle, WA, USA, 5–7 April 2009; pp. 946–955.
10. Lawson, R.; Ogden, R. Sustainability and process benefits of modular construction. In Proceedings of the TG57-Special Track 18th CIB World Building Congress, Salford, UK, 10–13 May 2010; p. 38.
11. Lopez, D.; Froese, T.M. Analysis of costs and benefits of panelized and modular prefabricated homes. *Proc. Eng.* **2016**, *145*, 1291–1297. [CrossRef]
12. Knyziak, P. The impact of construction quality on the safety of prefabricated multi-family dwellings. *Eng. Fail. Anal.* **2019**, *100*, 37–48. [CrossRef]
13. Aye, L.; Ngo, T.; Crawford, R.H.; Gammampila, R.; Mendis, P. Life cycle greenhouse gas emissions and energy analysis of prefabricated reusable building modules. *Energy Build.* **2012**, *47*, 159–168. [CrossRef]
14. Hong, J.; Shen, G.Q.; Mao, C.; Li, Z.; Li, K. Life-cycle energy analysis of prefabricated building components: An input–output-based hybrid model. *J. Clean. Prod.* **2016**, *112*, 2198–2207. [CrossRef]

15. Tumminia, G.; Guarino, F.; Longo, S.; Ferraro, M.; Cellura, M.; Antonucci, V. Life cycle energy performances and environmental impacts of a prefabricated building module. *Renew. Sust. Energy Rev.* **2018**, *92*, 272–283. [CrossRef]

16. Silva, P.C.; Almeida, M.; Bragança, L.; Mesquita, V. Development of prefabricated retrofit module towards nearly zero energy buildings. *Energy Build.* **2013**, *56*, 115–125. [CrossRef]

17. Gunawardena, T.; Ngo, T.; Aye, L.; Mendis, P. Innovative Prefabricated Modular Structures–An Overview and Life Cycle Energy Analysis. In Proceedings of the International Conference on Structural Engineering Construction and Management, Kandy, Central, Sri Lanka, 16–18 December 2011.

18. Yin, X.; Dong, Q.; Zhou, S.; Yu, J.; Huang, L.; Sun, C. Energy-Saving Potential of Applying Prefabricated Straw Bale Construction (PSBC) in Domestic Buildings in Northern China. *Sustainability* **2020**, *12*, 3464. [CrossRef]

19. Bruno, R.; Bevilacqua, P.; Cuconati, T.; Arcuri, N. Energy evaluations of an innovative multi-storey wooden near Zero Energy Building designed for Mediterranean areas. *Appl. Energy* **2019**, *238*, 929–941. [CrossRef]

20. Da Cunha, S.R.L.; de Aguiar, J.L.B. Phase change materials and energy efficiency of buildings: A review of knowledge. *J. Energy Storage* **2020**, *27*, 101083. [CrossRef]

21. Nghana, B.; Tariku, F. Phase change material's (PCM) impacts on the energy performance and thermal comfort of buildings in a mild climate. *Build. Environ.* **2016**, *99*, 221–238. [CrossRef]

22. Ascione, F.; Bianco, N.; De Masi, R.F.; Mastellone, M.; Vanoli, G.P. Phase change materials for reducing cooling energy demand and improving indoor comfort: A step-by-step retrofit of a Mediterranean educational building. *Energies* **2019**, *12*, 3661. [CrossRef]

23. Tenpierik, M.; Wattez, Y.; Turrin, M.; Cosmatu, T.; Tsafou, S. Temperature Control in (Translucent) Phase Change Materials Applied in Facades: A Numerical Study. *Energies* **2019**, *12*, 3286. [CrossRef]

24. Saafi, K.; Daouas, N. Energy and cost efficiency of phase change materials integrated in building envelopes under Tunisia Mediterranean climate. *Energy* **2019**, *187*, 115987. [CrossRef]

25. Guarino, F.; Longo, S.; Cellura, M.; Mistretta, M.; La Rocca, V. Phase change materials applications to optimize cooling performance of buildings in the Mediterranean area: A parametric analysis. *Energy Proc.* **2015**, *78*, 1708–1713. [CrossRef]

26. Costanzo, V.; Evola, G.; Marletta, L.; Nocera, F. The effectiveness of phase change materials in relation to summer thermal comfort in air-conditioned office buildings. In Proceedings of the Building Simulation, Cambridge, UK, 11–12 September 2018; pp. 1145–1161.

27. Karaoulis, A. Investigation of energy performance in conventional and lightweight building components with the use of phase change materials (PCMs): Energy savings in summer season. *Proc. Environ. Sci.* **2017**, *38*, 796–803. [CrossRef]

28. SUPRIM. Available online: https://suprim.gr/?lang=en (accessed on 29 July 2020).

29. Soares, N.; Gaspar, A.; Santos, P.; Costa, J. Multi-dimensional optimization of the incorporation of PCM-drywalls in lightweight steel-framed residential buildings in different climates. *Energy Build.* **2014**, *70*, 411–421. [CrossRef]

30. Ascione, F.; Bianco, N.; De Masi, R.F.; de'Rossi, F.; Vanoli, G.P. Energy refurbishment of existing buildings through the use of phase change materials: Energy savings and indoor comfort in the cooling season. *Appl. Energy* **2014**, *113*, 990–1007. [CrossRef]

31. Kośny, J. Short history of PCM applications in building envelopes. In *PCM-Enhanced Building Components*; Springer: Berlin/Heidelberg, Germany, 2015; pp. 21–59.

32. Kosny, J.; Yarbrough, D.W.; Miller, W.A.; Petrie, T.; Childs, P.W.; Syed, A.M. *2006/07 Field Testing of Cellulose Fiber Insulation Enhanced with Phase Change Material*; Oak Ridge National Lab.(ORNL), Building Technologies Research and Integration Center: Oak Ridge, TN, USA, 2008.

33. Zhang, M.; Medina, M.A.; King, J.B. Development of a thermally enhanced frame wall with phase-change materials for on-peak air conditioning demand reduction and energy savings in residential buildings. *Int. J. Energy Res.* **2005**, *29*, 795–809. [CrossRef]

34. Pomianowski, M.; Heiselberg, P.; Zhang, Y. Review of thermal energy storage technologies based on PCM application in buildings. *Energy Build.* **2013**, *67*, 56–69. [CrossRef]

35. Cabeza, L.F.; Castell, A.; Barreneche, C.d.; De Gracia, A.; Fernández, A. Materials used as PCM in thermal energy storage in buildings: A review. *Renew. Sustain. Energy Rev.* **2011**, *15*, 1675–1695. [CrossRef]

36. Al-Absi, Z.A.; Mohd Isa, M.H.; Ismail, M. Phase Change Materials (PCMs) and Their Optimum Position in Building Walls. *Sustainability* **2020**, *12*, 1294. [CrossRef]
37. Panayiotou, G.; Kalogirou, S.A.; Tassou, S.A. Evaluation of the application of Phase Change Materials (PCM) on the envelope of a typical dwelling in the Mediterranean region. *Renew. Energy* **2016**, *97*, 24–32. [CrossRef]
38. Ozdenefe, M.; Dewsbury, J. Thermal performance of a typical residential Cyprus building with phase change materials. *Build. Serv. Eng. Res. Technol.* **2016**, *37*, 85–102.
39. IEEE STANDARD. IEEE Std 442-IEEE Guide for Soil Thermal Resistivity Measurements. *EUA IEEE* **1981**. [CrossRef]
40. ASTM. Standard test method for determination of thermal conductivity of soil and soft rock by thermal needle probe procedure. *ASTM Data Ser. Publ.* **2008**, *5334*, 1–8.
41. TOTEE20701-1/2017. Technical Guides of the recast of the Hellenic Thermal Regulation of the Energy Assessment of Buildings. 2017. Available online: http://portal.tee.gr/portal/page/portal/SCIENTIFIC_WORK/GR_ENERGEIAS/kenak/files/TOTEE_20701-1_2017_TEE_1st_Edition.pdf (accessed on 10 July 2020). (In Greek).
42. Tabares-Velasco, P.C.; Christensen, C.; Bianchi, M. Verification and validation of EnergyPlus phase change material model for opaque wall assemblies. *Build. Environ.* **2012**, *54*, 186–196. [CrossRef]
43. Yu, S.; Cui, Y.; Shao, Y.; Han, F. Research on the Comprehensive Performance of Hygroscopic Materials in an Office Building Based on EnergyPlus. *Energies* **2019**, *12*, 191. [CrossRef]
44. Shrestha, S.; Maxwell, G. Empirical validation of building energy simulation software: EnergyPlus. In Proceedings of the Building Simulation; Iowa State University: Ames, IA, USA, 2006; pp. 2935–2942.
45. Goia, F.; Chaudhary, G.; Fantucci, S. Modelling and experimental validation of an algorithm for simulation of hysteresis effects in phase change materials for building components. *Energy Build.* **2018**, *174*, 54–67. [CrossRef]
46. Shabunko, V.; Lim, C.; Mathew, S. EnergyPlus models for the benchmarking of residential buildings in Brunei Darussalam. *Energy Build.* **2018**, *169*, 507–516. [CrossRef]
47. Bojić, M.; Yik, F. Application of advanced glazing to high-rise residential buildings in Hong Kong. *Build. Environ.* **2007**, *42*, 820–828.
48. Dávi, G.A.; Caamaño-Martín, E.; Rüther, R.; Solano, J. Energy performance evaluation of a net plus-energy residential building with grid-connected photovoltaic system in Brazil. *Energy Build.* **2016**, *120*, 19–29. [CrossRef]
49. Boyano, A.; Hernandez, P.; Wolf, O. Energy demands and potential savings in European office buildings: Case studies based on EnergyPlus simulations. *Energy Build.* **2013**, *65*, 19–28. [CrossRef]
50. Korolija, I.; Zhang, Y.; Marjanovic-Halburd, L.; Hanby, V.I. Regression models for predicting UK office building energy consumption from heating and cooling demands. *Energy Build.* **2013**, *59*, 214–227.
51. Vujošević, M.; Krstić-Furundžić, A. The influence of atrium on energy performance of hotel building. *Energy Build.* **2017**, *156*, 140–150. [CrossRef]
52. Hong, S.H.; Lee, J.M.; Moon, J.W.; Lee, K.H. Thermal comfort, energy and cost impacts of PMV control considering individual metabolic rate variations in residential building. *Energies* **2018**, *11*, 1767. [CrossRef]
53. Yao, J.; Chow, D.H.C.; Zheng, R.-Y.; Yan, C.-W. Occupants' impact on indoor thermal comfort: A co-simulation study on stochastic control of solar shades. *J. Build. Perform. Simul.* **2016**, *9*, 272–287. [CrossRef]
54. Ramos, G.; Ghisi, E. Analysis of daylight calculated using the EnergyPlus programme. *Renew. Sust. Energy Rev.* **2010**, *14*, 1948–1958. [CrossRef]
55. Motamedi, S.; Liedl, P. Integrative algorithm to optimize skylights considering fully impacts of daylight on energy. *Energy Build.* **2017**, *138*, 655–665. [CrossRef]
56. Váz Sá, A.; Almeida, R.; Sousa, H.; Delgado, J. Numerical analysis of the energy improvement of plastering mortars with phase change materials. *Adv. Mater. Sci. Eng.* **2014**, *2014*. [CrossRef]
57. Zastawna-Rumin, A.; Kisilewicz, T.; Berardi, U. Novel Simulation Algorithm for Modeling the Hysteresis of Phase Change Materials. *Energies* **2020**, *13*, 1200. [CrossRef]
58. Crawley, D.B.; Lawrie, L.K.; Winkelmann, F.C.; Buhl, W.F.; Huang, Y.J.; Pedersen, C.O.; Strand, R.K.; Liesen, R.J.; Fisher, D.E.; Witte, M.J. EnergyPlus: Creating a new-generation building energy simulation program. *Energy Build.* **2001**, *33*, 319–331. [CrossRef]

59. Tabares-Velasco, P.C.; Griffith, B. Diagnostic test cases for verifying surface heat transfer algorithms and boundary conditions in building energy simulation programs. *J. Build. Perform. Simul.* **2012**, *5*, 329–346. [CrossRef]

60. Jin, X.; Xu, X.; Zhang, X.; Yin, Y. Determination of the PCM melting temperature range using DSC. *Thermochim. Acta* **2014**, *595*, 17–21. [CrossRef]

61. Castellón, C.; Günther, E.; Mehling, H.; Hiebler, S.; Cabeza, L.F. Determination of the enthalpy of PCM as a function of temperature using a heat-flux DSC—A study of different measurement procedures and their accuracy. *Int. J. Energy Res.* **2008**, *32*, 1258–1265. [CrossRef]

Article

NZEB Analyses by Means of Dynamic Simulation and Experimental Monitoring in Mediterranean Climate

Anna Magrini * and Giorgia Lentini

Department of Civil Engineering and Architecture, University of Pavia, 27100 Pavia, Italy;
giorgia.lentini01@universitadipavia.it
* Correspondence: magrini@unipv.it

Received: 20 July 2020; Accepted: 11 September 2020; Published: 14 September 2020

Abstract: The reduction of energy consumption in the building sector has promoted the spread of the NZEB (Nearly Zero Energy Building) model. A future target is represented by positive-energy buildings (PEB), which produce more energy than they consume. The study is centred on the examination of some peculiarities of NZEB through a case study and on the analysis of opportunities for further increase in energy performance, to trace the road that each designer should take, through an extensive evaluation of the potentials variations on the project that could lead to better results. The project assessments are developed through a dynamic simulation model and the data from the monitoring of the building's performance are used to evaluate the actual energy saving conditions. The analyses demonstrate the importance of an accurate design of the envelope and technical building systems associated with a smart management of the control systems and the setting of the set points, for the optimal operation of the systems. Ambitious but feasible design choices and an accurate analysis of the possibility of increasing the energy performance of a NZEB can lead to reaching the PEB target and energy independence, enhancing the production of energy from renewable sources.

Keywords: energy efficiency policy; nearly zero energy building; Positive Energy Building; energy performance of buildings; thermal behaviour; thermal dynamic simulation

---

1. Introduction

This research aims at enriching the knowledge of some of the most useful actions in building design to reach the best energy performance towards the NZEB (Nearly Zero Energy Building) target, outlining some steps of the building design, providing examples of the assessments that should be developed, to support the best choices, depending on the specific context in which the building is located (Section 2). The focus is on new buildings, even if most of the aspects here considered can be applied also to existing buildings retrofit.

Through the analysis of an exemplary case study, a good combination of envelope-system design and smart management is analysed, simulation results and monitored data are considered, and some variations are proposed and evaluated, to show advantages and problems, depending on a cost analysis. The main characteristics of the building envelope and the technical systems of the case study are typical of a single family house in an urban context not densely populated (Section 3).

The energy analysis is carried out by means of dynamic simulation to highlight the aspects relating to real-time control combined with smart management of the temperature set point, and the systems operating times (Section 4). Firstly, a comparison between this approach and the results of a semi-stationary model in winter conditions is presented, and the energy performance data are compared with the real consumption to critically analyse the different methods.

The results of the dynamic simulation method allow to put in evidence different effects of solar energy, such as the contribution of the solar greenhouse whose behaviour is analysed in detail during the winter season. In the summer season, the possibility of using the greenhouse structure to protect the façade with reflective white curtains and the possibility of opening the vertical glazed sections of the greenhouse (sliding) cancels any potentially negative effects of overheating.

The heat pump operation, supported by the PV system, is analysed in detail in relation to the internal/external temperatures of the months of July and August. The analysis of the data collected by the photovoltaic system allows to confirm, as foreseen by the project, the total coverage of the building's energy needs at an annual cycle and to highlight the surplus of energy produced which would allow the building to be classified as PEB to represent a pole in the distribution of renewable energy produced in excess.

Finally, some possibilities are described for realizing a better project or any interventions on the current one, through better management (Section 5). Potential design and technical systems management variants are therefore analysed. Among these, only the most relevant and promising results for a potential increase in energy performance or for an increase in internal comfort are highlighted, obtained by considering the following aspects:

- At the plant management level, actions on the regulation of the internal temperature, with possible variations to the current setting of the room temperature set points.
- At the design level, evaluation of the actual need of the solar greenhouse, which, analysed through dynamic simulation, turns out to be a very important element in terms of energy savings in winter conditions, also in relation to the intelligent management of the air flow rates from it to internal environments.
- Evaluation of the possibility of energy storage in batteries.
- Opportunity to use a biomass heat generator to support the heat pump, in periods when the PV system does not produce enough energy to meet the needs.

2. Literature Review

Some aspects related to the current European NZEB energy model towards its future positive energy implementation, its concrete methods of application through technologies and studies for new optimized materials, and the importance of an integrated design between building envelope and systems are presented briefly.

2.1. The New Energy Performance Target

Some notes on the NZEB model are here summarised as it currently represents the target to reach for the new buildings and for the restoration, when possible. The introduction of the NZEB (Nearly Zero Energy Building) model in 2010 represents a further development of the particular focus on energy consumption of the building stock since 2002 [1]. The construction sector continues to be under observation, as underlined by the European Commission Recommendation (2019) [2], which indicates the buildings, responsible for almost the 40% of final energy consumption, at the heart of the Union's energy efficiency policy. Therefore, the NZEB objective, initially conceived for new buildings, begins to be extended to existing buildings, in a long-term renovation strategy (Directive 2018/844/EU [3]).

The peculiarity of a NZEB compared to other buildings is the zero (or almost) energy balance between energy demand compared to the generation of energy from renewable sources [4]. Attention is also paid to the quality conditions of the indoor environments (IEQ-Indoor Environmental Quality). In fact, a NZEB should guarantee not only a very low energy consumption, largely covered by RES, but also a good level in terms of thermohygrometric comfort and air healthiness, as well as visual and acoustic comfort.

Another important aspect is the efficient integration of the building into a "smart grid". This transition can be supported by the insertion of control mechanisms (BCS—building control system) and the management (BMS—building management system) of the plant subsystems, for which the building can interact with the occupants and the electricity grid, for keeping the required internal conditions unchanged and limiting energy consumption. The smart grid would allow the integrated management of individual NZEBs, to realize an energy distribution network and, therefore, the organization of entire neighbourhoods in which some buildings could be identified as PEB (Positive Energy Building), buildings designed and sized to produce more energy than that necessary for their sustenance [5,6]. The current limited diffusion of this model could be attributed to two main reasons: absence of indications on the PEB targets, which should be formulated by the European Commission, and need for extensive adaptation of existing urban networks, to be ready to accept exchanges between the energy produced by the PEB and the user request. Anyway, in recent years, the PEB model has begun to be considered as a possible evolution of the NZEB target, jointly with the PED (Positive Energy District) one. With this last model, energy-independent districts could be built, where energy flows are exchanged between high-performance buildings and others unable to autonomously provide for their own energy support.

2.2. Technologies and Strategies to Support the NZEB Target

The most recent studies take into account both consolidate and innovative solutions for building envelope and systems: heavy structures, materials, plant systems already on the market used in an innovative way, increase in the use of plant regulation. However, it is difficult to generalize methods and approaches to outline indications on the use of advanced materials and of sustainable energy generation systems: indications both on materials and systems must be carefully evaluated according to the climate and the architectural characteristics of the building linked to the urban context in which it is located [5].

Often, well-known technologies are re-proposed in an innovative version and adapted to new needs, as they can be very useful for making the most appropriate choices. For example, a well-known technology, such as that of the Trombe wall, represents a passive solution for building envelopes that can lead to high-level energy performance [7].

Innovative wall configurations can limit the thermal losses whilst increasing solar heat gains to the heated spaces, also in cold climates [8]. The combination between the Trombe wall and the PCM thermal storage properties can allow obtaining good results to reduce thermal losses effectively.

The possibility of using sustainable materials in new buildings is particularly interesting. For example, the use of wood could find wider diffusion. However, technologies that provide for an important use of wood are not very well established, especially in areas where the wall thermal inertia is important for reducing the effects of solar radiation. Although the requirements of new buildings in the Mediterranean area appear particularly binding for multi-storey buildings with a massive employment of wooden materials, this technology can help to reach good results [9]. From dynamic simulation, it was assessed that a rational exploitation of the building thermal mass can be obtained by the use of wide dead band thermostats, in addition with triple pane glass windows, to obtain a good compromise between thermal losses and solar gains, and heat pumps coupled with a PV generator.

In all cases where a more accurate and in-depth analysis of energy performance is required, the use of dynamic simulation software appears indispensable, as it allows the analysis, in detail, of the behaviour of the structures and the possibility of heat storage in the daily cycle or in wider intervals. Furthermore, for a more adequate formulation and an accurate knowledge of the choices to be made to obtain the best energy performance of a building, it is necessary to proceed with cost assessments in parallel with the energy calculations and to perform combined analyses on energy-economic aspects.

The selection of the most appropriate simulation method is essential to limit uncertainty in the results and guarantee their accuracy. Usually the use of simplified models, instead of dynamic simulation, does not allow the analysis, in detail, of changes in internal conditions and to describe the thermal comfort conditions.

This problem becomes of particular importance when the aim is the management potential of the energy produced for achieving positive energy performance. In this case, parametric analyses can be useful [10] to identify the optimum combination of technologies through suitable technical and financial criteria, for example, based on:

- Fabric energy efficiency, by means of wall insulation and high performance windows;
- Low temperature heating (district heating and cooling network);
- Heat pumps, mainly geothermal units; and
- PV panels and battery storage

The technical analysis must be performed by means of the most appropriate software to determine the energy flows at the building and district level and key performance indicators, such as energy self-supply level by RES, emissions reduction and payback period can be calculated.

An element of comparison that still seems problematic is the value of the reference parameters for NZEB. From a wide investigation on 34 NZEB case study (with only two residential buildings) in hot and humid climates [11], the problem of the heterogeneous definition of the references for NZEB was highlighted: in some cases, the NZEB target can be achieved even just by a large use of renewable energy resources, despite a high EUI (energy use intensity) value. In fact, in a few cases the EUI was found higher than 200 kWh/(m^{2}y), even if a reference value, used for the comparisons, was the one considered by the New Building Institute (NBI) in the United States, EUI = 56.8 kWh/(m^{2}y). However, in general, energy efficiency should be a priority for a true NZEB, therefore referred to a low EUI, properly supported by renewable energy.

To deepen the knowledge of the NZEB effective functioning, extensive experimentation is required. In the literature, there are still few cases monitored, at least in the Mediterranean climate. The monitoring carried out on case studies usually is mainly used for dynamic simulation tools validation rather than for comparing design and real energy performance results.

A contribution to the energy data monitoring on district scale was given by the ECO-Life Project, cofounded by the EU Commission in the period 2010–2016, referred to urban districts, including new and existing refurbished buildings. From the real consumption data, in this case, a difference attributed to the occupants' behaviour was put into evidence [12,13]. The obtained data were normalized in order to compare different projects, at different periods, with different weather conditions, etc., and to analyse variations in a building's energy use. The calculations and results were based on monthly data, referring to the simplified quasi-steady state approach of the EPBD-calculation method. From the analyses, a large energy saving potential was put in evidence, concerning occupant behaviour and average energy use in new dwellings higher than expected from calculations. The use of a good management of technical systems can effectively solve some of the causes of high energy use.

2.3. Contributions to the NZEB Design

Among the different elements to consider in the design and management of a NZEB, the present research intends to highlight two important aspects, represented by both the microclimatic control, and the regulation of the building technical systems. It is important that, in the design phase, utmost care is taken to respect the objectives not only of energy saving and high energy performance but also of environmental comfort. Therefore, the project, although accurate and respectful of the NZEB objectives, should put the greatest attention to the continuous control that supports or replaces the user's behaviour and its building management.

An accurate design of the envelope and technical building systems associated with a smart management of the control systems and the setting of the set points, for the optimal operation of the systems, represent elements that need to be considered closely related.

Such integrated design can lead to a good coverage of the energy use for system operation by renewable energy that can be higher than 80%, as tested in the Czech Republic. An energy system for space heating and domestic hot water preparation in a family house, supported by a combination of PV photovoltaic system with a heat pump and a ground heat storage under the house, was designed and monitored for two years [14]: its management results support the belief that the NZEB target is not that distant and ambitious.

Combined energy systems seem promising to provide complete, optimized and achievable, cost-effective and sustainable solutions that will be increasingly needed in the near future. It is important to mention the importance of the design optimization that requires suitable methods to solve the design problems by means of multi-objective analyses [15].

The design of the energy systems should be analysed through a wide numbers of point of views, including the grid interaction, and different design alternatives such as fuel cells and renewable energy generation systems. The collected information can help for similar buildings [16]. In most cases, even if the overall PV energy yearly generation surpasses the electricity consumption and is fed into the grid, the building has to import from the grid a large energy amount and, therefore, it is not independent from the grid. Therefore, the evaluation of on-site energy storage systems is useful to improve the flexibility of the building to adapt to the load, reducing the dependence on the grid in low-generation times.

3. Materials and Methods

The software tools used for the analyses, the characteristics of the case study in relation to the envelope and systems and the climatic context in which it is located, and the input data of the calculations are described below.

3.1. Energy Simulation Tools

The dynamic energy simulation allows to acquire greater awareness of the detailed behaviour of the building structures, of the possibility of exploiting the thermal accumulation, depending on the climatic conditions and of the impact of solar radiation, all elements variable in the day-night, weekly, monthly cycle, seasonal. Above all, it is thus possible to obtain more in-depth information on the plant regulation and on the definition of the set points for their operation, in relation to internal needs and external climatic conditions. Moreover, contributions of elements such as the solar greenhouse can be evaluated taking into account the air flow and its temperature, from this space to the indoor environment.

Usually, the building design process uses simulation tools based on quasi-stationary models to determine the energy performance in summer and winter conditions and, therefore, to evaluate the energy needs of the building in the annual cycle and to verify the compliance with national regulations. The temperatures, the internal contributions, the expected consumptions refer to monthly balance and the calculations are therefore simplified.

It may be useful here to compare the results of the building simulation conducted during the design phase with these simplified models in a semi-stationary regime, with the one used for this study, which is based on a dynamic simulation model. Differences are expected, as indicated in several researches. Comparisons between steady-state calculation and dynamic simulation for energy efficiency have evidenced differences mainly due to temperature and solar radiation variations that can vary from 16 to 44% in cold zones [17].

The limits of the quasi-steady-state method, represented by the overestimation of thermal losses, are a recognized problem affecting the low energy consumption buildings and particularly NZEBs. In studies developed on buildings located in Mediterranean areas [18], the gap between models was found to be around 10% for heating energy needs, and lower for cooling energy needs.

The energy performance assessment is here developed following national and international indications, as required by the European Directive [1] and summarised in [19,20]. The software used for the dynamic hourly simulation of buildings sets the calculations in accordance with the EN ISO 52016-1: 2018 standard [21,22], and allows calculations for:

- Dynamic energy needs of the building for heating and cooling services;

- Analysis of summer comfort in free cooling conditions or with the systems support;
- Evaluation of the energy required to maintain the design temperatures for heating and cooling;
- Study of the contribution of a solar greenhouse for the energy needs reduction;
- Analysis of the night ventilation effectiveness; and
- Design and validation of fixed shields.

The results of the simulations will also be compared with the data collected from long-term monitoring both to confirm the possibility of describing thermal behaviour by means of simulation methods and to verify whether the regulation techniques adopted in practice in building energy management allow better exploitation of energy resources.

3.2. Case Study—Building Features

The case study is represented by a single-family residential building, built in 2013 in Northern Italy, characterized by a traditional architectural form, congruent with the hilly context in which it is inserted (Figure 1).

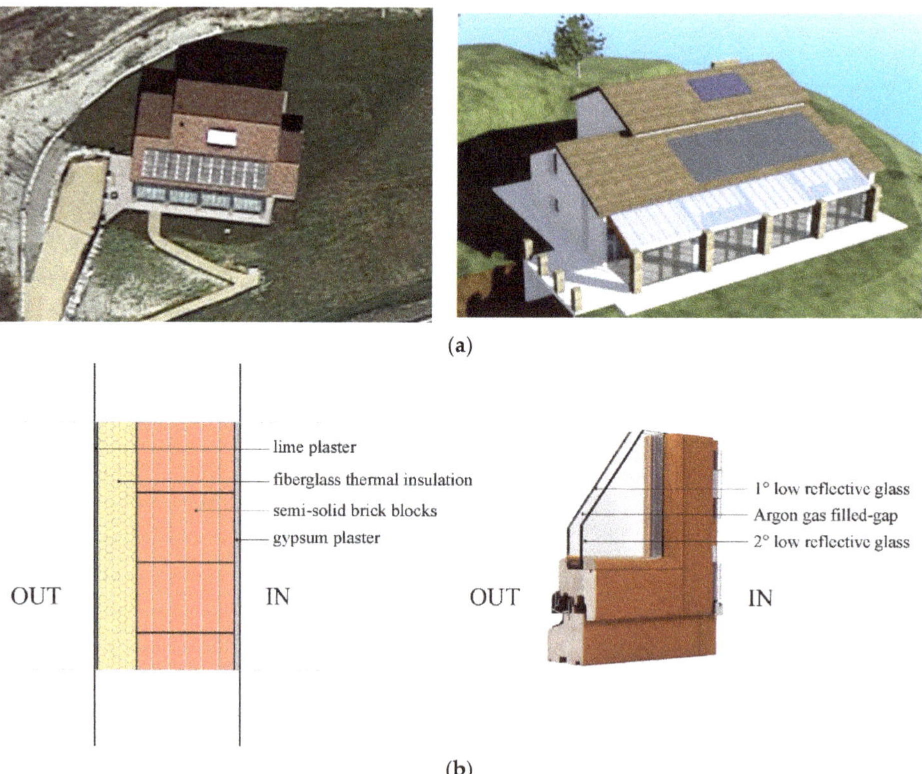

Figure 1. (**a**) Investigated building: overview from Google Maps (left) and lateral view (right); (**b**) stratigraphy of the main building envelope: perimeter wall (left) and low-emission double glazing (right).

The building, chosen as case-study, has attracted attention, not only at a regional level, by winning a regional call for its construction, but also nationally. It has been included in the census of the national NZEB through a summary and descriptive card produced by the "National Observatory of almost zero energy buildings" [23], promoted by the National Agency for New Technologies, Energy and

Sustainable Economic Development (ENEA). The building, together with 25 others located in 10 Italian regions, represents an example of good NZEB practices at national level [24], in particular for the energy performance of the building envelope (external insulation, wooden window frames with thermal break and double glazing), the high efficiency of the plant system (heat pump, CMV, air intake system from the solar greenhouse), and for the almost total coverage of consumption from renewable sources (94%).

Two particular aspects can be highlighted: the first concerns the fact that, despite being a high efficiency building, the construction is typical of the rural context in which it is built and is not characterized by modern lines, which would cause great visual impact on the context urban and landscape; the second relates to the fact that the project is prior to the national directives and laws that defined the NZEB model and its performance. However, the building respects this target and has even far better performances that could define it as PEB (positive-energy building).

The building has a useful area of 322.4 m^2 divided on three levels:

1. Basement (cellars, garage, bathroom, the only heated room on the floor, the thermal power plant);
2. Ground floor (living area); a mezzanine floor with two bathrooms and two bedrooms; and
3. First floor (main bedroom, bathroom and dressing room) and a mezzanine.

3.2.1. Opaque Elements

The structures used for the case study are mainly represented by 12 elements (structural or dividing walls, floor slabs, inter-floor and roofing). The design thermal transmittance U-values are compared with the reference U-values for NZEB, defined by the DM 26/06/2015 [25] (Table 1). As regards the periodic transmittance Y_{IE}, verification of the limit value is not necessary as the project site is characterized by an average irradiance value less than 290 W/m^2, in the month of maximum insolation.

Table 1. List of stratigraphic packages with thermal transmittance checks.

	Description	Udesign	U rif
		$(W/(m^2K))$	$(W/(m^2K))$
1	Earth retaining wall (thickness 35 cm)	0.68	0.30
2	Perimeter wall (thickness 52 cm)	0.20	0.30
3	Mezzanine floor partition (thickness 22 cm)	0.58	0.80
4	Partition heated rooms in the basement (thickness 28 cm)	0.31	0.30
5	Internal partition (thickness 10 cm)	2.05	0.80
6	Heated basement slab (thickness 81.3 cm)	0.12	0.30
7	Unheated basement slab (thickness 80.1 cm)	0.12	0.30
8	Ground floor slab (thickness 57 cm)	0.31	0.80
9	Mezzanine floor slab (thickness 52 cm)	2.90	0.80
10	First floor slab (thickness 50 cm)	0.26	0.80
11	Attic slab (thickness 38 cm)	0.24	0.25
12	Roof slab (thickness 33.3 cm)	0.23	0.80

From the comparison of the U values, the only opaque elements that do not meet the transmittance verification are the ones between an unheated environment and the outside (uninsulated masonry against the ground) or vertical and horizontal dividing elements between heated internal environments. Therefore, this failure does not lead to losses for the energy requirement calculations.

3.2.2. Transparent Elements

The highly efficient windows are chosen in line with the local architecture: they are composed of low-emission double glazing, with argon gas in the cavity, and thermal break oak frames.

The preferred orientation for natural lighting is the South; from the four large windows of the living area on the ground floor, it is possible to access the solar greenhouse, characterized by sliding doors in tempered glass with a folding system (open in summer and closed in winter) and a 45% inclined coverage (laminated glass), shielded by white roller blinds, to reduce the risk of summer overheating.

The three openings on the east side are not shielded, as they are protected by an overhang with an extension of 4 m; to the north, the rooms have four openings, which do not require shielding, given the lack of direct solar input; finally, the West side is characterized by three windows, one of which is double height, for which a mobile shielding system with white wooden venetian blinds has been provided, since it is affected by solar radiation in the second part of the day.

All the windows of the building envelope are verified as characterized by a thermal transmittance with a mean value of U_w = 1.35 W/(m^{2}K), lower than the limit equal to 1.80 W/(m^{2}K) in accordance with the Ministerial Decree 26/06/2015 [25].

A large glazed element built on the south-facing façade on the ground floor is represented by a large solar greenhouse. During the heating period, the white roller blinds, which shield the roof of the greenhouse, are rewound to promote maximum irradiation and the sliding windows, made of tempered glass, are closed to allow airtightness. The environment represents a passive indirect-gain solar system: the accumulated heat is used for heating the internal rooms through an air intake system from the greenhouse, which, by means of two 100 m^3/h fans, introduces into the area the air of the greenhouse when its temperature is 5 °C higher than the temperature of the indoor environment.

3.3. Case Study—Technical Systems

In the thermal power plant of the basement, there are almost all the systems, represented by:

- An air-water electric heat pump, with a useful thermal power of 13.1 kW and ON-OFF operation, which covers the role of generation system both for the radiant heating and cooling system and for production domestic hot water;
- A controlled mechanical ventilation (CMV) system, equipped with a heat recovery unit (90% efficiency), represented by an enthalpy exchanger operating continuously with an air flow that varies between 50 m^3/h and 550 m^3/h;
- A dehumidifier, which starts if the internal humidity conditions are such as to create discomfort (set-up value RH = 50%); it happens especially in summer, due to the high vapour content of the air which is cooled by the radiant system;
- A water storage tank (500 L), used to collect the hot water from the solar thermal system: the domestic hot water is heated by the heat pump only in cases where the solar thermal collectors are not able to work, due to lack of solar radiation.

In addition, on the ground floor, there is a secondary ventilation system that allows to transfer an air flow of 200 m^3/h from the solar greenhouse to the internal environment. A summary of the plant components, in a schematic representation of the most significant sections of the plant is reported in Figure 2. Although the solar greenhouse is not actually part of the plant system, it is included, as it plays a role in supporting the radiant heating system. The only system for generating the entire system is the heat pump, which exchange energy with the other components. For calculating the energy requirement in a semi-stationary regime, the building has been divided into five heated thermal zones, and three unheated zones (Figure 3, Table 2). Table 3 summarizes the characteristics of the radiant system (heating period: 15 October–15 April, 8.00–16.00, cooling period: 1 July–31 August, 8.00–16.00, CMV active 24 h a day throughout the year).

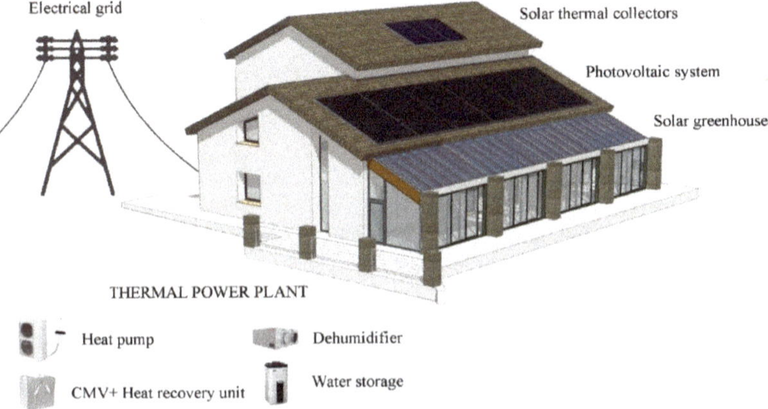

Figure 2. Schematic representation of the case study's systems.

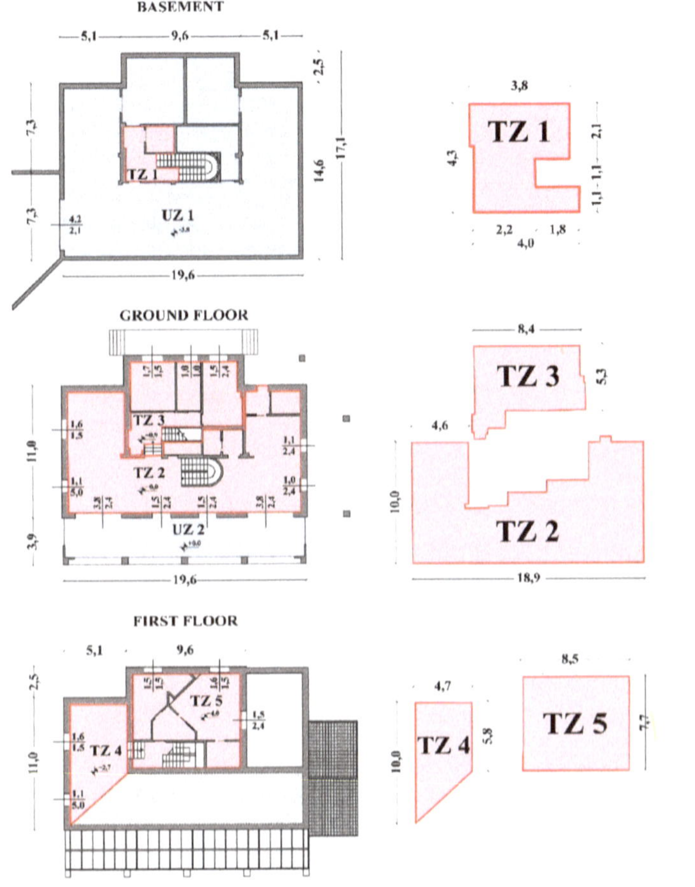

Figure 3. Heated thermal zones (TZ) (in red) and unheated zones (UZ) (in blue) (left: external opaque wall dimensions and windows; right: internal dimensions of the thermal zones).

Table 2. Subdivision of the building into heated thermal zones and unheated zones. (A = net surface; V = net volume).

Code	Heated Thermal Zones	A (m^2)	V (m^3)	Code	Unheated Zones	Area (m^2)	V (m^3)
TZ1	Bathroom (basement)	16.1	53.1	UZ1	Garage (basement)	252	875
TZ2	Living area (ground floor)	132.8	543.5	UZ2	Solar greenhouse (ground floor)	63.4	190.2
TZ3	Sleeping area (mezzanine floor)	38.8	104.8	UZ3	Attic (second floor)	68.3	93.6
TZ4	Loft (first floor)	36.6	84.6				
TZ5	Sleeping area (first floor)	51.7	139.6				

Table 3. Plants serving heated thermal zones (TZ) and location of radiant panels (floor/ceiling); CMV: I = immission, E = extraction.

	Heating-Cooling with Radiant Panels	DHW	CMV	CMV Flow (m^3/h)
TZ 1	1.3 kW (ceiling)	yes	E	50
TZ 2	6.6 kW (floor)	yes	I/E	200
TZ 3	3.1 kW (ceiling)	yes	I/E	100
TZ 4	4.0 kW (ceiling)	no	I/E	100
TZ 5	4.1 kW (ceiling)	yes	I	100 (I), 150 (E)

The contribution of renewable sources is represented by a photovoltaic system (about 79.5 m^2) and a solar thermal system (6.6 m^2), placed on the 45% inclined surface with respect to the horizontal, with a South orientation. The LED lighting is used both for external and internal use. A recovery rainwater system helps to cover the need for the irrigation of the surrounding garden (2800 m^3/year).

The design of the building envelope and the technical systems sizing have been set synergistically: the management of the building is entrusted to an appropriate control system (BCS) and management system (BMS), capable of maintaining the required conditions of internal comfort, with reduced energy consumption. A climatic control unit acquires the temperature and humidity data in each room and the ones coming from an external climate probe, located on the north side of the building.

Furthermore, the free-cooling technique is used to control energy consumption in summer: it permits a reduction in the summer night temperature of the rooms. A fan is activated for external temperatures of 24 °C or lower. The external air intake is located near the northeast corner of the building, and the air distribution takes place separately for each environment, thus promoting a decrease in the internal temperature of up to 4 °C.

3.4. Case Study—Environmental Design Data

The following data were used in the calculation:

- Degree days: 2631 DD;
- Italian climatic zone E (heating period. 183 days);
- Minimum design temperature for heating system: −9.2 °C;
- Internal conditions—heating period: Temperature = 20 °C, Relative Humidity = 50%; and
- Internal conditions—cooling period: Temperature = 26 °C, Relative Humidity = 50%.

Monthly mean temperature (T), relative humidity (RH) and irradiation values for each orientation are resumed in Table 4.

Table 4. Climatic data: solar irradiation.

	T (°C)	RH (%)	Monthly Mean Irradiation (MJ/day)								
			N	NE	E	SE	S	SW	W	NW	Oriz.
January	6.5	79.4	1.7	1.9	4.1	7.1	9.1	7.1	4.1	1.9	5
February	7.5	79.2	2.6	3.2	6.1	9.1	10.9	9.1	6.1	3.2	7.9
March	10.2	71.2	3.7	5.4	8.8	11.1	11.7	11.1	8.8	5.4	12.1
April	13.3	69.9	5.3	8.1	11.3	11.9	10.7	11.9	11.3	8.1	16.4
May	16.4	71	7.5	10.4	12.9	11.9	9.6	11.9	12.9	10.4	19.6
June	20.5	68.4	9.1	11.9	14	12.1	9.4	12.1	14	11.9	21.7
July	23.1	65.6	9	12.7	15.6	13.7	10.5	13.7	15.6	12.7	23.8
August	23.2	66.7	6.3	9.8	13.2	13.1	11.1	13.1	13.2	9.8	19.4
September	20.9	74	4.2	6.7	10.4	12.2	12.1	12.2	10.4	6.7	14.5
October	15.7	77.6	2.9	3.9	7.3	10.4	12.1	10.4	7.3	3.9	9.6
November	11.5	80.5	1.9	2.2	4.5	7.5	9.4	7.5	4.5	2.2	5.6
December	7.9	82.5	1.5	1.7	3.9	7.2	9.4	7.2	3.9	1.7	4.6

4. Results

4.1. Heating: Quasi-Static and Dynamic Simulation vs. Real Energy Consumption

Figure 4 shows the difference between the primary energy requirement for heating, calculated using the semi-stationary method with monthly balances, deriving from the calculation in the hourly dynamic regime. The energy consumption data obtained from the monitored data are also reported. These are, overall, 35% lower than the estimate carried out by hourly dynamic simulation: the dynamic model was calibrated on the basis of the ventilation data recorded by the building monitoring system.

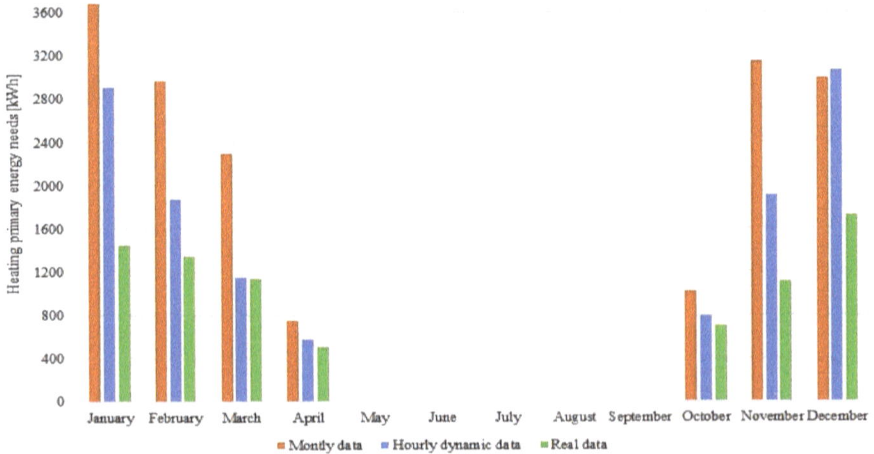

Figure 4. Hourly dynamic, monthly semi-stationary calculated data and real primary energy needs.

As already highlighted in several studies, the results obtained through quasi-stationary simulation models are different from those that can be obtained from a more detailed study through dynamic simulation models. The difference may be due to a more or less accurate knowledge of the characteristics of materials, construction techniques and systems. Furthermore, it is usually difficult to provide a precise description of the operating and management conditions of the system set points.

A comparison between different models [26] has shown that the quasi-steady state approach and the dynamic simulation methods can give differences in energy need calculations for heating in winter months up to 25.8%. Reasons of these discrepancies can be found considering some elements that are not always well known, such as occupant behaviour, ventilation, temperature set point, but also calculation methods and simplifications. Depending on climatic conditions, the differences between methods and real consumption data can be positive or negative: solar gains, for example, can determine opposite effects causing underestimation or overestimation of the energy needs [27].

The results of the semi-stationary simulation (which significantly overestimates real consumption) of the case-study, compared with the reference values by national rules, are indicated in Tables 5 and 6. The comparison between design values and reference values highlight the excellent performance of the building, better than the NZEB target. In the Energy Performance Certificate, energy class A4 (the maximum rate) and the NZEB status are attributed to the building.

Table 5. Geometric parameters and energy performance indices.

Parameters	Design Values	Reference Values [25]	Difference
Gross dispersing surface S	370.8 m^2		
Building heated volume V	1450.5 m^3		
Shape ratio S/V	0.26 m^{-1}		
Global average heat transfer coefficent H'$_T$	0.25 W/(m^{2}K)	0.75 W/(m^{2}K)	67%
Equivalent solar area/Floor area A_{sol}/A_f	0.029	0.030	3%
Energy performance indicator for heating EP$_{H,nd}$	46.32 kWh/(m^{2}y)	50.37 kWh/(m^{2}y)	8%
Energy performance indicator for cooling EP$_{C,nd}$	12.56 kWh/(m^{2}y)	16.34 kWh/m^2 y)	23%
Global non-renewable energy performance index of the building EP$_{gl,nren}$	1.00 kWh/(m^{2}y)	36.12 kWh/m^2 y)	97%
Total energy performance index of the building EP$_{gl,tot}$	47.69 kWh/(m^{2}y)	100.05 kWh/m^2 y)	52%

Table 6. Efficiency parameters.

Parameters—Renewable Energy	Design Values	Reference Values
Share of total renewable energy QR total	98%	50%
Share of renewable energy destined for the production of domestic hot water QR DHW	99.4%	50%
Photovoltaic system power	10.33 kW	7.61 kW
Parameters—Global average seasonal performance	**Design Values**	**Reference Values**
Heating performance $\eta_{gl,H}$	91.9%	70.6%
Cooling performance $\eta_{gl,C}$	325.4%	202.5%
DHW performance $\eta_{gl,W}$	96.7%	68.5%

4.2. Dynamic Simulation—Energy Performance

The main data used in the model are the following: set point temperatures for the heating and cooling periods, geometric parameters, an accurate description of the shading of the windows and the definition of hourly profiles for internal contributions, the ventilation flows, and the powers provided by the heating and cooling.

During the heating period, and even more for the cooling period, only a small amount of the power available from the radiant system is actually used, as it is not necessary for reaching the set point temperature. The living area (TZ2) is the only thermal zone for which up to 92% of the available thermal power has been used, given its large volume and useful surface.

It may be useful to highlight the percentage distribution of power (Figure 5), used for heating (in red) and cooling (in blue): the heating system mainly operates for powers ranging from 20% to 70%, while it never works for 100% of the available power, as the minimum site design temperature (−9.2 °C) is rarely reached.

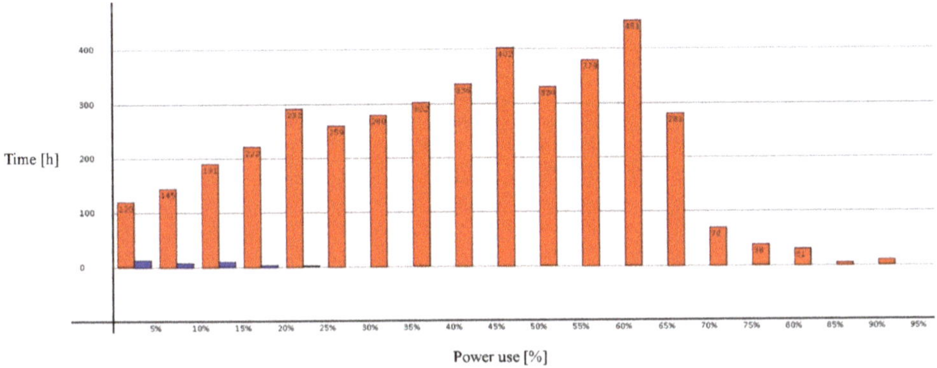

Figure 5. Percentage distribution of power used by the heating and cooling systems in the living area.

The heating system works with the available power of 65% (which corresponds to the maximum number of operating hours, or 451) only for 19 days out of the 183 that make up the heating season. The cooling system, on the other hand, operates with available powers of the order of 10%, confirming the fact that the thermal zone does not require much cooling energy to reach or maintain the summer set point temperature.

The sleeping area on the first floor (TZ5) requires much less energy than the living area by using about 10% of the available power most of the time (Figure 6).

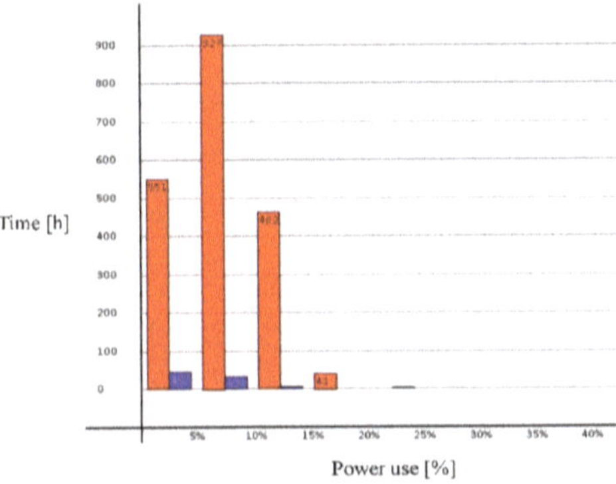

Figure 6. Percentage distribution of the power used for heating and cooling systems in the sleeping area on the first floor.

4.3. Solar Greenhouse: Simulation and Temperature Analysis

The solar greenhouse is a very important element in winter, as it allows a reduction in the thermal load of the space heating system. To maximize its efficiency, numerous elements must be considered in its design and management, such as exposure, optical properties and thermal capacity of the opaque surfaces, the amount of ventilation and of the shading devices [28]. The same elements are considered particularly important in the conclusions of a research developed through simulation models and experimentation on a prototype [29]. In fact, the need to avoid any summer overheating, to carry out in-depth calculations on the effects of thermal inertia and to provide for the control of microclimatic conditions through intelligent management of the CMV is highlighted.

The case study building uses the solar greenhouse during the heating period, from 15 October to 15 April, as support for the radiant heating system of the living area, with which it is in direct communication. The air intake system from the greenhouse to the internal environment is active from 11.00 to 14.00, when the temperature of the air inside the greenhouse reaches at least a temperature of 25 °C, i.e., when it exceeds by 5 °C the design temperature of the air inside the living area (winter set point of 20 °C).

It may be interesting to analyse the behaviour of the greenhouse as a function of time. Hourly data of indoor and greenhouse temperatures were simulated depending on external temperature. The results have been examined monthly. The detailed results for January are indicated in Figure 7, while a synthesis is presented for the other periods.

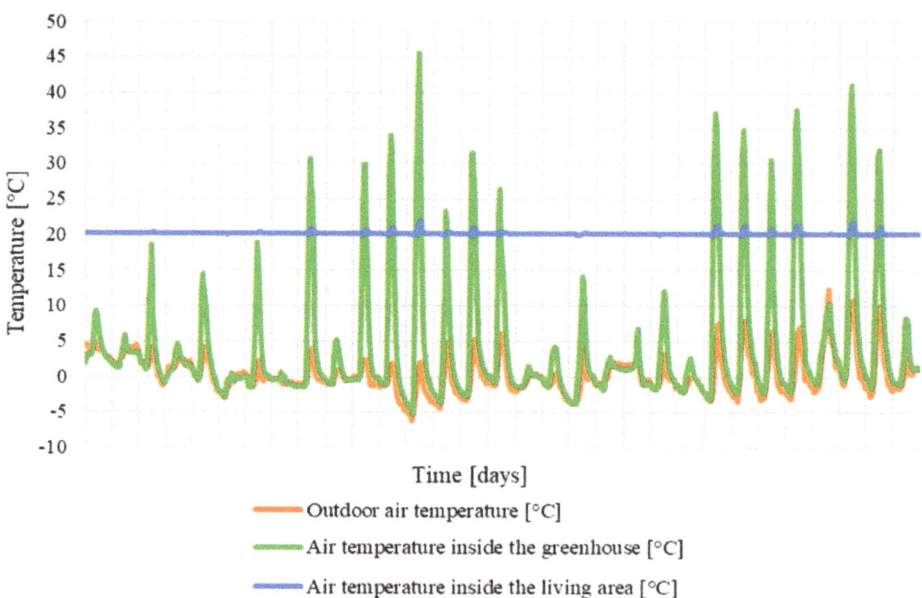

Figure 7. Trend of outdoor air temperatures (orange), inside the living area (blue) and the greenhouse (green) for the month of January.

In the hottest hours, indicatively from 11.00 to 13.00, solar radiation causes an increase in the temperature of the greenhouse with peaks up to 45 °C. The temperature of the internal environment (set point = 20 °C) remains constant for most of the time, allowing the reduction of the operating time of the heat pump and in some cases there is an increase that remains contained inside appropriate values to maintain comfort conditions in the indoor environment.

It may be interesting to observe how long the temperature remains around a certain value in the greenhouse: it is observed that the recurrent temperature is 0 °C (Figure 8), for 21% of the time. The fan that transfers the air from the solar greenhouse to the internal environment works when the temperature of the greenhouse reaches 25 °C and this occurs 15% of the time (37 h).

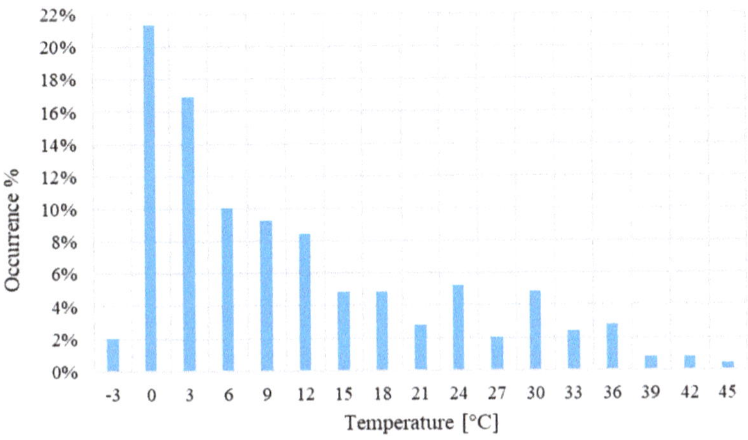

Figure 8. Percentage of time a certain temperature occurs in the greenhouse.

The results obtained for the complete heating period are summarized in Table 7, which shows the maximum and minimum values of the external temperature and of the greenhouse for each winter month. The following columns for each month indicate the percentage of time in which:

(A)   The temperature remains around the most recurrent value outside;
(B)   The most recurrent temperature value in the greenhouse occurs; and
(C)   The air intake system from the greenhouse is activated.

Table 7. Analysis of the contribution of the greenhouse during the heating season.

Month	Temperature (°C)				(A) Outdoor		(B) Greenhouse		(C) Greenhouse		(D) Greenhouse	
	Outdoor		Greenhouse									
	min	max	min	max	(°C)	(%)	(°C)	(%)	(h)	(%)	(h)	(%)
January	−6	12	−3	45.5	0	21	32	14	37	15	54	22
February	−6	15	−2	46.5	3	12	26	11	75	28	106	39
March	−3	21	0	47	18	10	34	9	151	42	193	54
April	0	20	2	42.5	24	11	39	10	82	44	111	59
October	1	21	4	46	18	12	27	11	62	37	89	54
November	−2	17	−1	39	9	18	32	17	36	15	56	23
December	−11	10	−7	38	6	16	32	16	31	13	46	20

(A) Outdoor: Temperature inside the greenhouse for the most recurrent interval [Tmin; Tmax]; (B) Greenhouse: Temperature inside the greenhouse for the most recurrent interval [26 °C; Tmax]; (C) Greenhouse: Operation of the air intake system from the greenhouse for the interval [26 °C; Tmax]; (D) Greenhouse: Operation of the air intake system from the greenhouse for the interval [21 °C; Tmax].

Furthermore, in the last section (D) the hypothesis of changing the set point for activating ventilation from the greenhouse is considered, bringing it to T = 20 °C. The suction of air from the greenhouse would therefore be activated when the temperature of the greenhouse reaches that of the internal environment. In this way, the hours in which the release of air from the greenhouse is activated would go from 474 to 655, with an increase of 38%. This solution could allow a further reduction in consumption due to the presence of the solar greenhouse and the circulation of heated air in the internal environment.

4.4. Analysis of the Heat Pump Operation

The operation of the heat pump, supported by the PV system, is analysed in relation to the internal/external temperatures of the summer months in which it is used for cooling (July/August). The heat pump allows cooling the water that is circulated in the radiant panels with a flow temperature of 18 °C. The return temperature is 25 °C. The goal of the system in the summer season is to maintain an air set point temperature in the environments of 26 °C.

The study focuses on the results of the hourly dynamic simulation of the living area (TZ2) and the sleeping area on the first floor (TZ5) during the two months of activation of the cooling system (July and August). In particular, the temperatures of the external, internal air and that of the external envelope surfaces (perimeter walls and windows) are compared.

From the trend of the external and internal temperatures of the two thermal zones for the month of July (Figure 9), it is noted that the external temperature exceeds the summer set point (26 °C) for 282 h out of 744 total hours, with peaks up to at 37.5 °C.

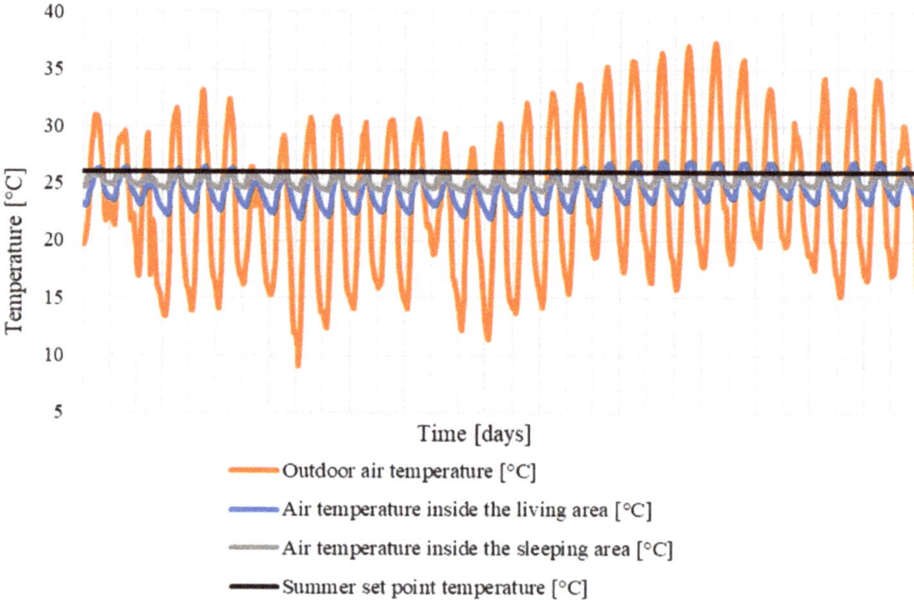

Figure 9. Trend in outdoor air temperatures, inside the living and sleeping areas in July.

Analysing in detail the temperature trends of the TZ2 and TZ5, it is observed that the set point temperature is exceeded in TZ2 for 111 h, while it is never exceeded in TZ5. Therefore, it can be said that the heat pump, which is activated for T ≥ 26 °C, worked for 111 h in July (about 5 h a day): in fact, it can be seen from the graph that the set temperature point is exceeded for 21 days out of 31 total. In this period, the cooling energy supplied by the heat pump is 350 kWh: the corresponding cooling power used (3.2 kW) can be compared with the useful cooling power of the heat pump (13.1 kW). From the comparison, it can be said that the cooling system in July needed only 24% of the useful power of the heat pump. With the same approach, for the month of August, the power used is 27% of the useful one.

4.5. Photovoltaic System

The availability of experimental data provided by the designer made it possible to conduct a comparison, extended for a period of five years (from June 2014 to June 2019), between the electricity consumed to meet the energy requirement and the production of electricity from the photovoltaic system. In addition, an economic calculation was developed to compare the economic cost of the energy required from the network with the profit obtained from the sale to the network of the energy surplus not intended for self-consumption.

Starting from 2011 with the publication of Legislative Decree n. 28 [30], attention was paid to promoting the use of energy from renewable sources. In fact, national indications require that at least 50% of the demand for domestic hot water, heating and cooling must be satisfied using renewable energy sources.

In compliance with the decree, the building under study has a photovoltaic system integrated into the lower pitch of the roof, with a net surface area occupied by the modules of 79.46 m^2, which produces an overall peak power of 10.33 kWp, 30% higher than the required reference power. From the comparison between electricity consumed and produced (Figure 10) it is clear that the electrical coverage by photovoltaics is guaranteed for almost the whole year, except for the winter months (December and January), for which it is necessary to obtain a modest amount of electricity to the grid (up to 320 kWh). The surplus of energy produced is transferred to the grid.

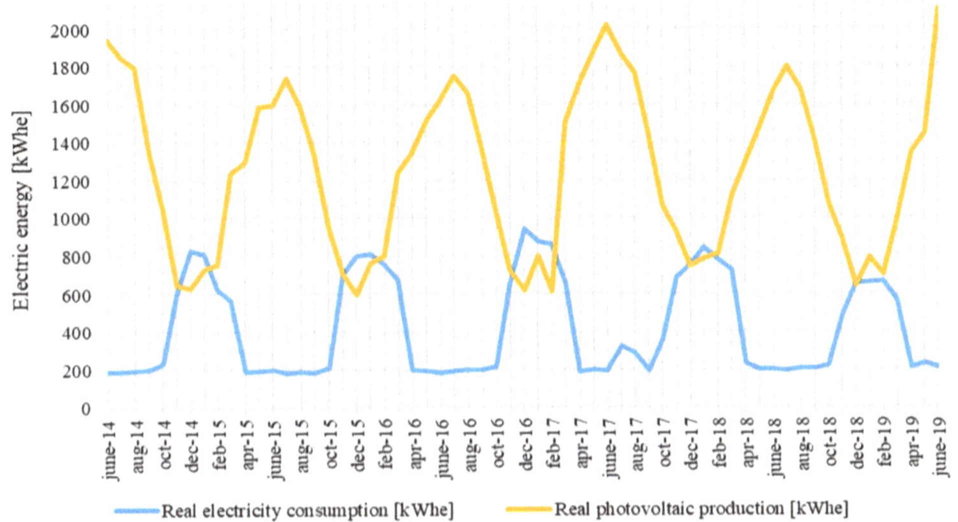

Figure 10. Comparison between real electricity consumption and photovoltaic production for the period June 2014–June 2019.

The comparison between the demand and the sale of energy is conducted considering the costs of electricity requested from the network in the winter months and the share sold to the network, when the plant produces more than it consumes. Reference is made to the following economic costs:

- Electricity drawn from the network: 0.1917 euro/kWh; and
- Electricity fed into the grid: 0.543 euro/kWh (all-inclusive tariff paid in 2013 to integrated photovoltaic systems with innovative features)

Table 8 shows the values of electricity consumed and produced for the five years analysed. From the comparison of the overall share of electricity consumed with that produced by the photovoltaic system, it can be said that only 34% of the energy available from photovoltaics was used for self-consumption.

Table 8. Energy data on electricity consumption and photovoltaic (PV) production for five years (June 2014–June 2019).

Year n.	Electricity Consumed	Electricity Produced by PV System	Surplus of Electricity Fed into the Grid		Electricity Purchased from the Grid	
	(kWh)	(kWh)	(kWh)	(€)	(kWh)	(€)
1	4.833	14.907	10.352	5.621	278	53
2	5.123	14.212	9.338	5.071	249	48
3	5.378	15.296	10.558	5.734	640	123
4	5.596	15.344	9.808	5.326	60	12
5	4.723	16.563	11.854	6.437	14	3
Tot.	25.653	76.322	51.910	28.188	1.241	239

Consequently, the amount of energy surplus is obtained from the difference between the third and second columns of the table. This surplus can also be estimated at an economic level: the economic return determined by the sale of electricity during the five years appears clearly higher than the small expenditure during the winter months in which the plant is not subject to adequate solar radiation to cover the energy needs.

5. Design Improvements Analyses

Building systems management and regulation have been analysed by means of simulation and monitoring data. The results highlight the good energy performance due to a smart design of both building envelope and systems and depending on an accurate management of the thermal systems operation. Some aspects of the project are evaluated below to highlight any further improvement opportunities. This procedure should be included in each project to evaluate the possibilities of obtaining more ambitious results than the target NZEB achieved.

Some features of the project are discussed below to highlight any further improvement opportunities. This procedure should be included in each project, to evaluate the possibilities of obtaining more ambitious results than the target NZEB achieved. Obviously, the peculiarities of each project mean that there are no indications generally applicable to everyone, but that more suitable alternatives must be identified case-by-case. In the analysed case study, different variants have been taken into consideration and only those considered most significant are discussed.

5.1. Improving Environmental Comfort

Three significant parameters for environmental comfort can be managed by a building technical system: temperature, humidity, and velocity of the air [31].

As indicated, the system has control of the temperature (t) and relative humidity (RH) in each zone with set points respectively set at t = 20 °C and RH = 50% in winter and t = 26 °C and RH = 50% in summer. The temperature in winter conditions may not be completely satisfactory in terms of comfort. For this reason, a comparison was made between the current conditions and those that would be achieved for a winter set point temperature set at 22 °C, both in terms of energy consumption and comfort. The evaluations were carried out assuming to maintain the constant temperature in the day-night cycle, and considering a maximum deviation of ±0.5 °C for the calculation of the deviation from the comfort conditions [32].

The comparison between the simulation results provides a 15% increase in the maximum power used which goes from 7.3 kW to 8.6 kW and a 36% increase in the energy used for heating, corresponding to approximately 7000 kWh. The monthly energy requirement that varies mainly as a function of the external temperature and solar contributions (Figure 11) shows a higher percentage increase in the less cold months.

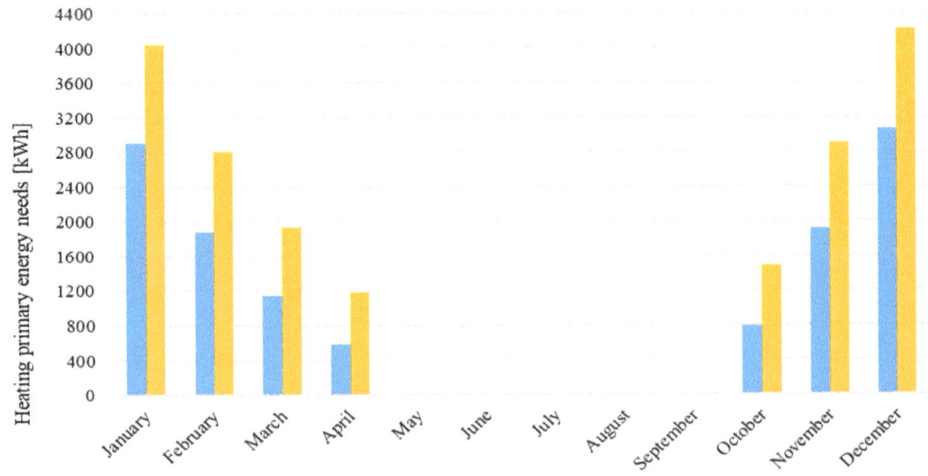

Figure 11. Comparison between the monthly heating primary energy needs for a set point of 20 °C (blue) and 22 °C (yellow).

The overproduction of energy from the photovoltaic system during these months, which are characterized by a greater availability of solar energy (March, April, October and November), can be used to power the heat pump and, therefore, to compensate for the greater energy requirement. During the other months of the heating season (January, February, and December), the photovoltaic system is not, however, able to cover the electricity needs, which, therefore, must be requested from the grid. The increase in demand entails an increase in costs, which, however, can be well compensated in the annual budget.

As regards the improvement of internal comfort, the difference between the two conditions in terms of deviation from the comfort condition can be assessed. From the dynamic energy simulation, the time in which the internal temperature underwent variations outside a range set at ±0.5 °C was calculated for each thermal zone (Table 9).

Table 9. Comfort hours for 20 °C and 22 °C temperature set point, in each thermal zone (TZ) referred to in Figure 3 (Δt = temperature difference).

Thermal Zone	Set Point	Time [h] Comfort		Time [h] Discomfort (Cold)		Mean Δt (Discomfort, Cold)	Time [h] Discomfort (Hot)		Mean Δt (Discomfort, Hot)
	(°C)	(h)	(%)	(h)	(%)	(°C)	(h)	(%)	(°C)
TZ1 Bathroom	20	6391	73	2363	27	1.4	0	0	0
	22	7176	82	1584	18	1.2	0	0	0
TZ2 Living area	20	723	8	8037	92	1.4	0	0	0
	22	6405	73	2355	27	0.8	0	0	0
TZ3 Sleeping area	20	4936	56	3823	43	0.8	1	0	0.3
	22	8246	94	490	6	0.5	24	0.3	0.3
TZ4 Loft	20	2714	31	6046	69	1.1	0	0	0
	22	7922	90	838	10	0.6	0	0	0
TZ5 Sleeping area	20	3497	40	5258	60	1.2	5	0.1	0.2
	22	8205	94	535	6	0.5	20	0.2	0.3

In the living area (TZ2), the hours of comfort go from 8% to 73%, while, for the two sleeping areas (TZ3 and TZ5), an even better condition is reached (94%) as regards the feeling of cold, while the number of hours in which the temperature is above the defined interval increases; even if in percentage terms it remains a negligible value.

The simulation was conducted assuming a fixed day-night set point and, from what has been observed in terms of comfort, the possibility of a different day-night set point could be considered, at least in areas TZ3 and TZ5 to improve comfort. This choice would also help to reduce the corresponding energy consumption.

5.2. Impact of the Solar Greenhouse

A peculiarity that makes the architectural form of the building homogeneous and uniform to the context is the solar greenhouse, a characteristic element of the biocompatible architecture, characterized by a glass envelope both on the wall and on the roof and by an optimal south exposure for capturing solar radiation.

The main goal of the solar greenhouse is to provide heat to the indoor environment in winter to reduce energy needs. The comparison between the energy requirements, calculated in the current configuration (with greenhouse) and without, highlights the importance of its contribution. In addition, the results are compared with those referring to the reference building corresponding to the NZEB standard for this building.

The presence of the solar greenhouse offers a positive contribution from an energy point of view (Figure 12), resulting in an overall decrease of 13% of the primary energy for heating compared to the configuration without the greenhouse, corresponding to a saving of 1550 kWh.

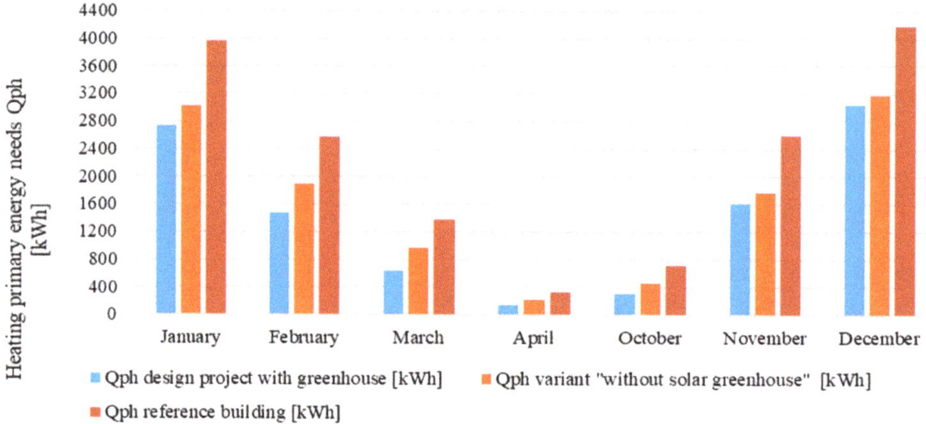

Figure 12. Comparison between the heating primary energy needs of the design building equipped with a solar greenhouse with the variant "without solar greenhouse" and the reference building.

For each month, the percentage reduction of the energy requirement of the project configuration compared to the variant is indicated. However, it can be observed that the energy requirement of the building without a greenhouse is always below the corresponding requirement of the reference building. Thus, even in the absence of a solar greenhouse the building would have been classified as NZEB.

The presence of the solar greenhouse leads to a decrease in the thermal performance index useful for heating, thanks to the free supply of heat offered by the thermal storage in winter: with reference to the main energy performance indices [25], the percentage difference between the configuration with the solar greenhouse and the one without can be calculated (Table 10).

Table 10. Energy performance indices of the configuration with solar greenhouse compared in percentage with those of the variant without solar greenhouse.

Parameters	% Difference between Configurations with and without Solar Greenhouse
Energy performance index of the envelope $EP_{H,nd}$	−11%
Global non-renewable energy performance index of the building $EP_{gl,nren}$	−17%
Total energy performance index of the building $EP_{gl,tot}$	−4%

5.3. PV Energy Storage System

Since, for a limited period of time in winter, the building needs electricity from the grid, the possibility of realizing the complete autonomy of the building and, therefore, making it stand-alone rather than grid-connected was evaluated through the use of an energy storage system.

It is assumed to use stored energy for an operating time of 3 h of the heat pump at full capacity during the winter months of photovoltaic underproduction (15 kWh). The cost of the system was estimated at around €10,000 in relation to the current market offer.

To assess how long the cost of the investment could be recovered, Year 3 (Table 8) is taken as a reference for consumption, as it is the period in which the greatest demand for electricity was detected. In December, January and February the electricity consumed (2682 kWh) exceeded the energy produced by the system (2042 kWh) for 640 kWh, which was taken from the network and which, in this case, is considered to be supplied from the energy storage system.

It is assumed that the storage batteries must supply energy for a period of 30 days (distributed 15 days in December 5 in January and 10 in February), for 3 h a day (from 19.00 to 21.00). This estimate was made considering in that time interval the absence of solar radiation, and the possibility of exploiting at other times both the contribution of the solar greenhouse and the energy from PV to recharge the accumulation for 749 h (248 in January, 269 in February and 232 in December).

The calculations show that the investment is not favourable in this context, due to the high energy performance of the building which ultimately requires energy from the network with an annual cost of approximately €120. The cost of storage does not seem reasonable, at least in the current context, also taking into account the battery life due to the charging and discharging cycles.

As an alternative to the energy storage system, it would be interesting to consider making high thermal capacity radiant screeds from the design stage: thanks to the high coefficient of thermal conductivity, this system could optimize the performance of the entire heating/cooling system with flooring favouring a homogeneous and gradual transmission of heat in the indoor environment.

5.4. Opportunity to Cover Part of the Winter Energy Needs with a Biomass Heat Generator

Another possibility of making the building autonomous from the network, the insertion of a solid biomass boiler (pellet), to support the heat pump, is analysed, referring to the coldest winter period, in which the PV system does not completely cover the need for hot water for both the radiant heating system and for sanitary use (Year 3).

The analysis of the heat generator systems led to the choice of an automatic loading biomass heat generator with a nominal useful power of 9 kW, to be installed in the basement, where the remaining part of the equipment is located.

For this type of heat generator, an automated system is provided, which allows the continuous and autonomous supply of fuel without the intervention of an operator. Indicatively, it can be estimated that the cost of a boiler of this type is around €4000, excluding costs for the chimney and for the electrical connections, which may be absorbed in construction costs.

The demand for electric energy from the network, equal to 640 kWh, assuming a performance coefficient of 2.5, corresponds to a need for thermal energy equal to 1600 kWh. In the event of a generation efficiency of 0.75, and the calorific value of the pellet equal to 4.9 kWh/kg, 435 kg of biomass (pellets) would be required, corresponding to an indicative cost of €116.

This estimate shows that the annual fuel expenses would be slightly lower than the annual cost of electricity required on the grid during the months of underproduction of the photovoltaic system. However, it is necessary to take into account the initial cost of the generator, the periodic maintenance costs (cleaning, smoke analysis), which have not been analysed as in any case they would not be recovered in an adequate time.

Although it is not convenient, from the economic point of view, to face the investment in a biomass generation system, in support of the heat pump, this solution would still be more advantageous than the previous one, as it is economically more convenient and better manageable, even in conditions other than those assumed.

The integration would be interesting to pursue the building's objective of energy autonomy from existing urban networks, with a better management possibility than that assessed for the electricity storage system. The system would, in any case, receive a contribution from renewable sources and, therefore, have minimal impact on the environment.

6. Discussion

Some considerations have been summarized here, deriving both from the study of the current conditions of the building, through simulation and monitoring, and from the evaluation of the proposed design alternatives:

1. The calculations carried out through semi-stationary simulation, according to national and international regulations, have shown that the building has the NZEB energy classification. These data, compared with those obtained from the dynamic simulation, highlighted, as expected, differences that in some months were also significant (about 45%). Only in December, the dynamic simulation gave higher consumption, probably caused by the hourly variation of temperatures. The comparison with the actual consumption data indicates that, although the model was calibrated according to the actual ventilation rates, the dynamic calculation was able to accurately describe the situation with an error more evident in the two coldest months. Probably the thermal inertia of the structures and the contribution of the solar greenhouse played more important roles in keeping the internal environment at the set temperature.

2. The use of dynamic simulation made it possible to analyse in detail the power used over time and verify that the system has been correctly sized for each thermal zone. Furthermore, the hourly description of the behaviour of the solar greenhouse made it possible to evaluate the opportunity to change the temperature set point for activating the fan between the greenhouse and the internal environment. In this way it is possible to obtain greater advantages in terms of the solar energy use for heating indoor air in the winter months. The importance of the solar greenhouse is also demonstrated through the comparison between the configuration without and with its presence.

3. The comparison between the data of energy consumption and energy produced by the photovoltaic system shows that the electrical coverage by photovoltaics is guaranteed for almost the whole year. Only in two months of the year is it necessary to use electricity from the grid, but overall electricity consumption represents only 34% of the energy produced over the whole year. Currently, in the absence of an energy storage system, the energy surplus by the photovoltaic system is reintroduced into existing urban networks. It would be interesting in the future to consider the possibility of connecting the building to a local smart grid, capable of redistributing the surplus of energy to other buildings connected to the system, to cover an energy requirement corresponding to about four times that of the case study.

4. The analysis of the internal thermohygrometric conditions indicates that the internal comfort can be improved in winter by setting an internal set point temperature of 22 °C instead of 20 °C. Even if the increase in energy consumption is equal to 36% for heating, it can be largely covered

by the summer overproduction in the annual budget. A good level of comfort is maintained for a much longer time, especially in the living area, but also in the two sleeping areas.

5.   In the analysis of the building project, two aspects that deserve further attention emerged and, therefore, were subject to verification: the possibility of energy storage was calculated by demonstrating that in the specific context the solution is not economically convenient. The support of a biomass generator to cover the winter needs, when the heat pump powered by photovoltaic energy is not sufficient, is equally not convenient from an economic point of view.

7. Conclusions

Some peculiar elements in the design and management of a NZEB have been analysed, mainly represented by microclimatic control, technical systems management, and regulation. Through the dynamic simulation, some exemplary results have been highlighted, considering the following aspects:

- The management system of the solar greenhouse and its energy contribution in winter;
- the operation of the heat pump, supported by the PV system, in summer;
- the almost total coverage of the building's energy needs in the annual cycle, thanks to the production of energy from the photovoltaic system;
- the potential PEB qualification of the building thanks to the surplus of energy produced which is fed into the network;
- the effective advantages, in terms of environmental comfort, due to the regulation of the internal set point temperature, with a consequent extra consumption, limited and covered by the overproduction of PV energy;
- the winter energy supply of the solar greenhouse, in relation to a smart management of the air flows from it to the internal environment and possible operating alternatives;
- the use of storage batteries for the energy produced by the PV system; and
- the use of a biomass heat generator to support the heat pump during the winter season to recover energy from the electricity grid.

The analysis of the case study is emblematic of an integrated approach, to be applied for all new buildings, which must take into account multiple aspects and alternatives to allow maximum effort to reduce energy consumption, avoiding unjustified extra costs and optimizing the integration of RES. It should represent the method to apply in every NZEB project to obtain optimal results.

The NZEB design, however, should not represent the goal, but the starting point: in fact, this model tends to create highly efficient buildings, but isolated from the urban context in which they are located, mainly characterized by existing buildings with a dispersing building envelope and an inefficient plant system. The further step forward, represented by PEB, could be evaluated technically and economically, to share its energy surplus to other buildings through the connection to a distribution network of a smart energy grid.

Author Contributions: Conceptualization, A.M. and G.L.; methodology, A.M.; software, validation, and analyses G.L.; discussion, A.M. and G.L.; writing—original draft preparation, G.L.; writing—review and editing, A.M. All authors have read and agreed to the published version of the manuscript.

Funding: This research received no external funding.

Acknowledgments: Many thanks go to Eng. Ludovica Marenco (IRE SpA, Regional Infrastructure for Energy Recovery), Genoa, Italy and Eng. Alberto Bodrato, designer of the case study building (Stecher Srl, Engineering Services, Ovada, Italy), which made available building project data and energy consumption monitoring.

Conflicts of Interest: The authors declare no conflict of interest.

References

1.   European Union. Directive 2010/31/EU of the European Parliament and of Council of 19 May 2010 on The Energy Performance of Building (EPBD Recast). *Off. J. Eur. Union* **2010**, *18*, 2010.

2. European Union. Commission Recommendation (EU) 2019/786 of 8 May 2019 on Building Renovation, C (2019) 3352, 16.5.2019. *Off. J. Eur. Union* **2019**, *127*, 37–79.

3. European Union. Directive (EU) 2018/844 of the European Parliament and of the Council of 30 May 2018 amending Directive 2010/31/EU on the Energy Performance of Buildings and Directive 2012/27/EU on Energy Efficiency, 19.6.2018. *Off. J. Eur. Union* **2018**, *156*, 75–91.

4. Cao, X.; Dai, X.; Liu, J. Building energy-consumption status worldwide and the state-of-the-art technologies for zero-energy buildings during the past decade. *Energy Build.* **2016**, *128*, 198–213. [CrossRef]

5. Magrini, A.; Lentini, G.; Cuman, S.; Bodrato, A.; Marenco, L. From nearly zero energy buildings (NZEB) to positive energy buildings (PEB): The next challenge—The most recent European trends with some notes on the energy analysis of a forerunner PEB example. *Dev. Built Environ.* **2020**, *3*, 100019. [CrossRef]

6. Kolokotsa, D.; Rovas, D.; Kalaitzakis, K.; Kosmatopoulos, E. A roadmap towards intelligent net zero- and positive-energy buildings. *Sol. Energy* **2011**, *85*, 3067–3084. [CrossRef]

7. Bevilacqua, P.; Benevento, F.; Bruno, R.; Arcuri, N. Are Trombe walls suitable passive systems for the reduction of the yearly building energy requirements? *Energy* **2019**, *185*, 554–566. [CrossRef]

8. Szyszka, J.; Bevilacqua, P.; Bruno, R. An Innovative Trombe Wall for Winter Use: The Thermo-Diode Trombe Wall. *Energies* **2020**, *13*, 2188. [CrossRef]

9. Bruno, R.; Bevilacqua, P.; Cuconati, T.; Arcuri, N. Energy evaluations of an innovative multi-storey wooden near Zero Energy Building designed for Mediterranean areas. *Appl. Energy* **2019**, *238*, 929–941. [CrossRef]

10. Sougkakis, V.; Lymperopoulos, K.; Nikolopoulos, N.; Margaritis, N.; Giourka, P.; Angelakoglou, K. An Investigation on the Feasibility of Near-Zero and Positive Energy Communities in the Greek Context. *Smart Cities* **2020**, *3*, 362–384. [CrossRef]

11. Feng, W.; Zhang, Q.; Ji, H.; Wang, R.; Zhou, N.; Ye, Q.; Hao, B.; Li, Y.; Luo, D.; Lau, S.S.Y. A review of net zero energy buildings in hot and humid climates: Experience learned from 34 case study buildings, Renewable and Sustainable. *Energy Rev.* **2019**, *114*, 109303. [CrossRef]

12. Himpe, E.; Janssens, A.; Vaillant Rebollar, J.E. Energy and comfort performance assessment of monitored low energy buildings connected to low-temperature district heating. *Energy Procedia* **2015**, *78*, 3465–3470. [CrossRef]

13. Van de Putte, S.; Bracke, W.; Delghust, M.; Steeman, M.; Janssens, A. Comparison of the actual and theoretical energy use in nZEB renovations of multi-family buildings using in situ monitoring. *E3S Web Conf.* **2020**, *172*, 22007. [CrossRef]

14. Matuska, T.; Sourek, B.; Broum, M. Energy system for nearly zero energy family buildings—Experience from operation. *Energy Rep.* **2020**, *6* (Suppl. S3), 117–123. [CrossRef]

15. Xu, Y.; Yan, C.; Liu, H.; Wang, J.; Yang, Z.; Jiang, Y. Smart energy systems: A critical review on design and operation optimization. *Sustain. Cities Soc.* **2020**, *62*, 102369. [CrossRef]

16. Tumminia, G.; Guarino, F.; Longo, S.; Aloisio, D.; Cellura, S.; Sergi, F.; Brunaccini, G.; Antonucci, V.; Ferraro, M. Grid interaction and environmental impact of a net zero energy building. *Energy Convers. Manag.* **2020**, *203*, 112228. [CrossRef]

17. Zhang, S.; Liu, Y.; Yang, L.; Liu, J.; Hou, L. Applicability of different energy efficiency calculation methods of residential buildings in severe cold and cold zones of China. *E ES* **2019**, *238*, 012029. [CrossRef]

18. Ballarini, I.; Primo, E.; Corrado, V. On the limits of the quasi-steady-state method to predict the energy performance of low-energy buildings. *Therm. Sci.* **2018**, *22* (Suppl. S4), 1117–1127. [CrossRef]

19. Magrini, A. *Building Refurbishment for Energy Performance—A Global Approach*; Springer: Cham, Switzerland, 2014; ISBN 978-3-319-03073-9.

20. Magrini, A.; Lazzari, S.; Marenco, L. Energy retrofitting of buildings and hygrothermal performance of building components. Application of the assessment methodology to a case study of social housing. *Int. J. Heat Technol.* **2017**, *35*, S205–S213. [CrossRef]

21. Available online: https://www.anit.it/software-anit/icaro/ (accessed on 31 August 2020). (In Italian)

22. European Committee for Standardization. *Energy Performance of Buildings—Energy Needs for Heating and Cooling, Internal Temperatures and Sensible and Latent Heat Loads—Part 1: Calculation Procedures (ISO 52016-1:2017)*; European Committee for Standardization: Brussels, Belgium, 2017.

23. ENEA. *Annual Report 2019—Energy Efficiency—Analysis and Results of Our Country's Energy Efficiency Policies*; ENEA: Rome, Italy, 2019; p. 312, ISBN 978-88-8286-382-1. Available online: https://www.enea.it/it/seguici/pubblicazioni/edizioni-enea/2019/rapporto-annuale-efficienza-energetica-2019 (accessed on 31 August 2020).

24. Mutani, G.; Pascali, F.; Martino, M.; Nuvoli, G. Nearly zero energy buildings: Analysis on monitoring energy consumptions for residential buildings in Piedmont Region (IT). In Proceedings of the 2017 IEEE International Telecommunications Energy Conference (INTELEC), Broadbeach, Australia, 22–26 October 2017; pp. 233–240, ISBN 978-1-5386-1019-0.

25. Ministry of Economic Development. *Applicazione delle Metodologie di Calcolo delle Prestazioni Energetiche e Definizione delle Prescrizioni e dei Requisiti Minimi Degli Edifici (Application of the Energy Performance Calculation Methodologies and Definition of Dispositions and Minimum Requirements for Buildings)*; OJ of The Italian Republic: Rome, Italy, 2015. (In Italian)

26. Zavrl, M.S.; Stegnar, G. Comparison of Simulated and Monitored Energy Performance. Indicators on NZEB case study Eco Silver House. *Procedia Environ. Sci.* **2017**, *38*, 52–59. [CrossRef]

27. Di Giuseppe, E.; Ulpiani, G.; Summa, S.; Tarabelli, L.; Di Perna, C.; D'Orazio, M. Hourly dynamic and monthly semi-stationary calculation methods applied to nZEBs: Impacts on energy and comfort. *IOP Conf. Ser. Mater. Sci. Eng.* **2019**, *609*, 072008. [CrossRef]

28. Oliveti, G.; Arcuri, N.; De Simone, M.; Bruno, R. Solar heat gains and operative temperature in attached sunspaces. *Renew. Energy* **2012**, *39*, 241–249. [CrossRef]

29. Ulpiani, G.; Giuliani, D.; Romagnoli, A.; di Perna, C. Experimental monitoring of a sunspace applied to a NZEB mock-up: Assessing and comparing the energy benefits of different configurations. *Energy Build.* **2017**, *152*, 194–215. [CrossRef]

30. Italian Government. *Legislative Decree 3 March 2011, n.28. Attuazione della Direttiva 2009/28/CE sulla Promozione Dell'uso Dell'energia da Fonti Rinnovabili. (Implementation of Directive 2009/28/EC on the Promotion of the Use of Energy from Renewable Sources)*; OJ of The Italian Republic: Rome, Italy, 2011. (In Italian)

31. European Committee for Standardization. *Ergonomics of the Thermal Environment—Analytical Determination and Interpretation of Thermal Comfort Using Calculation of the PMV and PPD Indices and Local Thermal Comfort*; (CEN) EN ISO 7730; ISO: Rome, Italy, 2005.

32. European Committee for Standardization. *Indoor Environmental Input Parameters for Design and Assessment of Energy Performance of Buildings Addressing Indoor Air Quality, Thermal Environment, Lighting and Acoustics*; (CEN) EN 15251; European Committee for Standardization: Brussels, Belgium, 2007.

Article

The Control of Venetian Blinds: A Solution for Reduction of Energy Consumption Preserving Visual Comfort

Francesco Nicoletti *, Cristina Carpino, Mario A. Cucumo and Natale Arcuri

Mechanical, Energy and Management Engineering Department (DIMEG), University of Calabria, 87036 Arcavacata di Rende (CS), Italy; cristina.carpino@unical.it (C.C.); mario.cucumo@unical.it (M.A.C.); natale.arcuri@unical.it (N.A.)
* Correspondence: francesco.nicoletti@unical.it

Received: 6 March 2020; Accepted: 1 April 2020; Published: 5 April 2020

Abstract: Glazing surfaces strongly affect the building energy balance considering heat losses, solar gains and daylighting. Appropriate operation of the screens is required to control the transmitted solar radiation, preventing internal overheating while assuring visual comfort. Consequently, in the building design phase, solar control systems have become crucial devices to achieve high energy standards. An operation based on well-defined control strategies can help to reduce cooling energy consumption and ensure appropriate levels of natural lighting. The present study aims at investigating the effect of smart screening strategies on the energy consumption of a test building designed in the Mediterranean climate. With the aim of automatically setting the inclination of venetian blind slats, the necessary equations are analytically found out and applied. Equations obtained are based on the position of the sun with respect to the wall orientation. In the case of a cloudy day or an unlit surface, empirical laws are determined to optimize the shielding. These are extrapolated through energy simulations conducted with the EnergyPlus software. Finally, using the same software, the actual benefits obtained by the method used are assessed, in terms of energy and CO_2 emissions saved in a test environment.

Keywords: smart solar shading; energy saving; venetian blinds

1. Introduction

Due to the extreme urgency of tackling climate change, it is necessary that buildings are as energy independent as possible [1]. It is important to implement measures for reducing energy consumption trying to take advantage of solar radiation [2,3]. European policies have identified energy efficiency in buildings as one of the key actions to limit greenhouse gas emissions [4]. The use of shielding devices is among the suggested interventions for energy conservation in buildings. An effective design of solar-control systems allows to reduce both cooling energy by preventing the transmission of direct-solar radiation in summer and heating energy maximizing solar gains in winter. Moreover, solar-control systems can help to reduce electricity for artificial lighting and ensure adequate visual comfort trying to take the most advantage of the healthy natural light. In addition to users' well-being, well-designed shielding devices can improve the architectural quality of the building. The type of shading used produces a different impact on the level of natural lighting, thermal comfort and visual comfort. Shielding systems can be divided into two categories, which are fixed and mobile systems [5]. The first category includes overhangs, external horizontal and vertical louvers, and egg-crates. The second category comprises venetian blinds, vertical blinds, and roller shades. In addition, according to the type of control, mobile systems can be furtherly distinguished in manual control, central up-down commands and fully automated control [6].

Several studies have investigated the effect of shading systems on thermal and daylighting performance of buildings. For example, the effect of external louver shading devices applied to different exposures of a building located at different latitudes was analyzed in [7]. Heating and cooling energy demand was quantified for different window and shading surfaces. A horizontal louver layout was considered for the South façade and a vertical louver layout was used for the East and West facades. Shading configuration (number of slats, tilt angle, spacing, distance from the window, and shading area) were optimized for each location in order to ensure adequate solar shading in summer and solar incidence in winter. The results showed that the implemented solar shadings are essential to reach internal comfort conditions and allow the achievement of significant energy savings for space cooling in all the studied locations. Horizontal shading, vertical shading and egg-crate shading were analyzed by [8] for a high-rise office building in Malaysia alongside different glazing configurations for the transparent facades. The authors demonstrated that cooling energy savings ranging from 5.0% to 9.9% can be obtained by applying the shields to all the facades with low-e double-glazing. Cooling energy saving will further increase between 5.6% and 10.4% if the shadings are applied to all facades with single clear glazing. [9] analyzed the effect of different fixed shading on the control of air temperature and the improvement of illuminance level of an office building in Jordan. Three types of shading (vertical fins, egg-crates and diagonal fins) were installed in three identical office and real-time experiments combined with computer simulations were adopted for thermal and daylight analyses proving that all shading devices help to improve thermal and visual environment in the office. [10] proposed an experimental configuration of an external shading device for apartment houses in south Korea. Simulations revealed that the experimental shading device, based on various adjustments of the slat angle, can offer the most efficient performance in terms of thermal and visual comfort for the occupants, compared to overhang, blind system, and light shelf, providing an energy saving of at least 11% during the cooling season. [11] carried out field measurements and simulations in order to assess the benefits of movable solar shadings on the energy, indoor thermal and visual comfort of a retrofitted residential building in Ningbo city in China. Results revealed that movable shading devices used for South facing windows can reduce building energy demand by about 34%, improve thermal comfort in summer and visual comfort by 20% and drastically reduce risks of uncomfortable conditions by 80%.

However, as observed by [12], manual or motorized blinds are limited in their ability to reduce energy consumption and to provide internal comfort because an operation of the blinds by the occupants themselves is expected to block direct solar radiation. On the contrary, the use of automated systems allows to more fully exploit the benefits of the blinds. The authors carried out thermal and visual experiments in a real scale test room and collected reports by the occupants of the dwelling in summer confirming the potential energy savings and the comfort enhancement when using automated blinds. [13] tested the applicability and effectiveness of an automated reflective shutter, which can work both as a sunshade and a daylight system. According to the results, the tested system was able to significantly improve daylight distribution and reduce energy consumption for artificial light by 60%. [14] used simulation to investigate the potential savings in energy demand achievable through the installation of blind shadings and application of shading and lighting controls in an office building with fully-glazed façade in Qatar. Results have shown that by applying shading controls the space total energy demand is reduced by 11.6% in north-oriented facades and by 24.8% in east-oriented facades. If lighting controls are added, energy savings will increase to 14.1% in north-oriented offices and 28.3% in east-oriented offices.

Several authors have analyzed the interaction of the occupants with shields and artificial light trying to understand the influence of the automation level and the possibility of manual operation on the environmental and energy performance and the degree of user satisfaction. A field study of human interactions with motorized roller shades and dimmable electric lights in a high-performance office building is presented in [15]. Four different control set-ups were explored, namely (1) Manual control with low-level of accessibility (wall switches); (2) Manual control with high level of accessibility (web interface); (3) Fully automated control; (4) Automated control with manual overrides. Indoor environmental variables were monitored through sensors and a feedback by occupants on their indoor perception was obtained with a survey. The analysis revealed a preference for customized indoor climate with consequently different energy impacts. [16] reported two experiments investigating the effect of the level automation and the type of system expressiveness on users' satisfaction with an automated blinds system. They found out that the use of an expressive interface communicating the status and intentions of the blinds system could favor users' acceptance and satisfaction level, reducing the sense of losing control when decisions on environmental control are made by technology.

The analyzed literature highlighted the benefits in terms of energy saving and internal comfort achievable with the application of solar shades [17]. Moreover, the improvement of the shielding operation can also involve advantages from an economic point of view, since it reduces the waste of energy for air conditioning [18]. However, the activation and setting strategies of these systems require a more in-depth analysis in order to identify the most effective solutions that can lead to maximize their performance. The objective of the present work is to investigate the effect of a smart-control system applied to venetian blinds on the building energy use. In particular, an innovative control strategy is designed based on the identification of the optimal slat angle that is progressively adjusted following the sun position in order to take the best advantage of the available solar radiation. Therefore, the transmitted solar radiation is reduced when it generates an undesired thermal load while it is allowed to enter when there is a heating demand. The occupants' comfort is considered as well in the smart operation, seeking to guarantee a suitable minimum level of natural lighting and avoid direct exposure to solar radiation. The control strategy is based on the IoT technology because it requires the use of "cognitive" objects consisting of sensors and actuators, which are able to detect environmental data, process information, and implement the most appropriate actions needed to achieve the desired conditions. The information relating to the external and internal climatic conditions is acquired and used by the venetian blind actuators to orientate the slats. In a future development, technology can also be integrated with artificial intelligence networks in order to manage any preferences of the occupant. The training, in that case, can be conducted by interrogating the user so that he can communicate if he is not satisfied with the level of illumination of the environment.

It is a solution achievable thanks to home automation, useful for improving the quality of life and living comfort. The system is completed if it is integrated with other unplanned commands, such as turning the lights on or off according to the illuminance and the presence of the occupants.

2. Venetian Blinds and Sensors

Venetian blinds represent a common solution often adopted to allow the user to modify the incoming solar energy based on external weather conditions and his own sense of visual pleasure. They consist of small horizontal slats that can be manually oriented. The following characteristic parameters are defined for this type of shielding (Figure 1): distance between two consecutive slats d; slat depth L; slat inclination (angle between the normal to the surface of the slat and the horizontal plane) *Slat*.

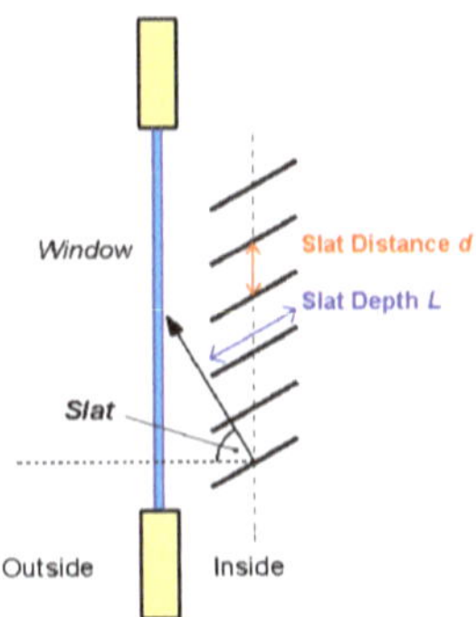

Figure 1. Schematization of venetian blinds.

In order to automate the opening of the blinds, it is necessary to modify the inclination angle by means of an actuator implementing the desired shielding. Information from indoor and outdoor environment are needed to evaluate the optimal inclination in every moment of the day. In particular, an internal temperature sensor T and a presence detection sensor are required for each room.

In correspondence of each wall containing a glazed surface, a solarimeter is placed on the outside with the task of measuring the global irradiance G incident on a vertical surface.

In addition to measuring the global irradiance value on the vertical surface, the solarimeter is used to understand if there is the presence of beam irradiance incident on the surface. The method to obtain this information consists in comparing the global irradiance measured by the solarimeter with the theoretical direct solar irradiance expected at that time of the day, estimated with the ASHRAE methodology [19]. Direct Normal irradiance is:

$$I_{bn} = A\, e^{-\frac{B}{sin\alpha}}$$ (1)

in which α is the solar altitude, A and B are listed according to the month of the year [19]. The direct incident irradiance G_b on a 90° inclined surface is:

$$G_b = I_{bn} \cos i$$ (2)

where i is the angle of incidence, obtained as a function of the solar altitude α, the azimuth of the surface γ_w and the solar azimuth γ:

$$\cos i = \cos(\gamma - \gamma_w) \cos \alpha$$ (3)

If the global measured irradiance is lower than the theoretical direct irradiance, then there is certainly no incidence of sunlight on the surface; otherwise, with good approximation, we can assume that there is direct irradiance. This hypothesis is a precaution in the moments in which users must be protected from glare. This method, therefore, allows to discern between clear and cloudy days. Through various simulations conducted in the EnergyPlus environment, it has been possible to observe how this condition is useful for the purpose.

3. Control Method

The shields are managed to adapt to the various operating conditions outlined by the sensor outputs. The goal of the following analysis is to identify the *Slat* angle to be taken for each operating condition. First of all, it is necessary to distinguish two different operating conditions, relative to the presence or absence of occupants inside the environment.

3.1. Absence of Occupants

If the room is unoccupied, the lamellas position depends on the internal temperature detected, in order to understand if it is necessary to favor or block solar contributions.

3.1.1. Winter Operation

In the case of temperatures below the set-point value of 21 °C, the maximum solar radiation must be introduced into the environment, without worrying about the glare of occupants. The optimal inclination angle varies if the surface is exposed or not to direct solar radiation. In the negative case, the *Slat* value is 110°. Numerous tests conducted within the EnergyPlus simulation environment have shown that this angle guarantees the transmission of the maximum diffuse radiation. In the positive case, in which there is direct radiation incident on the glass surface, the slat of the venetian blinds is adjusted in order to be in a direction parallel to the sun's rays. The slats' inclination is, therefore, defined by the projection of the sun on the vertical plane perpendicular to the surface. Figure 2 shows the sun profile angle β that the sun projection forms with the horizontal plane.

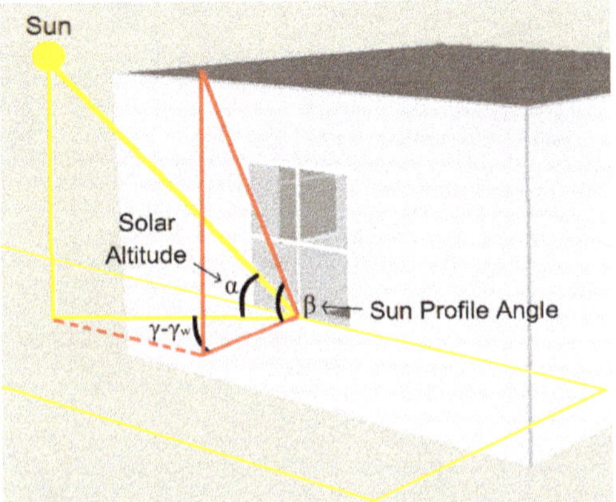

Figure 2. Relationship between sun profile angle, solar azimuth angle and solar altitude angle.

β is obtained according to the following relation [20]:

$$\beta = \arctan\left(\frac{\tan \alpha}{\cos(\gamma - \gamma_w)}\right) \tag{4}$$

The method defined in this paper is structured according to the following logic. The results shown by the simulations allowed to validate these conditions. If β is lower than 65°, the optimal slat angle is fixed equal to $\beta + 90°$, such as to keep the lamellas parallel to the solar rays.

Otherwise:

if G > 300 W/m^2 direct component is high and the optimal slat angle is still $\beta + 90°$;

if G < 300 W/m^2 direct radiation has a lower energy contribution than diffuse radiation, thus a Slat angle able to maximize the entry of the latter should be set. It is given by the following correlation:

$$Slat = 120° - 0.66\,\alpha \tag{5}$$

In the case of temperatures above the set-point value of 25 °C, shields must be completely shut since it is not necessary to guarantee natural illumination inside the environment.

3.1.2. Intermediate Operation

Finally, if the temperature is between 21–25 °C, *Slat* is fixed to 80° (there is no need to exploit the solar contribution or even to shield).

3.2. *Presence of Occupants*

In the presence of occupants, three different cases are distinguished again.

3.2.1. Winter Operation

If the temperature is lower than the set-point value of 21 °C, solar radiation, which in this case is a free contribution, must be used as much as possible.

- The presence of direct radiation could cause visual discomfort on the work surface to people. In this case, the optimal configuration consists in shielding sun's rays by keeping the lamellas as open as possible. The requirement that guarantees this condition is that the vertex A of the lower lamella, the vertex B of the upper lamella and the point representation of the sun are on the same line (Figure 3). The angle of inclination that identifies this position of the lamella is indicated in

Figure 3 with the name of *Slat**. In fact, if these points are aligned, certainly the sun's rays do not reach the work surface and the slats are the most open.

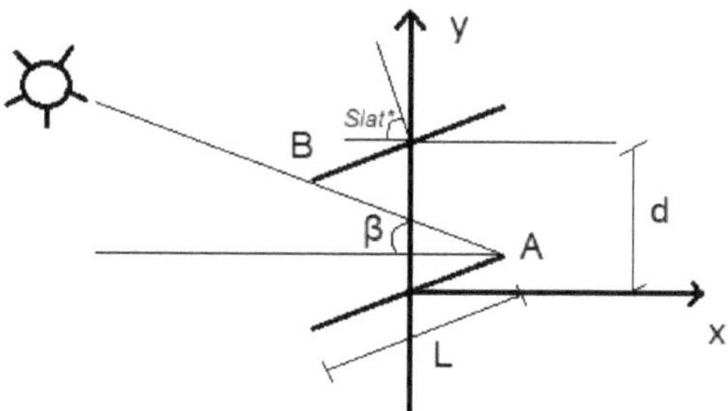

Figure 3. Limit inclination for direct radiation blocking.

with reference to Figure 3, the coordinates of points A and B are the following:

$$\begin{cases} x_A = \frac{L}{2}\sin(Slat^*) \\ y_A = \frac{L}{2}\cos(Slat^*) \end{cases} \qquad \begin{cases} x_B = -\frac{L}{2}\sin(Slat^*) \\ y_B = d - \frac{L}{2}\cos(Slat^*) \end{cases} \tag{6}$$

Therefore, it is necessary that points A and B form the angle β with respect to the horizontal:

$$-\tan\beta = \frac{y_B - y_A}{x_B - x_A} = \frac{d - \frac{L}{2}\cos(Slat^*) - \frac{L}{2}\cos(Slat^*)}{-\frac{L}{2}\sin(Slat^*) - \frac{L}{2}\sin(Slat^*)} \tag{7}$$

After some mathematical steps, *Slat** results:

$$Slat^* = 2\arctan\frac{\tan\beta + \sqrt{\tan^2\beta - \left(\frac{d}{L}\right)^2 + 1}}{1 + \frac{d}{L}} \tag{8}$$

- If the direct solar radiation does not affect the walls, then the configuration that allows the maximum diffuse solar radiation to enter is Slat equal to 110°.

3.2.2. Summer Operation

For temperatures above the set-point value of 25 °C, it is necessary to shield the solar radiation, but not completely, so that a minimum of natural illumination is guaranteed. This is possible with *Slat* equal to 45°.

3.2.3. Intermediate Operation

If the temperature is between 21 °C and 25 °C, the operation depends on the exposure to direct solar radiation: in the negative case, *Slat* is 80°; while in the case of exposure to direct solar radiation, the optimal angle is determined by equation (8) to avoid glare to the occupants.

4. Definition of Simulation Environment

The actual energy saving and operation of the smart-control strategy have been simulated in EnergyPlus. This software was selected for its renowned ability to accurately simulate and estimate the energy performance of buildings as well as its capability to properly integrate shading devices [14]. The building thermal zone calculation method in EnergyPlus is a heat balance model. The fundamental assumption of heat balance models is that air in each thermal zone can be modeled with uniform temperature. The fenestration module includes the ability to consider the angular dependence of transmission and absorption for both solar and visible radiation, and temperature-dependent U-value. In EnergyPlus, the calculation of diffuse solar radiation from the sky incident on an exterior surface takes into account the anisotropic radiance distribution of the sky. The sky model used for calculation is "CIE sunny clear day", with additional direct illumination from the sun. It has bright patches due to direct illumination from the sun with relatively dark areas where direct sunlight does not fall. The simulation tool allowed to evaluate the effectiveness of the system under dynamic conditions. Theoretical analysis and simulation are essential to test the operation of the devised smart system and to quantify its impact on energy consumption. The same approach has been adopted in other research works found in the literature [21–23], in which the effects of solar shading systems on energy demand and visual comfort were assessed by means of theoretical studies and simulations. However, this is intended to be a preliminary phase for the development of a prototype, which will be used to validate the conceived model with the aid of experimental data.

The problem was solved as a time-dependent model. The climatic parameters set refer to a year, obtained for the city of Cosenza (South Italy) from Meteonorm [24]. For the investigated location, the maximum value of external air temperature in summer is 36.4 °C and the minimum value in winter is 0.2 °C. The maximum value of the direct normal solar radiation is 997 W/m^2 while the maximum value of the diffuse solar radiation on a horizontal surface is 489 W/m^2.

The test building consists of a 25 m^2 square room designed to have a window for each exposure (Figure 4). The thermal transmittance values of external walls, floor and roof are equal to 0.38, 0.34, 0.27 W/m^{2}K, respectively. The stained glass windows represent the 15% of the total walls and are composed of double glass and air gap (4-12-4) with a thermal transmittance of 1.91 W/m^{2}K. The number of occupants is zero from 8.00 a.m. to 2.00 p.m. and it is two during the rest of the day. The lighting system, equipped with linear control, is active from 7.00 a.m. to 8.00 a.m. and from 2.00 p.m. to 11:00 p.m. and ensure 500 Lux on the work surface, placed at 0.75 m above the ground. Finally, temperature control is managed by a fan coil system, always on, set at 20 °C for winter days and 26 °C for summer days.

Figure 4. Test building.

5. Analysis of Results

In order to analyze the effects of smart shielding devices, the solar radiation transmitted with the dynamic control is compared with the solar radiation transmitted in the case of fixed shielding with an inclination angle of 80°.

Figure 5 shows the trends of three different contributes of irradiance *I* for the different exposures: the global external radiation incident on the vertical surface *Gv*, the solar radiation transmitted with the orientable shields *Tadj* and the solar radiation transmitted with the fixed shields *Tfixed*. In the figure, moreover, the lamellas inclination angle *Slat*, in case of activated control is reported. The simulations are conducted with reference to a typical day in the winter period (January 4th). In Figure 5a, relative to the East exposure, the gain deriving from the use of mobile shields emerges. In the early hours, they help to provide a greater solar contribution. The same considerations also apply to the Southern exposure (Figure 5d), in which *Slat* = β + 90°, with β determined by the Equation (7). In the morning hours, for the North and West exposures, not directly illuminated by the sun, *Slat* is equal to 110°, such as to maximize the incoming diffuse solar radiation. From the graph displaying the west exposure (Figure 5c) emerges, instead, how dynamic control is implemented to ensure visual comfort in the presence of the occupants. The solar radiation transmitted, therefore, is limited from 15:00 to 17:00 by the control described by Equation (8).

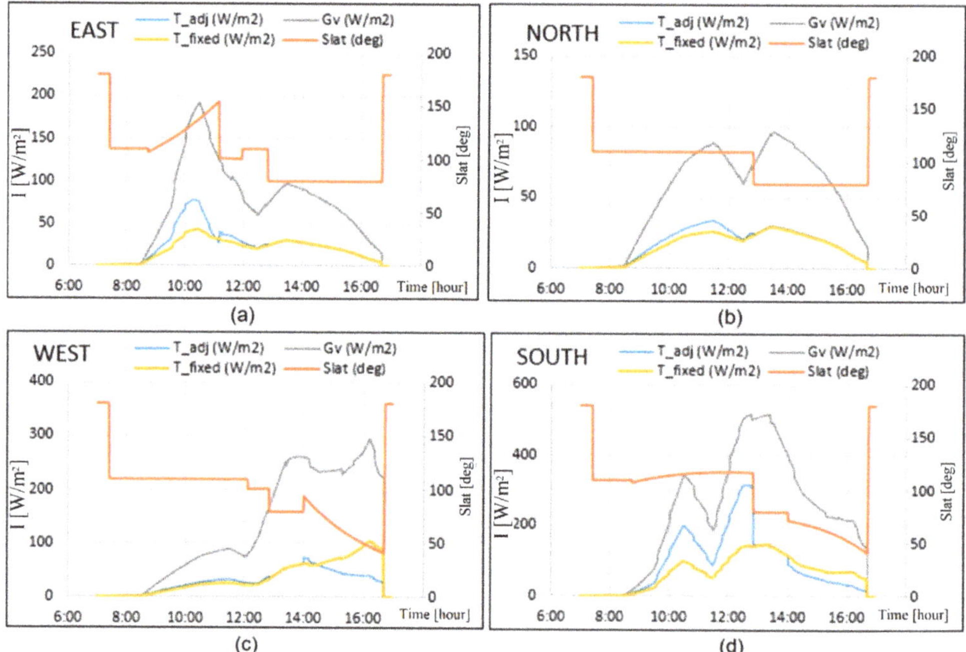

Figure 5. Irradiance and Slat angle in winter day for different exposure: (**a**) East (**b**) North (**c**) West (**d**) South.

For the south facade, which is exposed to solar radiation for long periods over the day, the double effect of the mobile shading systems emerges. In the first part of the day, they increase the incoming solar radiation while, in the second part of the day, they optimize the amount of solar radiation that can be transmitted in conjunction with the visual comfort of the occupants. Under the simulated conditions, the daily heating demand is 2.05 kWh with the smart logic control and 5.38 kWh with the fixed slat angle, obtaining a reduction of 62%.

In Figure 6, instead, the same quantities are shown with reference to a typical day in the summer period (July 18th).

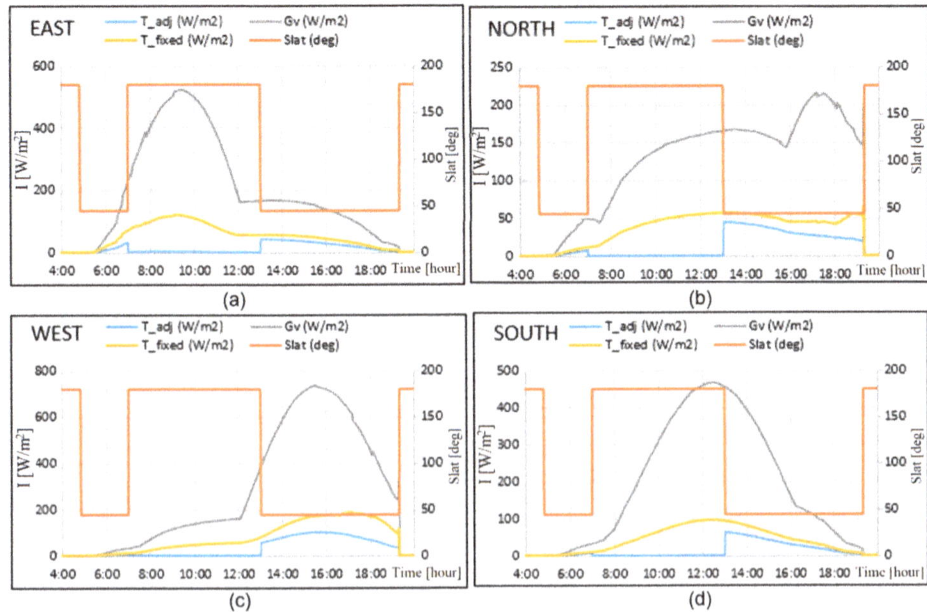

Figure 6. Irradiance and Slat angle in summer day for different exposure: (**a**) East (**b**) North (**c**) West (**d**) South.

In the conditions of a clear day and high outdoor temperature, the solar radiation transmitted through the windows must be shielded as much as possible and, therefore, the thermal load on the cooling system can be reduced. The graphs in Figure 6 show how the control, set up to keep the shields completely closed in the absence of occupants and at an angle of 45° in the presence of occupants, guarantees its effectiveness by introducing into the environment a much smaller amount of solar radiation compared to the case with fixed shielding. The daily cooling energy demand is 11.5 kWh with the smart logic control and 12.9 kWh with the fixed slat angle, achieving a reduction of 10.8%. Figure 7 shows the monthly requirements deriving from the sum of energy consumption related to heating, cooling and lighting. In every month there is a tangible reduction in consumption thanks to the application of intelligent control of shielding systems. The percentage saving is between 30% and 60% in the winter months and between 10% and 20% in the summer months. On an annual level, total saving is around 15%.

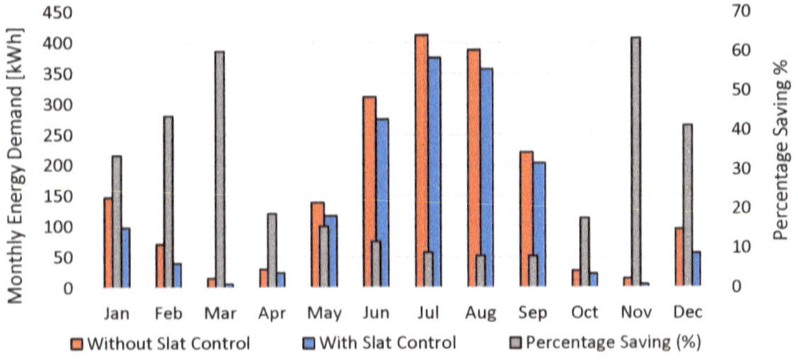

Figure 7. Monthly energy demand and percentage saving.

Considering the energy demand satisfied by means of an electric heat pump with average efficiencies COP = 3.5 and EER = 3.0, the values shown in Figure 8 are obtained for the annual supplied energy (electricity). The values are normalized with respect to the heated and cooled net area (19.36 m^2). The figure also shows the energy costs calculated assuming an average electricity cost of 0.25 €/kWh including Value-Added Tax [25]. The graph reveals that the application of the smart shading control allows to obtain savings in supplied electricity of about 5.0 kWh/m^2·year, and an economic saving in the bill of 1.20 €/m^2·year.

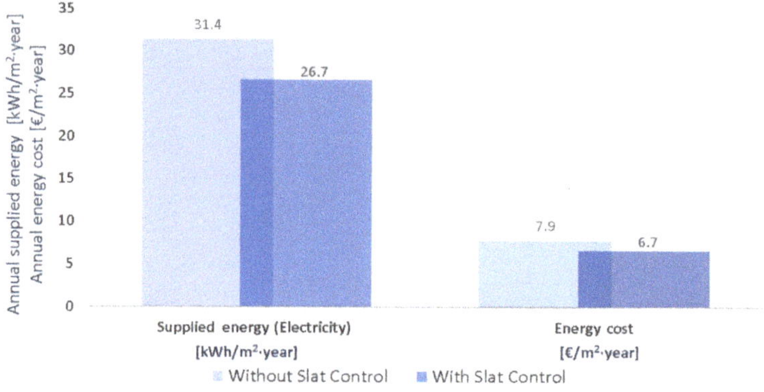

Figure 8. Savings of annual energy supply and correspondent reduction of the annual energy cost provided by the application of the smart shading control compared to the case without slats control.

By applying a nonrenewable primary energy factor of 1.95 [26] the annual source energy is calculated. Based on the results reported in Figure 9, the automated management strategy of the slats leads to a primary energy saving of 177.80 kWh/year and a reduction of CO2 emissions of 79.2 kgCO2/year [27].

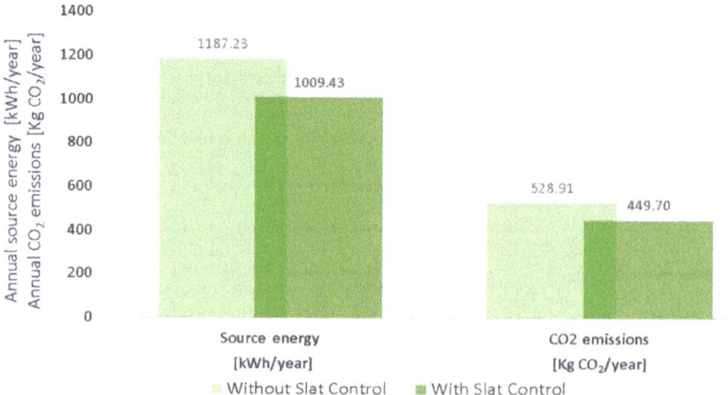

Figure 9. Saving of source energy and reduction of CO_2 emissions obtainable through smart control of the venetian blinds.

It is worth noting that the evaluations are carried out on a simplified model of limited surface. By scaling up the problem to cases characterized by a larger building surface, the CO2 emissions reductions are even more significant, considering that savings of 4.10 (kg CO2/m^2·year) can be reached.

6. Conclusions

The study focuses on the energy effects derived from the use of a system of internal mobile shields controlled with a precise control method, in order to determine any benefits on the energy needs of a residential building model for a specific location.

The use of mobile shields leads to a reduction in the energy demand for cooling during the summer period and guarantees an aid to the energy needs in the winter period, when solar radiation can help in reducing the energy waste for heating and lighting. Furthermore, the best shielding solution must protect the visual comfort of the occupants, who could be exposed to glare.

The dynamic control of the shields allows obtaining a considerable energy saving, with significant effects especially in the winter period. In one year, the estimated reduction in energy consumption for the analyzed building was 15%.

The automation of venetian blinds can find wide diffusion and utility in domestic environments and offices. With reference to residential applications, clear advantages are constituted by energy saving and, therefore, by economic saving. Solar loads are managed automatically when the user is in another room in the house or away from home. This allows complete house management at all times. Another advantage is represented by the absence of visual discomfort caused by glare, an aspect that is also important in the workplace, in the office, to always maintain concentration.

Further developments and improvements on the control logic could concern the parametric study of the geometric characteristics of the slats, in order to identify the optimal configuration to be adopted during installation. It is also interesting to carry out the same study in other locations with different latitudes, in order to assess how much the location can influence the resulting benefits. A further development is represented by the implementation of the system in an experimental environment, in which it is supported by the machine learning logic to account for any human preferences.

Author Contributions: Conceptualization, F.N. and N.A.; methodology, F.N. and C.C.; software, F.N.; validation, C.C. and M.A.C.; formal analysis, N.A. and M.A.C.; investigation, F.N. and C.C.; resources, C.C.; data curation, F.N.; writing—original draft preparation, F.N. and C.C.; writing—review and editing, F.N. and C.C.; visualization, F.N. and C.C.; supervision, M.A.C. and N.A.; project administration, M.A.C. and N.A.; funding acquisition, F.N. and M.A.C. All authors have read and agreed to the published version of the manuscript.

Funding: This research was funded by Regione Calabria (PAC CALABRIA 2014-2020 - Asse Prioritario 12, Azione B) 10.5.12 - CUP H28D19000040006).

Conflicts of Interest: The authors declare no conflict of interest.

References

1. Bruno, R.; Bevilacqua, P.; Cuconati, T.; Arcuri, N. Energy evaluations of an innovative multi-storey wooden near Zero Energy Building designed for Mediterranean areas. *Appl. Energy* **2019**, *238*, 929–941. [CrossRef]

2. Bevilacqua, P.; Bruno, R.; Arcuri, N. Green roofs in a Mediterranean climate: Energy performances based on in-situ experimental data. *Renew. Energy* **2020**. in-press. [CrossRef]

3. Bevilacqua, P.; Benevento, F.; Bruno, R.; Arcuri, N. Are Trombe walls suitable passive systems for the reduction of the yearly building energy requirements? *Energy* **2019**, *185*, 554–566. [CrossRef]

4. Directive 2010/31/EU of the European Parliament and of the Council of 19 May 2010 on the energy performance of buildings (recast). *Off. J. Eur. Union 2010* **2010**, *3*, 124–146.

5. Kirimtat, A.; Koyunbaba, B.K.; Chatzikonstantinou, I.; Sariyildiz, S. Review of simulation modeling for shading devices in buildings. *Renew. Sustain. Energy Rev.* **2016**, *53*, 23–49. [CrossRef]

6. Kuhn, T.E. State of the art of advanced solar control devices for buildings. *Sol. Energy* **2017**, *154*, 112–133. [CrossRef]

7. Palmero-Marrero, A.I.; Oliveira, A.C. Effect of louver shading devices on building energy requirements. *Appl. Energy* **2010**, *87*, 2040–2049. [CrossRef]

8. Lau, A.K.K.; Salleh, E.; Lim, C.H.; Sulaiman, M.Y. Potential of shading devices and glazing configurations on cooling energy savings for high-rise office buildings in hot-humid climates: The case of Malaysia. *Int. J. Sustain. Built Environ.* **2016**, *5*, 387–399. [CrossRef]

9. Freewan, A.A.Y. Impact of external shading devices on thermal and daylighting performance of offices in hot climate regions. *Sol. Energy* **2014**, *102*, 14–30. [CrossRef]
10. Kim, G.; Lim, H.S.; Lim, T.S.; Schaefer, L.; Kim, J.T. Comparative advantage of an exterior shading device in thermal performance for residential buildings. *Energy Build.* **2012**, *46*, 105–111. [CrossRef]
11. Yao, J. An investigation into the impact of movable solar shades on energy, indoor thermal and visual comfort improvements. *Build. Environ.* **2014**, *71*, 24–32. [CrossRef]
12. Kim, J.H.; Park, Y.J.; Yeo, M.S.; Kim, K.W. An experimental study on the environmental performance of the automated blind in summer. *Build. Environ.* **2009**, *44*, 1517–1527. [CrossRef]
13. Hashemi, A. Daylighting and solar shading performances of an innovative automated reflective louvre system. *Energy Build.* **2014**, *82*, 607–620. [CrossRef]
14. Al Touma, A.; Ouahrani, D. Quantifying savings in spaces energy demands and CO2emissions by shading and lighting controls in the Arabian Gulf. *J. Build. Eng.* **2018**, *18*, 429–437. [CrossRef]
15. Sadeghi, S.A.; Karava, P.; Konstantzos, I.; Tzempelikos, A. Occupant interactions with shading and lighting systems using different control interfaces: A pilot field study. *Build. Environ.* **2016**, *97*, 177–195. [CrossRef]
16. Meerbeek, B.W.; de Bakker, C.; de Kort, Y.A.W.; van Loenen, E.J.; Bergman, T. Automated blinds with light feedback to increase occupant satisfaction and energy saving. *Build. Environ.* **2016**, *103*, 70–85. [CrossRef]
17. Bruno, R.; Pizzuti, G.; Arcuri, N. The Prediction of Thermal Loads in Building by Means of the en ISO 13790 Dynamic Model: A Comparison with TRNSYS. *Energy Procedia* **2016**, *101*, 192–199. [CrossRef]
18. Bruno, R.; Bevilacqua, P.; Arcuri, N. Assessing cooling energy demands with the EN ISO 52016-1 quasi-steady approach in the Mediterranean area. *J. Build. Eng.* **2019**, *24*. [CrossRef]
19. *ASHRAE Handbook, Fundamentals*; American Society of Heating, Refrigerating and Air-Conditioning Engineers: New York, NY, USA, 1989.
20. Hu, J.; Olbina, S. Illuminance-based slat angle selection model for automated control of split blinds. *Build. Environ.* **2011**, *46*, 786–796. [CrossRef]
21. Kim, D.W.; Park, C.S. Comparative control strategies of exterior and interior blind systems. *Light. Res. Technol.* **2012**, *44*, 291–308. [CrossRef]
22. Tzempelikos, A.; Athienitis, A.K. The impact of shading design and control on building cooling and lighting demand. *Sol. Energy* **2007**, *81*, 369–382. [CrossRef]
23. Yun, G.; Yoon, K.C.; Kim, K.S. The influence of shading control strategies on the visual comfort and energy demand of office buildings. *Energy Build.* **2014**, *84*, 70–85. [CrossRef]
24. Meteonorm. Meteonorm Global Meteorogical Database Version 7.1.8. 2017. Available online: https://meteonorm.com/en/ (accessed on 26 March 2020).
25. Available online: https://www.arera.it/it/dati/eep35.htm (accessed on 26 March 2020).
26. Repubblica Italiana, D.M. *26 Giugno 2015—Parte 1. Applicazione Delle Metodologie di Calcolo Delle Prestazioni Energetiche e Definizione Delle Prescrizioni e dei Requisiti Minimi Degli Edifici*; Gazzetta Ufficiale Serie Generale n.162 del 15-07-2015—Suppl. Ordinario n. 39; Istituto Poligrafico e Zecca dello Stato: Rome, Italy, 2015.
27. ISPRA—Istituto Superiore per la Protezione e la Ricerca Ambientale. *Rapporti 303/2019*; ISPRA: Rome, Italy, 2019; ISBN 978-88-448-0945-4.

Article

On the Implementation of the Nearly Zero Energy Building Concept for Jointly Acting Renewables Self-Consumers in Mediterranean Climate Conditions

Faidra Kotarela [1], Anastasios Kyritsis [2] and Nick Papanikolaou [1,*

[1] Department of Electrical & Computer Engineering, Democritus University of Thrace, 67100 Xanthi, Greece; freikota1@ee.duth.gr

[2] Photovoltaic Systems and Distributed Generation Department, Centre for Renewable Energy Sources and Saving, 19009 Athens, Greece; kyritsis@cres.gr

* Correspondence: npapanik@ee.duth.gr; Tel.: +30-2541-079-921

Received: 4 February 2020; Accepted: 23 February 2020; Published: 25 February 2020

Abstract: Cost-effective energy saving in the building sector is a high priority in Europe; The European Union has set ambitious targets for buildings' energy performance in order to convert old energy-intensive ones into nearly zero energy buildings (nZEBs). This study focuses on the implementation of a collective self-consumption nZEB concept in Mediterranean climate conditions, considering a typical multi-family building (or apartment block) in the urban environment. The aggregated use of PVs, geothermal and energy storage systems allow the self-production and self-consumption of energy, in a way that the independence from fossil fuels and the reliability of the electricity grid are enhanced. The proposed nZEB implementation scheme will be analyzed from techno-economical perspective, presenting detailed calculations regarding the components dimensioning and costs-giving emphasis on life cycle cost analysis (LCCA) indexes—as well as the energy transactions between the building and the electricity grid. The main outcomes of this work are that the proposed nZEB implementation is a sustainable solution for the Mediterranean area, whereas the incorporation of electrical energy storage units—though beneficial for the reliability of the grid—calls for the implementation of positive policies regarding the reduction of their payback period.

Keywords: nZEB; photovoltaics; geothermal energy system; energy storage units; energy transactions; life cycle cost assessment; payback period

1. Introduction

Buildings in EU countries account for approximately 40% of the total primary energy consumption and 36% of greenhouse emissions [1]. EU climate change objectives for 2020, such as a 20% increase of RES in gross final consumption of energy, 20% reduction of buildings primary energy consumption and 20% reduction of greenhouse gas emissions compared to 1990 levels are key issues nowadays, whereas new more ambitious targets were set in 2030 climate and energy framework [2].

In this context, the concept of highly efficient buildings has been perceived. Specifically, the term "nearly Zero Energy Building" (nZEB) has been introduced, referring to a building of high energy performance which takes advantage of various types of RES in order to service a significant number of its total annual energy needs [3,4]. The nearly zero or very low amount of energy required can be produced in site (under the net metering scheme) or in the near area (under the virtual net metering scheme). Therefore, nZEBs are interconnected to the electricity distribution networks, using them as a backup system for their power balance. Analytically, the excess electricity that is injected into the grid (during low building energy consumption intervals) can be used during low energy production intervals. It is worth noting that the majority of residential buildings under the net metering scheme receive almost 70% of the energy required to maintain power balance from the grid, in real time. Indeed, by studying self-consumption ratio of southern European residential buildings (buildings with PV installations under the net metering scheme) it is concluded that the mean self-consumption ratio of such buildings ranges between 30 and 35%. Hence, nearly zero or very low energy consumption should be studied in regard to energy transactions with the electricity networks [5]. Considering the energy consumption and production (where applicable) profiles of residential buildings, it is generally accepted that demand side management techniques and energy storage systems installation seem to be the most suitable smart-grid services in such applications. As far as demand side management is concerned, demand and response techniques could provide flexibility on electrical system operation either by reducing residential consumers' electricity consumption during electrical load peak periods, or by motivating buildings owners to modify their electrical consumption profile according to RES availability [6–9]. Demand side management may be more efficient when implemented at the level of final consumers' aggregators, which have direct control on air conditioning, heat pump units and water heaters during periods of peak demand or peak RES production. Building owners which provide demand side management services could be remunerated either for their demand and response availability (following contracted time-based rates and critical peak rebates) or for the procurement of the service (following variable peak pricing or real-time pricing and time-of-use pricing).

Additionally, it is worth mentioning that unlike detached houses where the installation of small RES systems (mainly PVs) is usually simple, in the case of apartment buildings the installation of many small and independent PV systems is less efficient and calls for a much more complicated design. Today, however, the implementation of nZEB concept in apartment buildings is facilitated by the scheme of energy communities [10,11]. The aggregated use of RES from a group of self-producers who are located in the same building or multi-apartment block allows the production of clean energy at the level of a whole building and not an individual property.

nZEBs consist of low thermal transmittance building materials with high energy standards and technical specifications, in order to increase the insulation levels of building envelope (walls, roof, floor and windows), achieving so significant reduction on thermal losses. Furthermore, highly efficient electromechanical installations for heating and cooling are employed in conjunction with bioclimatic architecture rules and passive solar design techniques for natural cooling and ventilation. Energy consumption defines the energy classification of each building to specific grades (i.e., A, B, C, D, E, F or G). For the development and widespread of nZEB concept, EU Commission has published the Directive 2010/31/EU which regards the residential and tertiary sectors (offices, public buildings etc.). As for the targets determined for new buildings, this Directive imposes all new buildings to become nZEBs by 31 December 2020, whereas the deadline for new buildings occupied and owned by public authorities was 31 December 2018. Additionally, the European Directive states that the upgrade of buildings energy performances should be pursued to the extent that this is technically, economically and functionally feasible.

The main issues raised by nZEB concept are the achievement of nearly zero energy transactions with the public grid on annual basis (and the certification of this condition), the type of energy used, the accepted renewable energy supply options and requirements in terms of energy efficiency, the indoor/outdoor climate conditions and the connection with the electricity grid [12]. State-of-the-art strategies for nZEB technologies are synopsized in three main categories: passive energy-saving technologies, energy-efficient building service systems and RES integration technologies. The implementation of these interventions is reviewed in [13]. Renovating constructions are a part of the building conversion into a nZEB, especially in old buildings, in order to attain high energy efficiency. In [14] it is demonstrated that deep energy renovation of existing buildings (especially if it is combined with the overall energy improvement of the building which also involves non-energy related aspects) is one of the major opportunities to reduce energy consumption and to improve internal comfort conditions for the residents. In a case study referred in Mediterranean climate conditions it is mentioned that nZEB mainly depends on a synthesis of existing technologies and know-how of consolidated or traditional building principles [15]. Reference [16] provides a brief overview of existing nZEBs in Mediterranean countries, by noting the main characteristics for microclimatic design and energy conversion from RES systems. The main part of [16] focuses on an existing single-family nZEB, which is equipped with various energy conversion systems from RES, such as solar thermal collectors, PVs and GHEs, aerothermal and multi-split systems as well. EESUs and thermal storage for DHW tank were also considered.

Many scientific papers perform the energy consumption analysis of the building sector. In [17] the energy consumption, the technical and the environmental characteristics of residential buildings in Mediterranean are presented, as well as their potential for energy saving, based on their actual energy consumption profiles. The technical characteristics of residential buildings related to their energy performance (thermal insulation, envelope specifications, heating and cooling energy system, etc.), the actual energy use and the environmental factors related to energy consumption, are also described. The authors of [18] have proposed a new methodology called BEA, in order to examine the energy classification of buildings; various aspects of BEA such as, heating/cooling and electrical load demand, energy consumption and CO_2 emissions were analyzed. In [19] the energy savings of multi-family residential building thermal systems, such as hybrid biomass-solar ones, is examined. The required measurements conducted taking into consideration parameters such as energy consumption and CO_2 emissions reduction. In view of developing energy-efficient structures, references [20,21] provide an overview of building design criteria that can reduce the thermal and cooling load demand of residential buildings. These criteria are based on buildings orientation, shape, envelope system, passive heating and cooling mechanisms, shading and glazing. These papers investigate the optimal sustainable and energy-efficient building design options. Energy efficiency measures and installation of RES technologies for buildings designed to become nZEBs are investigated in [22], with the use of quasi-dynamic simulation tools and models. In [23] the principles of integrated designed procedure and smart building technologies (along with energy efficiency methodologies and innovative techniques) are presented, highlighting that the integration of smart technologies requires a holistic approach that takes into account all aspects of sustainability from the early design phases of the building, in order to take full advantage of the benefits of the process and the opportunities that smart grids offer.

As for the installed RES technologies on buildings, there are many studies that propose various combinations of RES systems considering the nZEB concept. Some of them focus on the implementation of PVs and solar-thermal collectors as combined heat and power systems, in order to cover both electrical and thermal load demand [24,25]. In [26], a comparison among the installation of building-integrated PV/T air collectors and side-by-side PV modules and solar thermal collectors is presented, whilst in [27] a BIPV/T has been used as the roof top of a building in order to increase the electrical energy per unit area figure and to meet thermal demands. Additionally, BIPV systems are noted as an interesting approach for newly built and refurbished residences, because they operate as multifunctional building construction materials (they produce energy and serve as part of buildings envelope) [28–32]. BIPVs are usually used as glazing and windows for transparent openings; moreover, they can be used instead of ceramic tiles on the roofs of buildings. Furthermore, BAPV in the form of façade components can be used as curtain walls, or as shadings and balcony barriers. Both types of building PV products do not require the existence of free building surfaces or free land space for installation, compared to usual residential PV plants. In addition, from an architectural perspective, the ability to add color to PV cells (with the addition of suitable coatings) as well as to PV modules frames, reduces any optic nuisance, allowing the aesthetic integration of PV technology into the shell of buildings [29–34]. Additionally, lightweight fibre reinforced composite BIPV and BAPV modules are able to be curved (maintaining high mechanical strength) and thus various surface finishes are possible [28].

In [35] the optimal integration of a CHP to provide district-level cooling, heating, and electrical energy to a residential area and the potential for combination of the CHP with PV system has been investigated. In this context, authors in [36] presented an innovative technological concept for energy supply (electrical and thermal energy) consisting of CHP and WPP. This proposed combination is able to manage the energy demand in a small-scale area, like buildings. Finally, a combination of a solar thermal and a cogeneration system for district heating with seasonal energy storage installed on building-level is represented in [37].

In [38] a Swedish case study presents an analysis on how EAHPs and GSHPs in combination with PV-systems affect the specific energy demand of buildings, considering the nZEB concept. In more details, this energy production combination (enhanced with heat recovery ventilation) leads to decreased specific energy demand. A hybrid system installation in Bahrain that consists of PVs, a wind turbine and hydrogen fuel cells is analyzed in [39]. These building integrated RES are proved to be efficient as well as of a significant environmental impact.

Considering that nZEBs usually operate under the net metering scheme, their design principles are associated with the on-site production from RES and the self-consumption of locally generated clean electricity. Self-consumption can be enhanced by installing ESUs-sized for the amounts of produced energy. In [40] a PV system with two types of energy storage is analyzed, i.e., lead acid batteries and water heat tank storage, where a part of the electricity produced by the PVs is stored as heat. Comparing these two types of energy storage, the water heat tank is competitive to the ESS when it comes to the level of self-consumption. On the other hand, lead-acid battery life is too short in compared to battery investment cost, and therefore not considered to be profitable. Thus, more suitable battery technologies should be incorporated in the nZEB concept, such as Li-ion batteries which is the case in the present work. Another proposed energy system is presented in [41], using hybrid ESS including both batteries and electric water heaters in residential nZEBs, in order to store the PV energy production. The outcomes of this work highlight that the proposed hybrid PV ESS can mitigate the issue of high load variation that is raised at the utility side by the communities with high PV penetration levels. A thorough economic analysis for a residential building with the integration of PVs and battery ESSs is presented in [42]. This work considers various tariff structures of the electrical market, as well as various energy savings and CO_2 emissions reduction scenarios. The reduction of CO_2 emissions per kWh produced from GSHP systems is investigated in [43,44]. Moreover, Lo Russo et al. [45] consider a very low-enthalpy geothermal plant, using GWHPs, in order to evaluate the sustainability and the benefits of the concept under-study by means of greenhouse gasses reduction. Additionally, in [46] a survey on various cases regarding the sustainable design of buildings is presented, whereas the reduction of CO_2 levels in national and global scale is discussed.

As for the installation of geothermal energy systems on buildings, there are many case studies that focus on the separate exploitation of this type of RES technology. References [47–49] present a comprehensive review on the geothermal energy exploitation in buildings, taking into consideration various restrictions that play a vital role in designing small- or large-scale systems, such as the length and diameter of the GHE tube, the space between the GHEs and their technical specifications. In addition, Niemelä et al. [50] present a method for efficient heating of DHW by using a GHP; the water is heated from the inlet temperature of cold water to the target DHW. This system balances the thermal/cooling load demands of multi-family apartment buildings. Moreover, a case study of a GHE, designed for extraction or injection of thermal energy from/into ground, has shown that the double U-tube boreholes are superior to the ones of the single U-tube with reduced borehole resistance [51].

In [52] an optimization model for a PV/ST and a GSHP system for office building is presented. This model focuses on the optimal sizing of the PV/ST and the geothermal system, under the restriction of limited available area and practical heating and cooling loads, without considering the sizing of ESUs. Referring to the thermal load of nZEBs, the work in [53] compares two types of thermal storage for a GHP system, i.e., exhaust air- and solar-thermal storage; the analysis has highlighted the outperformance of the exhaust air-thermal storage option. In addition, a smart renewable energy system is introduced in [54], consisting of solar PVs, an ASHP and an ESU, in order to prove the significant decrease of the energy consumption for buildings and the accompanied reduction on greenhouse emissions. A comparison between the proposed system and the traditional energy supply system (electricity grid and natural gas) is also represented in this paper, where the surplus energy can be either delivered to the electricity grid or to the ESU. The performance of hybrid solar thermal and geothermal energy systems is discussed in [55–61]. In these studies, energy is supplied to the geothermal flow from wells in order to boost the power output. Next, Rad et al. [62] discuss the viability of hybrid GSHP systems combining with solar thermal collectors as the supplemental component in heating dominated buildings. This examination has shown that the installation of solar TES in the ground can considerably reduce the GHE length. As for the TESs design, a case study describes the installation of GSHPs fed by BHEs; in this framework, the installation of UTES for high temperature heat storage has been proven to be successful [63]. In this direction, it analyzes the operating performance of a PV/GHP system on a residential building that maximizes the self-consumption of the energy production [64]. The experimental analysis is based on real time data, regarding both the operation of the two systems and the interaction with the electricity grid. Finally, the case study in [65] presents a combined CCHP, i.e., a PV and a GSHP integrated system, which is designed for a large office.

As for the energy storage in smart grids, it emphasizes on the importance of the energy storage for the effective management of energy demand and supply, by analyzing the various types of ESUs (electrical, electrochemical, thermal, and mechanical ones) [66]. Also, in [67] the new types of energy storage being integrated into the grid are addressed. In this context, the design of smart grids will take advantage of the available storage capacity in dealing with more dynamic loads and sources.

Following the review on relevant scientific literature, we conclude that there are various combinations of RES systems in order to cover electrical and thermal (heating and cooling) load demands of nZEBs. As for ESUs, the existing installations incorporate either thermal or electrochemical storage units. Also, in many studies, the exploitation of the electricity grid for net metering application is presented. In conclusion, the outcomes of all these works regard the case of individual buildings, whereas there is a gap in the literature for the case of multi-family buildings or apartment blocks. Additionally, the main scope of the introduced RES and ESUs has been the decrease of CO_2 emissions and the increase of self-consumption ratio of the locally installed RES, at individual building/owner level. Therefore, the benefits of local installed RES and ESU units are limited from electricity grid perspective, because it is not manageable to handle numerous small systems at low voltage distribution level. In this context, it is worth noting that demand-response services are easier to be applied to aggregated low voltage prosumers (by using price signals to rearrange their consumption, production or storage services), than negotiating with numerus individual owners of small RES and ESU units at "stand–alone" buildings. In this way, demand-response can be a more credible and cost-effective solution for shifting energy demand (e.g., by reducing peak consumption) and avoiding high load fluctuations at grid side. Additionally, considering that more than half of buildings energy consumption corresponds to the heating and cooling needs (in order to achieve acceptable internal comfort conditions), it is concluded that a great share of demand-response lies in thermal appliances.

As for the LCCA of those installations, there are many case studies in the specific literature. Among them, in [68] the basic methodology of LCCA is presented, emphasizing on the analysis of the present value of system components, in order to calculate the total cost of ownership over its life time, including the costs to purchase, install, operate and maintain. In [69] a life cycle analysis of various renovation scenarios for multi-family buildings in Portugal is presented, by focusing on the goals of the nZEB and ZEB design process and on the contribution of solar systems, such as solar thermal collectors and PVs. According to [70], the LCCA is proposed for a PV and battery energy system in a residential building in India, by using analytical expressions for the payback period calculation. Also, Marszal andHeiselberg [71] use the LCCA method and considers a multi-family nZEB to indicate that in order to build a cost-effective nZEB, the energy use should be reduced to a small amount which should be covered by renewable energy generation. The work in [72] presents a comparison of the LCC of five alternative heating systems, namely GDHS, natural gas-fired boilers, pellet-fired boilers, GCHP and coal-fired boiler technologies, in four climate zones in Turkey. In addition, life cycle cost calculations for geothermal energy systems are presented in [73,74]. In case of battery ESSs, the LCCA method is used in [75–78].

The present paper focuses on the implementation of a collective self-consumption nZEB concept for multi-family building of B-energy class certification (which is imposed by the relevant legislation), that uses RES and ESUs [10]; the area of interest in this work is the Mediterranean, due to its excellent RES potential, its mild temperature conditions, and the relatively poor energy performance of its building sector (characterized by old buildings of high thermal losses and limited use of state of the art HVAC systems) [2–4]. The term collective self-consumption is used to describe a group of jointly acting renewables self-consumers who are located in the same building or multi-apartment block [10,11]. The joint use of RES technologies in conjunction with the installation of electrochemical and thermal ESUs, strengthens the scheme of energy communities; in terms of electricity grid operation, the installed energy storage not only increases the self-consumption ratio of the on-site installed RES (increasing the economic benefits of the users of the buildings) but it also relaxes load flow congestion and manages the evening peaks of electricity demand. These aspects are crucial for the secure operation of the electrical grids and will become more intensely as the electricity generation shifts from the centralized generation model to the distributed intermittent RES generation concept [5]. In this context, the adopted net metering schemes around Europe permit the residents of a multi-family building to establish an energy community and install a common PV system in order to serve their total annual electricity needs. The supplier offsets the energy bill by virtually sharing the production of the PV plant according to the rates given by the jointly acting renewables self-consumers. The design principles of this concept and its application are described in the following sections.

Last but not least, the estimation of the life cycle cost of the investment by using the LCCA method (for the proposed collective self-consumption nZEB scheme) is also performed, in order to evaluate its sustainability and describe the necessary incentives for its successful application.

2. Materials and Methods

2.1. Description of Collective Self-Consumption nZEB under Study

For the purposes of this study, a hypothetical multi-family building in the city of Athens (B climate zone area according to the National Regulation [79], which is characterized by mild temperatures during winter and dry weather conditions during summer period), with three floors and two dwellings per floor, is considered as the case study building. This type of building stock is common in many small and medium-sized cities in European Mediterranean countries. The Commission Recommendation 2016/1318/EU (as regards the application of the nZEB definition in practice) benchmarks for the energy performance of nZEBs the bespoken targets; those targets are mandatory for the new buildings in EU. Indicatively, for new households in Mediterranean area, the annual primary energy consumption ranges between 0–15 kWh$_{primary}$/m^2, according to [80]. However, for the purpose of this work, each dwelling is classified in B energy class (supposing that all necessary renovation measures have been already applied). In this regard, it is worth noting that at national level in Greece 65% of dwellings are classified in energy classes E–H, 32% in C–D and only 3% in A–B. Additionally, the national plan for Greece requires the energy improvement of refurbished buildings to B-class before their nZEB transformation. Therefore, in order to convert them into nZEBs, it is necessary first to reduce their energy demand by using energy efficient materials (such as low emission glazing for transparent openings and thermal envelope materials with appropriate U-Value for the insulation of roof, floor, thermal bridges and external walls) and upgrading the electromechanical equipment to the minimum energy efficiency standards, and afterwards to utilize RES to meet the extremely low energy demand. Although the upgrade of buildings thermal insulation is a key issue on their energy performance, as it has been thoroughly discussed in the scientific literature [81,82], the present work focuses on the effective energy supply of nZEBs (employing GHPs and PVs) and the optimal integration of EESUs. Any further discussion regarding renovation measures for enhancing the thermal insulation level of buildings envelope is out of the scope of this paper.

Last but not least, according to the Hellenic Statistic Authority [83], the average annual primary energy consumption in Greek households that belong to B energy class is roughly 98 kWh$_{primary}$/m^2 [78]. Thus, this primary energy consumption value was selected for, in order to formulate the energy consumption profiles of the building under study. The term "Primary Energy" refers to the energy production from RES and non-RES which has not undergone any conversion or transformation process and it is calculated by using the primary energy factors shown in Table 1 [3,4].

Table 1. Primary energy factors.

Energy Source	Primary Energy Factor	Emissions kg CO$_2$/kWh
Natural Gas	1.05	0.196
Heating Diesel	1.1	0.264
Electric Energy	2.9	0.989
LPG	1.05	0.238
Biomass	1	-
District heating by thermal stations	0.7	0.347
District heating by RES	0.5	-

As it will be further analyzed below, the present study focuses on the implementation of a collective self-consumption nZEB concept by exploiting PVs and geothermal energy systems on a multi-family residential building. The featured nZEB concept is based on the high energy productivity of PV systems in Mediterranean countries [84]. Additionally, a central geothermal energy system has been selected to serve both heating and cooling needs of the multi-family residential building, due to the technological maturity of these systems as well as their high efficiency [85]. In this case study, the calculation of the energy consumption of the rest appliances is also taken into consideration, i.e., the electrical consumption of the building before the installation of GHPs. As for the rest electrical appliances, it is mentioned that they account for approximately 20–25% of the annual energy consumption [86]. For this reason, this energy consumption component should not be disregarded from the rest analysis. The total area of each dwelling is considered to be 150 m^2. Having in mind that the ceiling height is roughly 3.2 m, the total heated and cooled volume of the building is equal to 2880 m^3. Moreover, the building is south-east oriented, and it is located in an urban area.

As an alternative solution, solar thermal panels could be installed instead of the geothermal energy system. Although solar thermal panels could reduce the required size of GHPs, for the sake of simplicity this scenario is not examined. Hence, in this paper a case study of a shallow geothermal energy system has been selected. It consists of GHEs in vertical layout, in order to occupy as less installation space as possible. The underground thermal energy is supplied to the building by using highly efficient GSHPs. Moreover, ESUs (electrical and thermal) are used to enhance nZEB performance. The schematic diagram of the proposed flexible nZEB system is presented in Figure 1. Additionally, the same figure illustrates the electrical energy transactions between the building under-study, the locally installed PV system and the electrical distribution network (along with the energy transactions with the EESU), as well as the thermal transactions between the building and the geothermal energy system.

The proposed combination of RES technologies comes up with certain restrictions for the necessary energy storage capacity (electrical and thermal), in order to avoid oversized systems and high initial investment costs that affect the pay-back period. The design aspects regarding the reduction of energy transactions with the electricity grid and the associated electrical/thermal energy storage capacity are discussed. Also, the optimal sizing of the ESUs manages to control demand-response and peak shavings. In order to achieve that, this study presents the calculations of the electrical PV production and consumption of the building for a 10-years period. This time period analysis is used for the following reasons: Firstly, the aim of this paper is the long-term design of the building energy system, taking into consideration the average life cycle of the ESU unit. Also, this time period analysis is quite enough for the reliable definition of the load demand and the excess energy or the energy losses on a monthly and on a yearly basis. Although (according to the nZEB concept) the energy transactions at the end of each year must be low, this is not a representative period of analysis, due to the weather conditions and especially the solar irradiation variations. Thus, the 10-years period enables the consideration of those variations on climatic conditions; apparently, by the end of the 10-years period the energy transactions must be low enough in order to meet the nZEB concept requirements. As for the economic parameters, the 10-year time frame is a usual pay-back period time for such investments. As regards the EESUs, their costs are a major part of this investment; hence their sizing should be carefully designed in order to come up with an acceptable payback period. In this study, the EESU is sized by means of an optimization process that takes into account the energy transactions among the nZEB and the electricity grid.

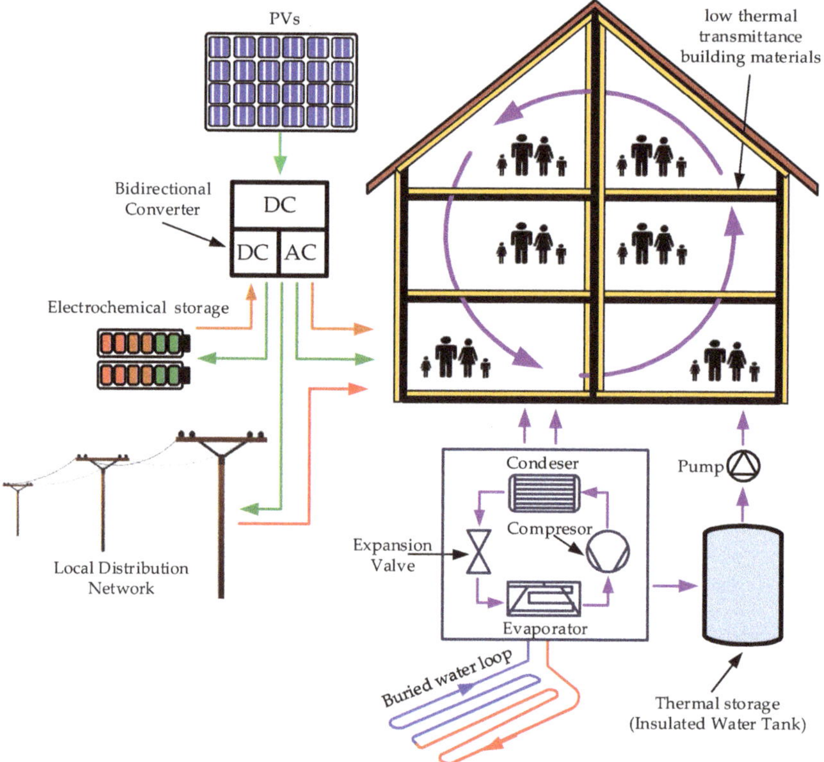

Figure 1. Schematic diagram of the proposed nZEB concept.

In next sections, the sizing of the collective self-consumption nZEB concept will be analyzed in detail. First, the energy consumption of the building is calculated on a yearly and a monthly basis. This is the fundamental step for the design of the PV unit and the geothermal system, as well as for the estimation of the energy transactions with the electricity grid. According to the previous section, the sizing of the ESUs should aim at minimizing those energy transactions. Finally, calculations on the LCCA of the proposed energy system, with and without using EESUs, are presented in Section 3.

In order to make the proposed case study more comprehensive, the flow diagram in Figure 2 describes the steps, the methods used and the aim of each step.

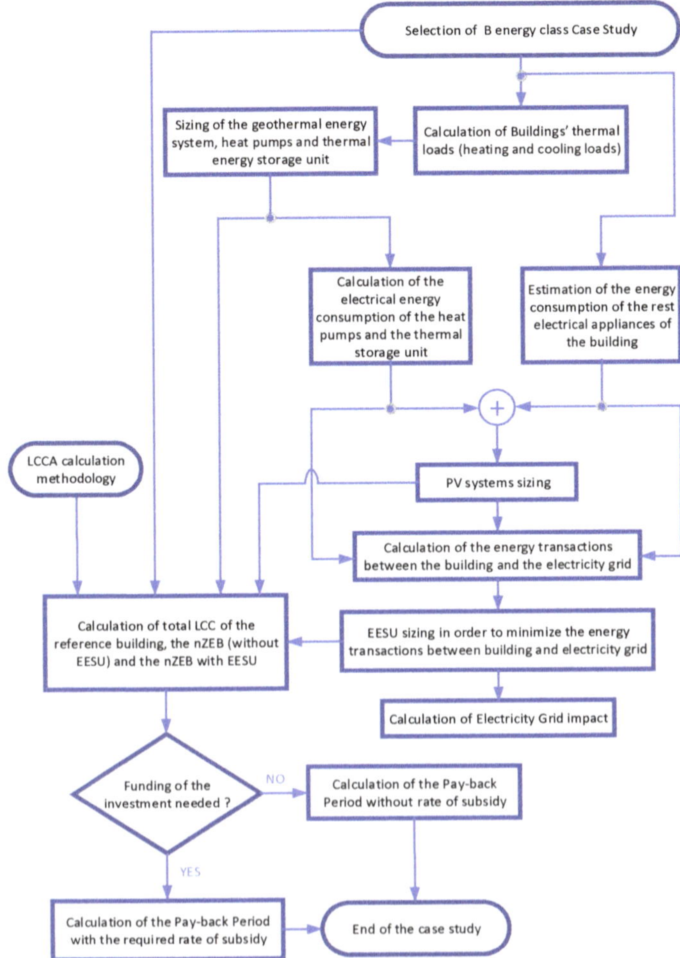

Figure 2. Flow diagram of the methodology and steps of this paper.

2.2. Definition of the Energy Consumption Profiles

As referred in the previous section, the average annual primary energy consumption in Greek households that are classified at B energy class is roughly 98 kWh$_{primary}$/m^2 [79,83]. In more details, the average annual thermal consumption, in terms of heating, of the aforementioned Greek households is 5244 kWh$_{th,h}$ and the average annual electrical consumption is 2400 kWh$_e$. In order to achieve the appropriate comfort indoor conditions during summer period, it is assumed that the need for cooling energy in each household is 2000 kWh$_{th,c}$, which is covered by GHPs. Hence, the distribution of heating and cooling load demands can be calculated on a monthly basis, during winter and summer period, as it is shown in Figure 2.

According to Table 1, those amounts of energy consumption are equal to 5506 kWh$_{th,h-primary}$, 6960 kWh$_{e,primary}$ and 2100 kWh$_{th,c-,primary}$, which leads to 14,566 kWh$_{primary}$ in total. This amount of energy consumption (calculated for the proposed building) is in line with the previously noted average total primary energy consumption per dwelling for the B energy class. Additionally, these energy needs must be covered by the utilization of GHPs and PVs in order to transform the building into a nZEB one. Next, the three types of energy load demand (heating, cooling and pure electric) of the building under study will be allocated on a monthly basis.

2.2.1. Thermal Energy Consumption (for Heating) Calculation

The definition of the thermal-heating load demand of the multi-family building is based on the average annual thermal consumption (for heating needs) that each dwelling has. Thus, according to the previous subsection, the total thermal energy load of the building is 31,464 $kWh_{th,h}$. This will be covered by the geothermal energy system; its annual operating hours (in order to cover the thermal energy demand for heating) regard the winter period and in general the cold days of the year. Those are estimated to 1700 h per year. This can be assumed by taking into consideration the mild weather conditions in Mediterranean climates during spring and autumn periods. Upon this parameter, the average annual heating load is 18.5 $kW_{th,h}$.

2.2.2. Thermal Energy Consumption (for Cooling) Calculation

As for the cooling load demand, this is similarly calculated to the heating one. Starting from the cooling load of each dwelling, the total need for cooling energy is 12,000 $kW_{hth,c}$. This amount of energy demand refers to the summer period and it is covered by using the same GHP (it performs both heating and cooling operation). The number of cooling operating hours is estimated to 990 h per year. So, the average annual cooling load is 12.13 $kW_{th,c}$.

2.2.3. Electrical Energy Consumption Calculation

The total electrical energy load is the sum of the pure electrical energy consumption of building electrical appliances and the electrical energy that heat pumps absorb during their operation, as it will be shown in the following sections. The second component of the electrical energy consumption can be calculated indirectly through the aforementioned heating and cooling loads and the efficiency of the heat pumps for those operational modes (heating and cooling). The average annual pure electrical energy consumption of the multi-family residential building is 14,400 kWh_e. Assuming a 24-h operation, the average annual electrical load of the building is 1.64 kW_e. Note that the total electrical energy consumption should be covered by the PV system. Having in mind the contracted electrical power capacity per dwelling, each dwelling may consume up to a maximum of 12 kW_e. Thus, according to the standardization of the Hellenic Electricity Distribution Network Operator S.A. (HEDNO S.A.), the contracted power capacity of the building is 85 kVA.

2.3. Geothermal Energy System Design

The shallow geothermal energy system is installed in the uncovered ground area of the land that the building is situated. According to Figure 1, the geothermal system is installed in a vertical layout and it consists of a GHP, the necessary GHEs, an insulated TES, a separate heat pump that charges/discharges the TES unit (HPSU) and the distribution network that transfers heat in the building.

GHPs perform heating and cooling operation and they must be able to cover the maximum thermal and cooling load demands. For the case under study, the total operating hours for the GHP are 2690 h per year in order to serve the average annual thermal (heating and cooling) energy demand. A significant parameter in the selection of the GHP is the *COP*, which is the ratio between the useful thermal energy and the absorbed electrical energy during its operation [87]. *COP* factor differs for heating and cooling operation. During cooling operation this factor is referred as *EER*, though it has the same meaning with *COP* factor. Hence, depending on the total operating hours, the *COP* and the *EER* factors determine the necessary electrical energy consumption of the heat pump. The technical data of the GHP and HPSU heat pumps, for the specific building, are shown in Table 2. It is noted that those are actual data from commercial heat pumps.

Table 2. Geothermal energy system technical data.

	GHP Technical Data	**HPSU Technical Data**
Heating capacity (kW$_{th,h}$)	22.26	71.8
Cooling capacity (kW$_{th,c}$)	24.9	50.5
COP	3.13	5.04
EER	4.4	3.46

Another important issue is the sizing of the TES unit, as it has to cover thermal energy demands deviations from the average annual value. The sizing of the TES is based on the estimation of the number of continuous hours per year that heating or cooling load demand is at its maximum value. Hence, considering a maximum of 12 h continuous maximum thermal load demand, a commercial TES unit of 40 m^3 volume has been selected (assuming a conservative capacity factor of 15 kWh/m^3 in case of water-tanks). This proposed heat volume is appropriate for underground installation, without calling for a lot of excavation work. Thus, GHP supplies the TES unit with thermal energy during low thermal load conditions, whereas the reverse thermal energy flow takes place during high thermal load conditions.

As a result, the electrical energy consumption of the building is higher due to heat pumps operation, comparing to the electrical energy consumption without using heat pumps (pure electrical energy), as shown in Figure 3. More specifically, taking into consideration the technical characteristics in Table 2, the total annual electrical energy that both pumps consume is calculated to be 18,668 kWh$_e$. Thus, the total annual electrical energy consumption becomes 33,068 kWh$_e$. The allocation of building energy consumption on a monthly basis (due to heat pumps operation) is illustrated in Figure 3 for a typical year. In more detail, this figure depicts the allocation of the thermal energy consumption of the building broken down by operation mode (cooling-kWh$_{th,c}$ and heating-kWh$_{th,h}$), as well as the monthly total (including the electrical consumption of the heat pumps), kWh$_{e,total}$, and the initial (prior to nZEB transformation) electrical consumption of the building, kWh$_e$.

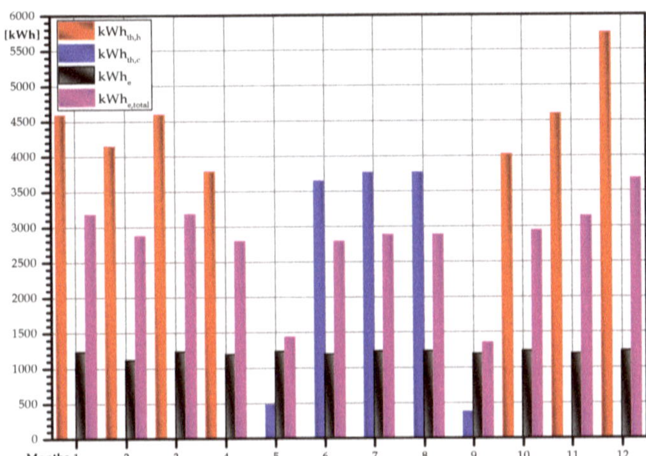

Figure 3. Allocation of the monthly energy consumption of the building for a typical year.

2.4. PV System Design

The sizing of the PV system is based on the total electrical energy consumption of the building, i.e., the electrical energy consumption of 33,068 kWh$_e$ according to Section 2.4. Considering the orientation of the building, the available meteorological data of the site, as well as the spatial planning restrictions, the proposed PV system nominal power is 21 kWp, and it is composed by three strings of

24 mainstream PV modules in series (considering PV panels of 295 Wp). The PV panels are installed on the rooftop of the building in tilt angle close to the latitude optimum inclination of 30° facing south.

It is noted that the relative production index for free standing-PV systems composed by crystalline silicon modules (placed in optimum tilt angle facing south) in the area of Athens, is roughly 1575 kWh$_e$/kWp. Additionally, the average monthly value of the capacity utilization index for PV systems in Greece is 17.73%, whereas the minimum and maximum recorded monthly values are 9.1% and 24.1% respectively. The minimum and maximum capacity utilization index values are recorded every year in July or August and in December or January respectively, highlighting the seasonal variation of PV electricity generation [85]. It is worth mentioning that the capacity utilization index is defined as the ratio of the final AC energy output (kWh) of a PV system over a specific period, to the AC rating (kW) of the plant, for the same time period, under STC.

The allocation of the monthly energy consumption presented in Figure 3 is the average energy consumption per month for a 10-years period (according to available climate data for the area of Athens). Indicatively, Figure 4 shows the monthly variation of solar irradiation, compared to the 10 years corresponding average value [88]. Similarly, Figure 5 shows the monthly variation of temperature; those data corroborate the proposed 10-years period of analysis, in order to take into account any climatic conditions deviations [88].

In order to calculate the PV energy production evolution over the period of analysis, the same climate data for the area of Athens are considered. In Figure 6, the 10-years period average energy production of the PV system per month in contrast to the monthly energy consumption of the building is presented.

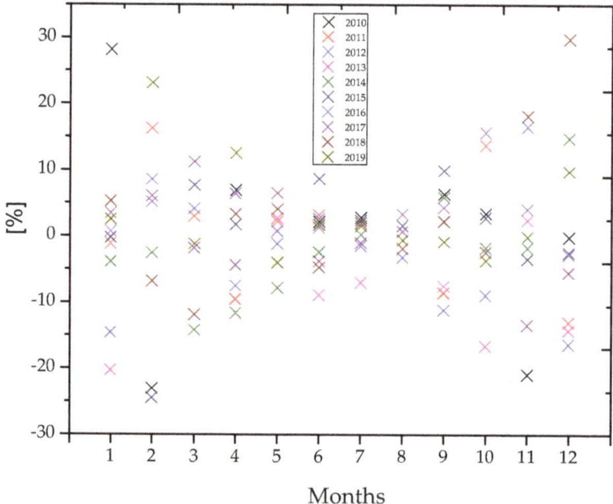

Figure 4. Monthly variation of solar irradiation for 10-years period (actual data for Athens area) [88].

As it is referred in Section 2.3, the electrical energy consumption of the building is higher due to the heat pumps operation. This is confirmed by Figure 6, comparing the electrical load of the building (referring strictly to building electrical appliances—pure electric energy) with the total electrical energy consumption during the heating/cooling operation of the heat pumps. Also, Figure 6 denotes the deviations between PV generation and load electrical demand on a monthly basis; during high electrical energy consumption intervals, the excessive energy needed is supplied by the electricity grid. The reverse electrical energy flow takes place during low load intervals, where the excessive PV energy production is supplied to the grid or to the EESU (as it is analyzed in Section 2.5), in order to balance the energy transactions between the building system and the electricity grid. In this context, it is noted that under the net-metering scheme the prosumers are allowed to inject the excessive amount of the produced PV electricity into the grid and use it at a later time to offset their consumption (when their renewable generation is absent or not sufficient).

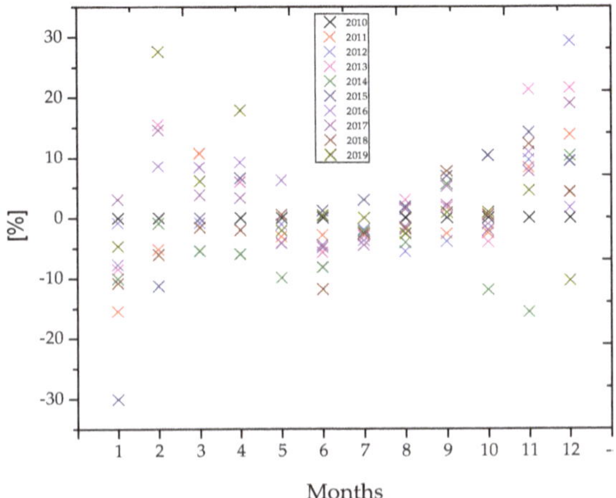

Figure 5. Monthly variation of temperature for 10-years period (actual data for Athens area) [88].

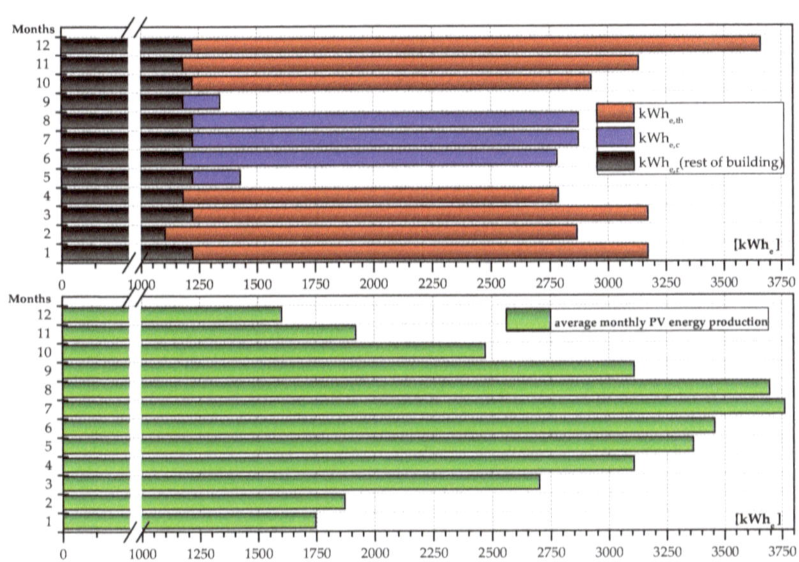

Figure 6. Total electrical energy consumption and average monthly PV production.

In any case, according to the above-mentioned data, by the end of the 10-years period the total electric energy supply by the grid is estimated to be 924 kWh$_e$, considering there is not any EESU installed. This amount of energy accounts for the 0.279% of the energy consumption of the building in the 10-years period. Having that in mind, the proposed residential building is characterized as a nZEB, indeed.

2.5. Sizing of the Electrical Energy Storage Unit

In the previous section, the amount of energy transactions considering the collective self-consumption nZEB concept without using EESUs was defined. In this section, the use of EESUs is presented, in order to reduce the energy transactions between the building and the electricity grid.

The EESU sizing is based on the energy fluctuations limitation (among the nZEB and the grid) per month. In this context, Figure 7 shows the energy exchanges per month (10-years window) for the EESU, as well as the energy transactions reduction (and the corresponding storage capacity). The maximum permitted monthly fluctuation in the EESU, $\Delta E_{storage,month}$, is set in the following scenarios, i.e., ± 100, ± 250, ± 400, ± 500, ± 750, ± 1000 and ± 2000 kWh$_e$.

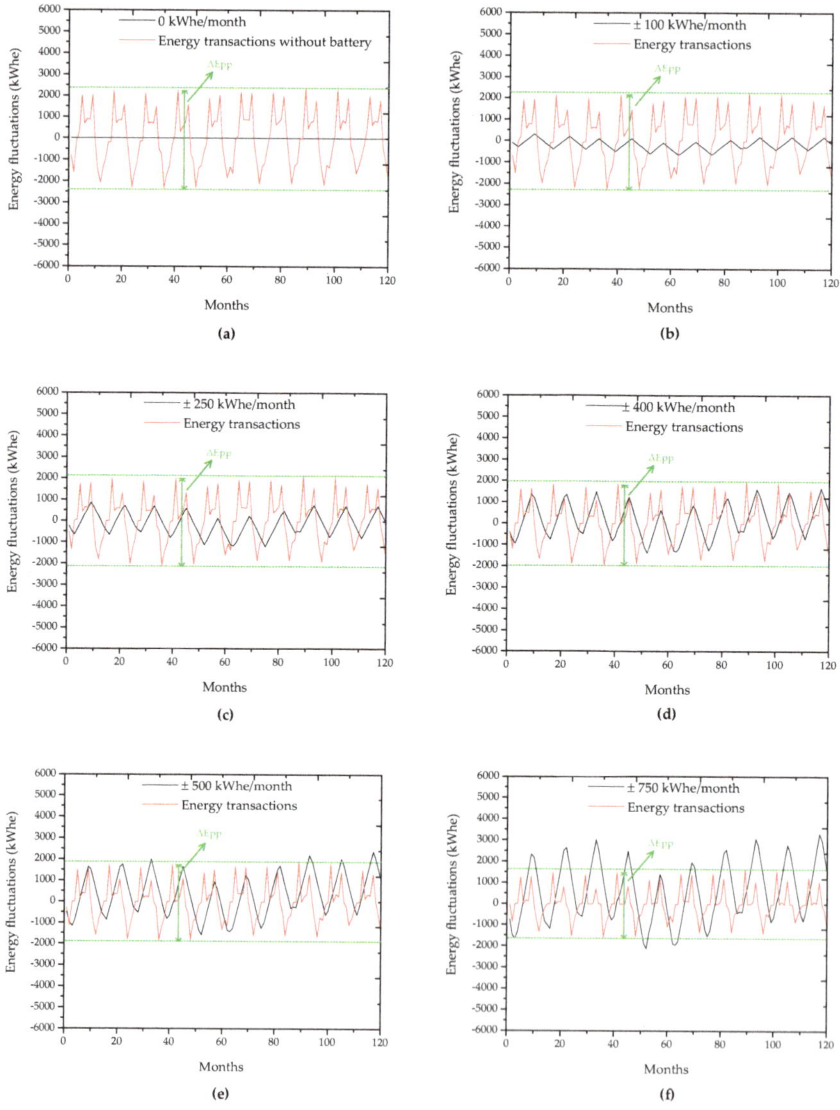

Figure 7. *Cont.*

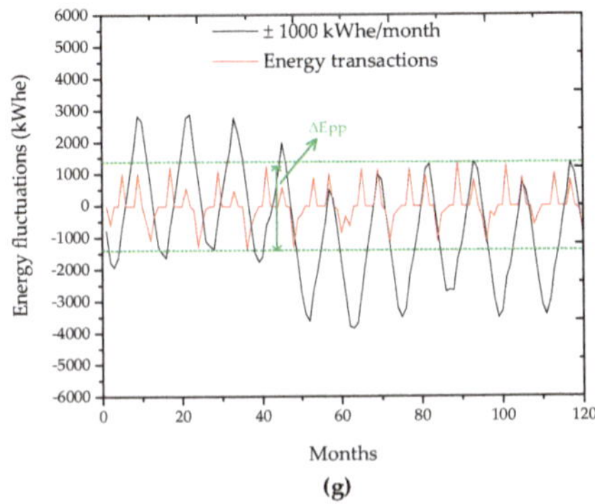
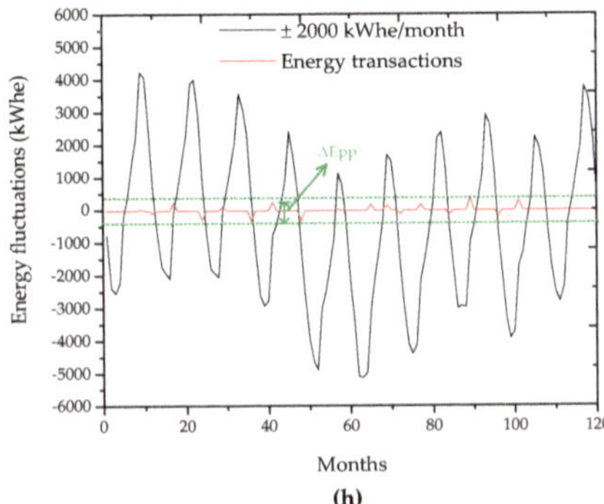

Figure 7. Monthly energy fluctuations in the EESU and energy transactions with the electricity grid for each scenario of EESU maximum permitted storage energy fluctuations. Maximum Permitted EESU Energy Fluctuations Per Month (kWh$_e$/month: (**a**) 0, (**b**) ±100, (**c**) ±250, (**d**) ±400, (**e**) ±500, (**f**) ±750, (**g**) ±1000, (**h**) ±2000.

The amount of energy that can be stored in the EESU depends on its acceptable maximum (SOC_{max}) and minimum (SOC_{min}) state of charge values. In this paper, it is assumed that SOC_{min} is 10% of the peak-to-peak stored energy fluctuation ($\Delta E_{peak-to-peak}$). Thus, the required storage capacity, $C_{storage}$, can be expressed as:

$$C_{storage} = SOC_{min} + \Delta E_{peak-to-peak} \tag{1}$$

During the EESU charging mode, the electrical energy is considered positive and vice versa. Table 3 shows the required storage capacity of each scenario, according to Equation (1) and the energy flows shown in Figure 7.

Table 3. Calculated electrical storage capacity for each scenario.

$C_{storage}$ (kWh)	Maximum Permitted EESU Energy Fluctuations Per Month (kWh$_e$/Month)
1040	±100
2264	±250
3377	±400
4342	±500
5988	±750
7400	±1000
13,080	±2000

The importance of sizing EESUs for the proposed collective self-consumption nZEB concept is related to the energy transactions with the electricity grid. This interrelation is presented in Figure 7; the more storage capacity is used (black line), the less energy transactions appear (red line). Of course, in case that not any EESU is used, the energy transactions with the grid are equal to the balance between the PV energy production and the energy consumption of the building. On the other hand, in case of 2000 kWh$_e$ per month permitted storage energy fluctuations, the energy transactions between the nZEB and the grid are mitigated. However, according to Table 3, this calls for unrealistic EESU capacities for building applications. Another deterrent for such an EESU scenario is its high cost and the danger for the residents and the building itself in case of damage. The economic analysis of the proposed system with and without an EESU is presented in the following sections.

Although all nZEBs do not present the same profile and size of power transactions with the electricity grid, their PV production is maximized at the same time interval (i.e., during midday). Having in mind the energy flows showed in Figure 7 (with and without EESUs) and considering a high penetration level of residential PV systems, it is concluded that various hazardous operating conditions (such as voltage rise, fault tripping and poor coordination of grid protection devices) could be emerged for the electricity grid [89]. Bearing in mind the above issues, the importance of EESU installation in nZEBs is highlighted—as a measure to secure electric grid operation.

2.6. Optimal EESU Sizing

In this section, an optimization sizing methodology for the EESU is proposed.

The optimal selection of the required storage capacity is based on the analysis presented in Section 2.5. In this context, the optimal storage capacity can be calculated using the following cost function:

$$e(K) = \left| \frac{C_{storage} - C_{storage,ref}}{C_{storage,ref}} \right| + \left| \frac{\Delta E_{grid,month} - \Delta E_{grid,month,ref}}{\Delta E_{grid,month,ref}} \right| \tag{2}$$

where $C_{storage}$ corresponds to the storage capacity of the EESU in kWh$_e$, derived by Equation (1), $\Delta E_{grid,month}$ is the absolute value of the maximum energy transaction between the building and the electricity grid during the 10-years period, in kWh$_e$. $C_{storage,ref}$ and $\Delta E_{grid,month,ref}$, are the reference values (weighting factors).

Also, the normalized variable K is defined by Equation (3):

$$K = \frac{\Delta E_{storage,month}}{\Delta E_{storage,month,ref}} \tag{3}$$

where $\Delta E_{storage,month}$ represents the maximum energy amount that can be stored in the EESU or it can be supplied to the building by the EESU, in kWh$_e$. In Equation (3), $\Delta E_{storage,month,ref}$ is arbitrary selected to be 250 kWh$_e$, and so variable K becomes an integer number which ranges between 0 and 8 (in regard to the scenarios discussed in Section 2.5).

In order to find the optimal storage capacity, the cost function must be minimized; according to the procedure described above, this means that the optimal EESU capacity is based on the energy transactions between the building and the electricity grid. Thus, in Table 4, the results of the optimization process regarding the EESU sizing is presented. Also, the reduction of the energy transactions for the optimum capacity of the EESU normalized in terms of building energy consumption (considering the nZEB concept), is illustrated in Figure 8.

Table 4. Optimization of electrical energy storage unit.

$C_{storage}$ (kWh)	$\Delta E_{grid,month}$ (kWh)	Reduction of Energy Transactions (%)	Optimum Storage Capacity Normalized in Terms of Building Energy Consumption (%)
0	2389	0	0
1040	2289	4.19	0.31
2264	2184	8.58	0.68
3377	1989	16.75	1.02
4342	1889	20.93	1.31
5988	1639	31.40	1.81
7400	1389	41.86	2.23
13,080	389	83.73	3.95

As shown in Figure 8, this study comes up to the conclusion that energy transactions between the building and the electricity grid reduce to a significant extent as the storage capacity increases (peak shavings). Moreover, it is estimated that the optimum EESU capacity is only a small ratio of the total energy consumption of the building during this 10-years period of analysis. The area below the red line shows that values from 0% to 0.68% refer to the practical application area, as higher energy storage values are not applicable at building level. Apparently, the increase of the EESU capacity enhances the independence from the electricity grid; however, as it will be discussed next, the investment cost is about to be extremely high. Last but not least, the results in Figure 8 highlight the effectiveness of the proposed collective self-consumption nZEB concept for multi-family buildings, as considerable EESU capacities can be exploited, reducing notably the energy transactions with the distribution grid.

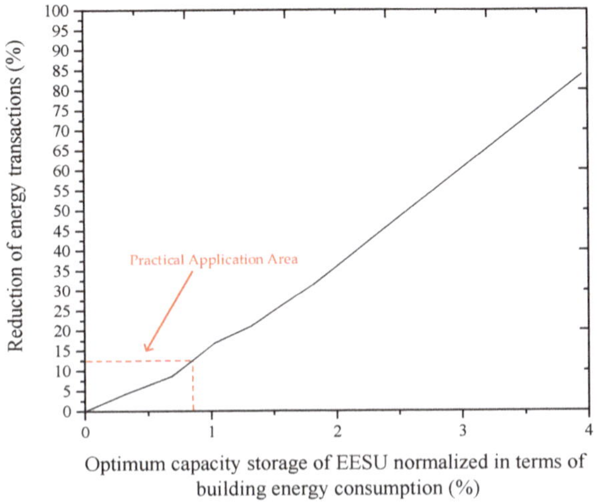

Figure 8. Interrelation between the energy transactions reduction and the optimum storage capacity.

3. Techno-Economic Analysis and Discussions

3.1. The LCCA Method

After the technical analysis of the energy system, in this section the cost-effectiveness of the proposed collective self-consumption nZEB is analyzed and discussed. As already noted, another important issue of this work is to calculate the total cost of the proposed investment, in order to estimate the payback period. The cost-effectiveness of the investment is the result of the LCCA method, by using the mathematical calculation procedure that will be fully analyzed in next sections. For the sake of calculations, LCCA is conducted for 25 years instead of 10 years. The option of the 25 years period analysis is related to the life cycle of the PV system.

In order to calculate the total life cycle cost of the abovementioned RES and EESU technologies combination, some economic and time parameters will be used, as presented in Table 5.

Table 5. Economic and time parameters for the calculation method of LCCA.

Economic Parameters	
Discount rate, d	0.25%, 0% (*)
Inflation rate, g	0.30%
Borrowing rate, i	0%
Down payment, D	0%
Escalation of energy costs, e	0.50%
Time Parameters	
Period of analysis, N	25 years
Borrowing period, N_Δ	10 years, 0 years (*)
(*) Values that correspond to the LCCA of the initial building	

The LCCA method calculates the total life cycle cost of the system, considering any future costs that are reduced to their present value (*PV*). In general, the reduction to present value of any investment, considering its increase due to the parameter of the inflation rate, is calculated as:

$$PV = X \times \frac{1}{d-g} \times \left[1 - \left(\frac{1+g}{1+d}\right)^N\right] = X \times PVF(d, g, N) \tag{4}$$

The purpose of the LCCA method used in this paper is to calculate:

(a) the Initial Cost (*IC*) of the investment, and
(b) the payback period of the proposed investment.

The total present value of the proposed investment (*PV*$_{TOT}$), is calculated by using Equation (5):

$$PV_{TOT} = PV_{SYS} + PV_{MISC} + PV_{REP} + PV_{ENER} - t \times PV_{INT} - PV_{ITC} - PV_{SV} \tag{5}$$

The parameters of Equation (4) will be described in next paragraphs, by giving their mathematical expressions:
(i) Present value of the total system costs (*PV*$_{SYS}$)

$$PV_{SYS} = D \times IC + PV_{LOAN} \tag{6}$$

PV$_{LOAN}$ is expressed by Equation (7):

$$PV_{LOAN} = (1 - D) \times IC \times \frac{PVF(d, 0, N_\Delta)}{PVF(i, 0, N_\Delta)}, \quad N \geq N \tag{7}$$

(ii) Present value of annual interest (*PV*$_{INT}$)

$$PV_{INT} = (1 - D) \times IC \times \left\{PVF(d, i, N_1) \times \left[i - \frac{1}{PVF(i, 0, N_\Delta)}\right] + \frac{PVF(d, 0, N_1)}{PVF(d, 0, N_\Delta)}\right\} \tag{8}$$

where: $N_1 = min(N_\Delta, N)$ and $D = 0$ (zero down payment is considered).
(iii) Present value of the funding of the investment (*PV*$_{ITC}$)

$$PV_{ITC} = \frac{I}{1+D} \tag{9}$$

(iv) Present value of operation and maintenance costs (*PV*$_{MISC}$)

$$PV_{MISC} = OM \times PVF(d, g, N) \tag{10}$$

OM costs are repetitive costs during the analysis period and can be calculated by using Equation (10):
(v) Present value of the energy cost that is supplied by the grid-according to the net metering scheme (*PV*$_{ENER}$)

$$PV_{ENER} = E \times PVF(d, e, N) \tag{11}$$

(vi) Present value of the replacement costs (*PV*$_{REP}$)
In many cases it may be necessary to replace some of the subsystems. The present value of the replacement cost is calculated by using Equation (12):

$$PV_{REP} = \frac{R}{(1+g)} \sum_{k=1}^{r} \left(\frac{1+g}{1+d}\right)^{\frac{Nk}{r+1}} \tag{12}$$

(vii) Present value of the remaining value at the end of the life cycle (PV_{SV})

The PV_{SV} calculation is related to the life cycle of the system. Given that PV_{SV} is difficult to be estimated, it is a common practice to extend the period of analysis up to the expected life period of the system. This is the reason for the LCCA extension to 25-years period, and so PV_{SV} becomes:

$$PV_{SV} = 0 \tag{13}$$

In next subsections, the LCCA of the building energy consumption and the proposed energy system is presented, according to the LCCA method analyzed above.

3.2. LCCA of the Proposed Energy System

In this section the results of the LCC calculations are presented. First, the total LCC of the building energy consumption before the installation of the proposed energy system is described. Then, the total LCC of the PV, the geothermal system and the EESU will be calculated.

3.2.1. LCCA of the Building Energy Consumption

As referred above, the annual electrical and thermal energy consumption is 26,400 kWh$_e$/y and 31,464 kWh$_{hth}$, respectively. Before the installation of the proposed energy system, the total energy consumption was covered by using the electricity grid and a diesel-boiler plant. According to the current rate prices of the electric energy and the diesel oil for heating purposes in Greece [90,91], the total annual costs of each consumption are 6391€ and 2260€, respectively. The rest costs of the building, according to the LCCA method, as well as some practical commercial data, are regarded as:

Initial Cost = 0€

Annual operation and maintenance costs = 700€

Replacement costs = 1500€ (i.e., replacement of the diesel boiler plant, $r = 1$)

Total energy cost = 8651€

Residual value at the end of the life cycle = 0€

Rate of subsidy = 0%

Thus, using Equations (5)–(13), as well as the data in Table 5, the total LCC of the building energy consumption is roughly 263k€. This total cost is essential in order to estimate the payback period and the energy and cost savings achieved by the proposed transformation to nZEB.

Next, the LCCA method is used for the proposed energy system, considering the net metering scheme; after the estimation of the total Initial Cost of the investment and the calculation of the total LCC of the system (LCC_{TOTAL}), the rate of the required subsidy is calculated, in order to make the whole investment sustainable.

3.2.2. LCCA of the PV System

As described in Section 2.4, the PV system is of 21 kWp nominal power in order to cover the 33,068 kWh$_e$ of electrical consumption. According to the LCCA method, as well as some practical commercial data, the costs of the PV system are estimated further down:

PV system cost = 21,600€

Annual operation and maintenance costs = 500 €/y

Other costs (insurance etc.) = 1200 €/y

Installation and interconnection costs (according to the Hellenic Electricity Distribution Network Operator) = 500€

Thus, the total Initial Cost of the PV system is roughly 22,100€. So, using Equation (5) and the data in Table 5, the total LCC of the PV system becomes 97,197€. Also, according to the LCC of the building energy consumption, the annual energy savings due to the PV energy production are equal to 6391 €/y.

3.2.3. LCCA of the Geothermal Energy System

The geothermal energy system has been designed to cover the 31,464 kWh$_{th}$ of thermal energy consumption. According to the LCC method, as well as some practical commercial data, the costs of the geothermal energy system are:

Initial Cost (construction costs of vertical loop system for 6 dwellings of 150 m^2) = 60,000€
Annual operation and maintenance costs = 4000 €/y
Replacement costs (one replacement of the heat pump unit) = 4500 €/y

Thus, the total LCC of the geothermal energy system is roughly 239 k€. Also, in comparison with the LCC of the building energy consumption, the annual energy savings due to the thermal production from the geothermal energy system are 2260 €/y.

3.2.4. LCCA of the TES Unit

As referred in Section 2.3, the TES unit is a 40 m^3 volume underground cylindrical water tank. The various costs of the TES system (according to some practical commercial data) are presented further down:

Tank cost (0.8€/lt) = 32,000€
Total Excavation costs (6 €/m^3) = 240€
Annual operation and maintenance costs = 640 €/y

Hence, the total Initial Cost of the TES is 32,240€, and the total LCC is 66,645€.

3.2.5. LCCA of the EESU

The EESU consists of a Li-ion battery bank. As discussed in Section 2.5, this study considers a number of alternative EESU capacity scenarios, in order to minimize the energy transactions between the building and the electricity grid on a monthly, yearly and 10-years basis. Thus, LCC is proceeded for all storage capacity scenarios and the LCC_{TOTAL} calculations of the whole investment are summarized in Table 6.

Table 6. LCCA of the whole investment for each scenario of EESU capacity.

$C_{storage}$ (kWh)	LCC_{TOTAL} (€)	Total Initial Cost (€)	Funding for the Investment (€)	Rate of Required Subsidy (%)
0	289k€	114k€	26k€	23.31
100	376k€	146k€	113k€	77.60
250	424k€	170k€	160k€	94.44
400	447k€	194k€	184k€	94.50
500	503k€	211k€	240k€	113.83
750	582k€	251k€	319k€	127.01
1000	662k€	292k€	398k€	136.5
2000	979k€	454k€	716k€	157.52

Several organizations have performed market analysis on Li-ion batteries (which is the selected EESU for our case too), providing forecasts regarding their price evolution over the next decades on the basis of some techno-economical hypotheses [92–95]. However, in the present work the considered future price formation of Li-ion batteries lies on the study of Bloomberg New Energy Finance Organization [91], due to the fact that these forecasts are the main scenario for the relevant market players. Thus, the present cost of Li-ion batteries is considered to be 162.5 €/kWh [92]. It is noted that at the 13th year (i.e., the year 2033) of the analysis period the EESU is going to be replaced. However, the cost of Li-ion batteries by this year is foreseen to be reduced to 74 €/kWh [92]. The various costs of the EESU system (taking into account some practical commercial data) are presented further down:

Annual operation and maintenance costs = 2% of the Initial Cost

Other costs (insurance etc.) = 1650 €/y
Cost of inverter unit = 150 €/kW
Cost of EESU activation = 150€
Cost for connection with the electricity grid = 300€.

In order to calculate the total LCC of the proposed investment, the regime of the net metering concept is also considered. According to the requirements of the Greek Net Metering scheme, prosumers avoid all charges (such as competitive charges, transmission and distribution grid charges, CO_2 emissions fees) for the electricity that they self-produce and consume simultaneously (self-utilization), with the sole exception of the social services fees. The excess energy (which is injected into the grid) is compensated with consumed energy; however, only competitive charges are subject to compensation.

As referred in Section 3.2.1, the LCC of the investment is calculated considering both self-consumed and injected PV electric energy. According to the household bills released by the HEDNO, the competitive charges for the electricity procurement are 0.04364 €/kWh$_e$. On the other hand, the surplus energy injected to the electricity grid is not credited.

The total amount of the excessive PV electricity which is injected into the grid and can be used at a later time on the 25-year basis, as well as the total cost that must be paid for this amount of PV energy, are presented in Table 7. Hereinafter and for the purposes of this study, the not directly consumed PV energy is referred as "absorbed energy".

Table 7. Cost of absorbed energy in each scenario of energy storage capacity.

$C_{storage}$ (kWh)	Total Absorbed Energy on 25 Year Basis (MWh$_e$)	Total Cost of Absorbed Energy (€)
0	159.2	6949
100	156.9	6848
250	152.6	6662
400	150.1	6554
500	146.5	6395
750	142.4	6215
1000	137.2	5991
2000	116.7	5097

The cost of the absorbed energy (under the net metering scheme), after the reduction to its present value ($PV_{ABSORBED}$), is calculated using Equation (14):

$$PV_{ABSORBED} = E_{ABSORBED} \times 0.04364 \times \frac{1}{d-e} \times \left[1 - \left(\frac{1+e}{1+d}\right)^N\right]$$
$$= E_{ABSORBED} \times 0.04364 \times PVF(d,e,N) \tag{14}$$

Next, the results of the total LCC of the proposed investment is presented in Table 6, by using Equation (15):

$$PV_{TOTAL} = PV_{SYS} + PV_{MISC} + PV_{ENER} + PV_{REP} + PV_{ABSORBED} - t \times PV_{INT} - PV_{ITC} - PV_{SV} \tag{15}$$

In addition, Table 7 depicts the necessary rate of subsidy for the investment depreciation within the 25-years period of analysis. According to these results, it seems that the proposed transformation to nZEB may remain sustainable as long as the installed EESU capacity remains below 400 kWh. This is an important outcome of this work, as it gives a concise sizing EESU capacity figure for the Mediterranean area, which can be exploited by the local subsidy policies that are about to be undertaken in order to transform the building sector into a nZEB one. Furthermore, the results in Table 7 highlight the fact that the installation of EESUs in an upper distribution level is more effective, in order to accommodate multiple nZEBs, e.g., in a common LVDS; hence, the mitigation of energy transactions with the electrical grid calls for the common share of larger-scale EESUs in the frame of LVDS and/or

Energy Communities [96,97]. Thus, Energy Communities can become beneficial for the reliability and the power quality of the grid.

4. Conclusions

This paper has proposed an efficient combination of RES technologies installed on a typical multi-family residential building in the urban environment, focusing on the Mediterranean area. This hybrid energy system consists of a PV and a geothermal unit, in order to meet electrical, thermal and cooling load demands. Furthermore, a methodology for sizing ESUs for thermal/cooling and electrical loads has been fully analyzed, so that the whole system has been designed considering the nZEB concept.

The technology that exploits geothermal energy is very efficient and environmentally friendly. However, an appropriate TES unit is needed, to meet thermal load deviations. In this paper, the TES has been sized so as to cover maximum load needs at any time of the year. This is a significant issue, as the main geothermal unit can balance only the average annual thermal/cooling load demand. The TES volume should be carefully designed, so that it could be undergroundly installed.

The analysis of the total life cycle cost of the proposed nZEB concept highlighted the fact that the proposed collective self-consumption nZEB concept for multi-family buildings is a sustainable way to transform the building sector of the Mediterranean area into a nZEB one; as for the installation of EESUs, this is facilitated by the collective self-consumption nZEB concept, according to the outcomes of this work. Nevertheless, as previously discussed, a great potential is foreseen in installing larger EESUs at LVDS level or in the frame of Energy Communities concept, enhancing so the reliability and the power quality of the electrical grid.

Moreover, the proposed collective self-consumption nZEB concept could be applicable even in large scale buildings, where there is no available space for RES systems to be installed, by exploiting the advantages of virtual net metering scheme.

As far as EESUs are concerned, an optimal methodology has been presented for their sizing, taking into account the corresponding energy transactions with the electricity grid. EESUs installation importance is given, as the more EESU capacity, the less energy transactions with the electricity grid are observed. In this context, the proposed dimensioning methodology contributes to avoid a negative impact on the operation of the electricity grid (such as reverse power flow, voltage rise, false tripping of protection measures, etc.) as a result of the increased penetration of residential PV systems into the same feeder.

Indeed, as we reach the 2030 target, the number of nZEB buildings and the capacity of intermittent RES systems (mainly PVs in densely populated areas) will be increasing, whilst electricity system inertia will be decreasing due to the replacement of synchronous generators with electronically coupled RES units. Thus, in order to secure the operation of the electricity grid and to avoid a curtailment operation for the RES units, mass EESUs installation either in the electricity grid or buildings is presumed. Therefore, the analysis on the energy transactions between nZEBs and the electricity grid as well as the optimal sizing of EESUs that have been presented in this work are of great interest for DSOs and policy makers, providing insightful assessments on the necessity of subsidy programs for nZEB transformation as well as on the grid impact due to the fluctuating production profile of PVs.

Author Contributions: Conceptualization, F.K., A.K. and N.P.; methodology, F.K., A.K. and N.P.; software, F.K. and A.K.; validation, A.K. and N.P.; formal analysis, F.K., A.K. and N.P.; investigation, F.K. and A.K.; data curation, N.P.; writing—original draft preparation, F.K.; writing—review and editing, A.K. and N.P.; visualization, A.K.; supervision, N.P.; funding acquisition, A.K. and N.P. All authors have read and agreed to the published version of the manuscript.

Funding: The research leading to these results has received partial funding from the European Union HORIZON 2020 Programme "EU HEROES" under grant agreement no 764805. The authors are solely responsible for the content of this publication.

Conflicts of Interest: The authors declare no conflict of interest.

Nomenclature

AC	Alternative Current
ASHP	Air-source heat pump
BAPV	Building Applied Photovoltaic System
BEA	Building Energy Analysis
BHEs	Borehole heat exchangers
BIPV	Building Integrated Photovoltaic System
BIPV/T	Building Integrated Photovoltaic/Thermal system
CCHP	Combined cooling, heating and power system
CHP	Cooling, heating and power system or cogeneration system
CO_2	Carbon dioxide
COP	Coefficient of performance
$C_{storage}$	Storage capacity (kWh$_e$)
$C_{storage,ref}$	Reference value of storage capacity (kWh$_e$)
d	Discount rate (%)
D	Down payment (%)
DHW	Domestic hot water
DSO	Distribution System Operator
e	Escalation of energy costs (%)
$e(K)$	Cost function of optimal storage capacity
E	Annual cost of the energy (€)
$E_{ABSORBED}$	Absorbed energy (€)
EAHP	Exhaust air heat pump
EER	Energy efficiency ratio
EESU	Electrical energy storage unit
ESS	Energy storage system
ESUs	Energy storage units
EU	European Union
g	Inflation rate (%)
GCHP	Gound-coupled heat pump
GDHS	Geothermal district heating system
GHEs	Geothermal heat exchangers or Ground heat exchangers
GHP	Geothermal heat pump
GSHP	Ground-source heat pump
GWHP	Ground-water heat pump
HEDNO	Hellenic Electricity Distribution Network Operator
HPSU	Heat pump storage unit
HVAC	Heating, ventilation and air-conditioning
i	Borrowing rate (%)
I	Funding of the investment (%)
IC	Initial Cost (€)
K	Normalized variable
LCC_{TOTAL}	Total life cycle cost of the building (€)
LCCA	Life Cycle Cost Assessment
Li-ion	Lithium iodide battery
LPG	Liquified Pretroleum Gas
LVDS	Low Voltage Distribution System
N	Period of analysis (years)
N_1	The minimum between N and N_1 (years)
N_Δ	Borrowing period (years)
nZEB	nearly Zero Energy Building
OM	Operation and Maintenace (€)
PV	Photovoltaic system
PV	Present Value (€)

$PV_{ABSORBED}$	Present value of absorbed energy (€)
PV_{ENER}	Present value of the energy cost that is supplied by the grid (€)
PV_{INT}	Present value of the annual interest (€)
PV_{ITC}	Present value of the funding of the investment (€)
PV_{LOAN}	Present value of the lean installment (€)
PV_{MISC}	Present value of the various costs during the life cycle, such as maintenance costs (€)
PV_{REP}	Present value of the replacement costs (€)
PV_{SYS}	Present value of total system cost (€)
PV_{SV}	Present value of the system remaining value at the end of the period of analysis (€)
PV_{TOTAL}	Total present value of the investment (€)
PVF	Present Value Function (€)
PV/GHP	Photovoltaic/Geothermal heat pump system
PV/ST	Photovoltaic/Solar thermal system
PV/T	Photovoltaic/Thermal system
r	The number of replacements during the analysis period
R	Replacement cost of a subsystem with reference to the first year of operation (€)
RES	Renewable Energy Sources
SOC_{max}	Maximum State of Charge value (%)
SOC_{min}	Mimimum State of Charge value (%)
STC	Standard Test Conditions
t	Income tax (%)
TES	Thermal energy storage
UTES	Underground thermal energy storage
WPP	Wind power plant
X	Any investment to be analyzed (€)
$\Delta E_{grid,\ month}$	Absolute value of the maximum energy transaction between the building and the electricity grid during the 10 – years period (kWh$_e$)
$\Delta E_{grid,\ month,ref}$	Reference value of maximum energy transaction between the building and the electricity grid during the 10 – years period (kWh$_e$)
$\Delta E_{peak\text{-}to\text{-}peak}$	Peak-to-peak stored energy fluctuation in EESU (kWh$_e$)
$\Delta E_{storage,\ month}$	Maximum permitted energy fluctuation in the EESU per month (kWh$_e$)
$\Delta E_{storage,\ month,ref}$	Reference value of maximum permitted energy fluctuation in the EESU per month (kWh$_e$)

References

1. D' Agostino, D.; Mazzarella, L. What is a Nearly zero energy building? Overview, implementation and comparison of definitions. *J. Build. Eng.* **2019**, *21*, 200–212. [CrossRef]
2. Green Paper, A 2030 Framework for Climate and Energy Policies COM 169. Available online: https://eur-lex.europa.eu/legal-content/EN/TXT/PDF/?uri=CELEX:52013DC0169&from=EN (accessed on 23 January 2020).
3. EU. Directive 2010/31/EU 2010 of the European Parliament and of the Council of 19 May 2010 on the Energy Performance of Buildings (recast). Available online: https://eur-lex.europa.eu/legal-content/EN/TXT/PDF/?uri=CELEX:32010L0031&from=el (accessed on 10 January 2020).
4. EU. Directive 2018/844 of the European Parliament and of the Council of 30 May 2018, Amending Directive 2010/31/EU on the Energy Performance of Buildings and Directive 2012/27/EU on Energy Efficiency. Available online: https://eur-lex.europa.eu/legal-content/EN/TXT/?uri=uriserv%3AOJ.L_.2018.156.01.0075.01.ENG (accessed on 7 February 2020).
5. Kyritsis, A.; Mathas, E.; Antonucci, D.; Grottke, M.; Tselepis, S. Energy improvement of office buildings in Southern Europe. *Energy Build.* **2016**, *123*, 17–33. [CrossRef]
6. Carli, R.; Dotoli, M. Decentralized control for residential energy management of a smart users ' microgrid with renewable energy exchange. *IEEE/CAA J. Autom. Sin.* **2019**, *6*, 641–656. [CrossRef]
7. Geidl, M.; Koeppel, G.; Favre-Perrod, P.; Klockl, B.; Andersson, G.; Frohlich, K. Energy hubs for the future. *IEEE Power Energy Mag.* **2007**, *5*, 24–30. [CrossRef]

8.   Hosseini, S.M.; Carli, R.; Dotoli, M. Robust Day-Ahead Energy Scheduling of a Smart Residential User Under Uncertainty. In Proceedings of the 2019 18th European Control Conference (ECC), Naples, Italy, 25–28 June 2019; pp. 935–940. [CrossRef]

9.   Park, S.; Salkuti, S.R. Optimal Energy Management of Railroad Electrical Systems with Renewable Energy and Energy Storage Systems. *Sustainability* **2019**, *11*, 6293. [CrossRef]

10.   EU. Directive 2018/2001 of the European Parliament and of the Council of 11 December 2018 on the Promotion of the Use of Energy from Renewable Sources (Recast). Available online: https://eur-lex.europa.eu/legalcontent/EN/TXT/PDF/?uri=CELEX:32018L2001&qid=1580204680261&from=EN (accessed on 10 January 2020).

11.   CEER Report C18-CRM9_DS7-05-03, 25. Regulatory Aspects of Self-Consumption and Energy Communities, Council of European Energy Regulators Asbl. Available online: https://www.ceer.eu/documents/104400/-/-/8ee38e61-a802-bd6f-db27-4fb61aa6eb6a (accessed on 10 January 2020).

12.   Marsza, A.J.; Heiselberg, P.; Bourrelle, J.S.; Musall, E.; Voss, K.; Sartori, I.; Napolitano, A. Zero Energy Building–A review of definitions and calculation methodologies. *Energy Build.* **2011**, *43*, 971–979. [CrossRef]

13.   Cao, X.; Dai, X.; Liu, J. Building energy-consumption status worldwide and the state of- the-art technologies for zero-energy buildings during the past decade. *Energy Build.* **2016**, *128*, 198–213. [CrossRef]

14.   Semprini, G.; Gulli, R.; Ferrante, A. Deep regeneration vs shallow renovation to achieve nearly Zero Energy in existing buildings: Energy saving and economic impact of design solutions in the housing stock of Bologna. *Energy Build.* **2017**, *156*, 327–342. [CrossRef]

15.   Ferrante, A.; Cascella, M.T. Zero energy balance and zero on-site CO_2 emission housing development in the Mediterranean climate. *Energy Build.* **2011**, *43*, 2002–2010. [CrossRef]

16.   Ascione, F.; Borrelli, M.; Francesca De Masi, R.; de Rossi, F.; Vanoli, G.P. A framework for nZEB design in Mediterranean climate: Design, building and set-up monitoring of a lab-small villa. *Sol. Energy* **2019**, *184*, 11–29. [CrossRef]

17.   Gaglia, A.G.; Dialynas, E.N.; Argiriou, A.A.; Kostopoulou, E.; Laskos, K.M. Energy performance of European residential buildings: Energy use, technical and environmental characteristics of the Greek residential sector–energy conservation and CO_2 reduction. *Energy Build.* **2019**, *183*, 86–104. [CrossRef]

18.   Rey, F.J.; Velasco, E.; Varela, F. Building energy analysis (BEA): A methodology to assess building energy labeling. *Energy Build.* **2007**, *39*, 709–716. [CrossRef]

19.   Calero, M.; Alameda-Hernandez, E.; Fernández-Serrano, M.; Ronda, A.; Martín-Lara, M.Á. Energy consumption reduction proposals for thermal systems in residential buildings. *Energy Build.* **2018**, *175*, 121–130. [CrossRef]

20.   Pacheco, R.; Ordóñez, J.; Martínez, G. Energy efficient design of building: A review. *Renew. Sustain. Energy Rev.* **2012**, *16*, 3559–3573. [CrossRef]

21.   Feng, Y. Thermal design standards for energy efficiency of residential buildings in hot summer/cold winter zones. *Energy Build.* **2004**, *36*, 1309–1312. [CrossRef]

22.   Kampelis, N.; Gobakis, K.; Vagias, V.; Kolokotsa, D.; Standardi, L.; Isidori, D.; Cristalli, C.; Montagninoc, F.M.; Paredes, F.; Muratore, P.; et al. Evaluation of the performance gap in industrial, residential & tertiary near-Zero energy buildings. *Energy Build.* **2017**, *148*, 58–73. [CrossRef]

23.   Karlessi, T.; Kampelis, N.; Kolokotsa, D.; Santamouris, M.; Standardi, L.; Isidori, D.; Cristalli, C. The concept of smart an nZEB buildings and the integrated design approach. *Procedia Eng.* **2017**, *180*, 1316–1325. [CrossRef]

24.   Kalogirou, S.A.A.; Tripanagnostopoulos, Y. Hybrid PV/T solar systems for domestic hot water and electricity production. *Energy Convers. Manag.* **2006**, *47*, 3368–3382. [CrossRef]

25.   Yin, H.M.; Yang, D.J.; Kelly, G.; Garant, J. Design and performance of a novel building integrated PV/thermal system for energy efficiency of buildings. *Sol. Energy* **2013**, *87*, 184–195. [CrossRef]

26.   Delisle, V.; Kummert, M. A novel approach to compare building-integrated photovoltaics/thermal air collectors to side-by-side PV modules and solar thermal collectors. *Sol. Energy* **2014**, *100*, 50–65. [CrossRef]

27.   Agrawal, B.; Tiwari, G.N. Optimizing the energy and exergy of building integrated photovoltaic thermal (BIPVT) systems under cold climatic conditions. *Appl. Energy* **2010**, *87*, 417–426. [CrossRef]

28.   Kyritsis, A.; Roman, E.; Kalogirou, S.A.; Nikoletatos, J.; Agathokleous, R.; Mathas, E.; Tselepis, S. Households with Fibre Reinforced Composite BIPV modules in Southern Europe under Net Metering Scheme. *Renew. Energy* **2019**, *137*, 167–176. [CrossRef]

29. López, P.C.S.; Frontini, F. Energy Efficiency and Renewable Solar Energy Integration in Heritage Historic Buildings. *Energy Procedia* **2014**, *48*, 1493–1502. [CrossRef]

30. López, P.C.S.; Lucchi, E.; Franco, G. Acceptance of Building Integrated Photovoltaic (BIPV) in heritage buildings and landscapes: Potentials, barriers and assessment criteria. In Proceedings of the Rehabend Congress 2020, Construction Pathology, Rehabilitation Technology and Heritage Management, Granada, Spain, 24–27 March 2020.

31. Jelle, B.P.; Breivik, C.; Drolsum Røkenes, H. Building integrated photovoltaic products: A state-of-the-art review and future research opportunities. *Sol. Energy Mater. Sol. Cells* **2012**, *100*, 69–96. [CrossRef]

32. Cerón, I.; Caamaño-Martín, E.; Javier Neila, F. 'State-of-the-art' of building integrated photovoltaic products. *Renew. Energy* **2013**, *58*, 127–133. [CrossRef]

33. Attoye, D.E.; Aoul, K.A.T.; Hassan, A. A Review on Building Integrated Photovoltaic Façade Customization Potentials. *Sustainability* **2017**, *9*, 1–24.

34. Sánchez-Pantoja, N.; Vidal, R.; Carmen Pastor, M. Aesthetic perception of photovoltaic integration within new proposals for ecological architecture. *Sustain. Cities Soc.* **2018**, *39*, 203–214. [CrossRef]

35. Ondeck, A.D.; Edgar, T.F.; Baldea, M. Optimal operation of a residential district-level combined photovoltaic/natural gas power and cooling system. *Appl. Energy* **2015**, *156*, 593–606. [CrossRef]

36. Postnikov, I.; Stennikov, V.; Penkovskii, A. Integrated energy supply schemes on basis of cogeneration Plants and wind power plants. *Energy Procedia* **2019**, *158*, 154–159. [CrossRef]

37. Lindenberger, D.; Bruckner, T.; Groscurth, H.M.; Kummel, R. Optimization of solar district heating systems: Seasonal storage, heat pumps, and cogeneration. *Energy* **2000**, *25*, 591–608. [CrossRef]

38. Thygesen, R.; Karlsson, B. An analysis on how proposed requirements for near zero energy buildings manages PV electricity in combination with two different types of heat pumps and its policy implications–A Swedish example. *Energy Policy* **2017**, *101*, 10–19. [CrossRef]

39. Alnaser, N.W. Building integrated renewable energy to achieve zero emission in Bahrain. *Energy Build.* **2015**, *93*, 32–39. [CrossRef]

40. Thygesen, R.; Karlsson, B. Simulation and analysis of a solar assisted heat pump system with two different storage types for high levels of PV electricity self-consumption. *Sol. Energy* **2014**, *103*, 19–27. [CrossRef]

41. Gong, H.; Rallabandi, V.; Ionel, D.M.; Colliver, D.; Duerr, S.; Ababei, C. Net Zero Energy Houses with Dispatchable Solar PV Power Supported by Electric Water Heater and Battery Energy Storage. In Proceedings of the 2018 IEEE Energy Conversion Congress and Exposition (ECCE), Portland, OR, USA, 23–27 September 2018; pp. 2498–2503. [CrossRef]

42. Akter, M.N.; Mahmud, M.A.; Amanullah, M.T.O. Comprehensive economic analysis of a residential building with solar photovoltaic and battery energy storage system An Australian case study. *Energy Build.* **2017**, *138*, 332–346. [CrossRef]

43. Blum, P.; Campillo, G.; Munch, W.; Kolbel, T. CO_2 savings of ground source heat pump systems–A regional analysis. *Renew. Energy* **2010**, *35*, 122–127. [CrossRef]

44. Sivasakthivel, T.; Murugesan, K.; Sahoo, P.K. Potential Reduction in CO_2 Emission and Saving in Electricity by Ground Source Heat Pump System for Space Heating Applications-A Study on Northern Part of India. International Conference on Modeling Optimization and Computing (ICMOC -2012), April 10 & 11. *Procedia Eng.* **2012**, *38*, 970–979. [CrossRef]

45. Lo Russo, S.; Boffa, C.; Civita, M.V. Low-enthalpy geothermal energy: An opportunity to meet increasing energy needs and reduce CO_2 and atmospheric pollutant emissions in Piemonte, Italy. *Geothermics* **2009**, *38*, 254–262. [CrossRef]

46. Knudstrup, M.A.; Ring-Hansen, H.T.; Brunsgaard, C. Approaches to the design of sustainable housing with low CO_2 emission in Denmark. *Renew. Energy* **2009**, *34*, 2007–2015. [CrossRef]

47. Kaushal, M. Geothermal cooling heating using GHE for various experimental and analytical studies: Comprehensive review. *Energy Build.* **2017**, *139*, 634–652. [CrossRef]

48. Florides, G.; Kalogirou, S. Ground heat exchangers-a review of systems, models and applications. *Renew. Energy* **2007**, *32*, 2461–2478. [CrossRef]

49. Self, S.J.; Reddy, B.V.; Rosen, M.A. Geothermal heat pump systems: Status review and comparison with other heating options. *Appl. Energy* **2013**, *101*, 341–348. [CrossRef]

50. Niemelä, T.; Manner, M.; Laitinen, A.; Sivula, T.M.; Kosonen, R. Computational and experimental performance analysis of a novel method for heating of domestic hot water with a ground source peat pump system. *Energy Build.* **2018**, *161*, 22–40. [CrossRef]
51. Zeng, H.Y.; Diao, N.R.; Fang, Z.H. Heat transfer analysis of boreholes in vertical ground heat exchangers. *Int. J. Heat Mass Transf.* **2003**, *46*, 4467–4481. [CrossRef]
52. Fu, Y.; Haiyang, L.; Sun, K.; Sun, Q.; Wnnersten, R. A multi objective optimization of PV/ST-GSHP system based on office buildings. *Energy Procedia* **2018**, *152*, 71–76. [CrossRef]
53. Fadejev, J.; Simson, R.; Kurnitski, J.; Kesti, J.; Mononen, T.; Lautso, P. Geothermal heat pump performance in a nearly zero-energy building. *Energy Procedia* **2016**, *96*, 489–502. [CrossRef]
54. Lv, Y.; Si, P.; Liu, J.; Ling, W.; Yan, J. Performance of a Hybrid Solar Photovoltaic - Air Source Heat Pump System with Energy Storage. *Energy Procedia* **2018**, *158*, 1311–1316. [CrossRef]
55. Lentz, A.; Almanza, R. Solar–geothermal hybrid system. *Appl. Therm. Eng.* **2006**, *26*, 1537–1544. [CrossRef]
56. Zhou, C.; Doroodchi, E.; Moghtaderi, B. An in-depth assessment of hybrid solar–geothermal power generation. *Energy Convers. Manag.* **2013**, *74*, 88–101. [CrossRef]
57. Li, H.; Xu, W.; Yu, Z.; Wu, J.; Yu, Z. Discussion of a combined solar thermal and ground source heat pump operation strategy for office heating. *Energy Build.* **2018**, *162*, 42–53. [CrossRef]
58. Dai, L.; Li, S.; DuanMu, L.; Li, X.; Shang, Y.; Dong, M. Experimental performance analysis of a solar assisted ground source heat pump system under different heating operation modes. *Appl. Therm. Eng.* **2015**, *75*, 325–333. [CrossRef]
59. Bi, Y.; Guo, T.; Zhang, L.; Chen, L. Solar and ground source heat-pump system. *Appl. Energy* **2004**, *78*, 231–245. [CrossRef]
60. Trillat-Berdal, V.; Souyri, B.; Fraisse, G. Experimental study of a ground-coupled heat pump combined with thermal solar collectors. *Energy Build.* **2006**, *38*, 1477–1484. [CrossRef]
61. Ozgener, O.; Hepbasli, A. Performance analysis of a solar assisted ground-source heat pump system for greenhouse heating: An experimental study. *Build. Environ.* **2005**, *40*, 1040–1050. [CrossRef]
62. Rad, F.M.; Fung, A.S.; Leong, W.H. Feasibility of combined solar thermal and ground source heat pump systems in cold climate, Canada. *Energy Build.* **2003**, *61*, 224–232. [CrossRef]
63. Sanner, B.; Karytsas, C.; Mendrinos, D.; Rybach, L. Current status of ground source heat pumps and underground thermal energy storage in Europe. *Geothermics* **2003**, *32*, 579–588. [CrossRef]
64. Franco, A.; Fabio, F. Experimental analysis of a self-consumption strategy for residential building: The integration of PV system and geothermal heat pump. *Renew. Energy* **2016**, *86*, 1075–1085. [CrossRef]
65. Ma, W.; Fang, S.; Liu, G. Hybrid optimization method and seasonal operation strategy for distributed energy system integrating CCHP, photovoltaic and ground source heat pump. *Energy* **2017**, *141*, 1439–1455. [CrossRef]
66. Kolokotsa, D.; Kampelis, N.; Mavrigiannaki, A.; Gentilozzi, M.; Paredes, F.; Montagnino, F.; Venezia, L. On the integration of the energy storage in smart grids: Technologies and applications. *Energy Storage* **2019**, *1*. [CrossRef]
67. Roberts, B.P.; Sandberg, C. The role of energy storage in development of smart grids. *Proc. IEEE* **2011**, *99*, 1139–1144. [CrossRef]
68. Li, H.; Qu, L.; Qiao, W. Life-cycle cost analysis for wind power converters. In Proceedings of the 2017 IEEE International Conference on Electro Information Technology (EIT), Lincoln, NE, USA, 14–17 May 2017; pp. 630–634. [CrossRef]
69. Silva, S.M.; Mateus, R.; Marques, L.; Ramos, M.; Almeida, M. Contribution of the solar systems to the nZEB and ZEB design concept in Portugal – Energy, economics and environmental life cycle analysis. *Sol. Energy Mater. Sol. Cells* **2016**, *156*, 59–74. [CrossRef]
70. Jacob, A.S.; Das, J.; Abraham, A.P.; Banerjee, R.; Ghosh, P.C. Cost and Energy Analysis of PV Battery Grid Backup System for a Residential Load in Urban India. *Energy Procedia* **2017**, *118*, 88–94. [CrossRef]
71. Marszal, A.J.; Heiselberg, P. Life cycle cost analysis of a multi-storey residential Net Zero Energy Building in Denmark. *Energy* **2011**, *36*, 5600–5609. [CrossRef]
72. Ates, S.A. Life cycle cost analysis: An evaluation of renewable heating systems in Turkey. *Energy Explor. Exploit.* **2015**, *33*, 621–638. [CrossRef]

73. Sullivan, J.L.; Clark, C.E.; Han, J.; Wang, M. Life-Cycle Analysis Results of Geothermal Systems in Comparison to Other Power Systems, Center for Transportation Research Energy Systems Division, Argonne National Laboratory August 2010. Available online: http://www.osti.gov/bridge (accessed on 10 January 2020).

74. Yilmaz, C. Life cycle cost assessment of a geothermal power assisted hydrogen energy system. *Geothermics* **2020**, *83*, 101737. [CrossRef]

75. Deng, Y.; Li, J.; Li, T.; Gao, X.; Yuan, C. Life cycle assessment of lithium sulfur battery for electric vehicles. *J. Power Sources* **2017**, *343*, 284–295. [CrossRef]

76. Schoenung, S.M.; Hassenzahl, W.V. Long- vs. Short-Term Energy Storage Technologies Analysis A Life-Cycle Cost Study A Study for the DOE Energy Storage Systems Program, August, 2003. Available online: https://prod-ng.sandia.gov/techlib-noauth/access-control.cgi/2003/032783.pdf (accessed on 10 January 2020).

77. Karanasios, E.; Ampatzis, M.; Nguyen, P.H.; Kling, W.L.; van Zwam, A. A model for the estimation of the cost of use of Li-Ion batteries in residential storage applications integrated with PV panels. In Proceedings of the 2014 49th International Universities Power Engineering Conference (UPEC), Cluj-Napoca, Romania, 2–5 September 2014; pp. 1–6. [CrossRef]

78. Mehdijev, S. Dimensioning and Life Cycle Costing of Battery Storage System in Residential Housing-A Case Study of Local System Operator Concept. Master of Science Thesis, Project in Electric Power Engineering, Stockholm, Sweden, June 2017. Available online: http://www.diva-portal.org/smash/get/diva2:1130036/FULLTEXT01.pdf (accessed on 10 January 2020).

79. Droutsa, P.; Kontogiannidis, S.; Dascalaki, E.; Blaras, C. Mapping the energy performance of hellenic residential buildings from EPC (energy performance certificate) data. *Energy* **2016**, *98*, 284–295. [CrossRef]

80. EU. Commission Recommendation 2016/1318/EU of 29 July 2016 on Guidelines for the Promotion of Nearly Zero-Energy Buildings and Best Practices to Ensure That, by 2020, All New Buildings are Nearly Zero-Energy Buildings. Available online: https://eur-lex.europa.eu/legal-content/EN/TXT/PDF/?uri=CELEX:32016H1318&from=EL (accessed on 7 February 2020).

81. Lucchi, E.; Tabak, M.; Troi, A. The "Cost Optimality" Approach for the Internal Insulation of Historic Buildings. *Energy Procedia* **2017**, *133*, 412–423. [CrossRef]

82. Paduos, S.; Corrado, V. Cost-optimal approach to transform the public buildings into nZEBs: A European cross-country comparison. *Energy Procedia* **2017**, *140*, 314–324. [CrossRef]

83. Hellenic Statistic Authority 2013. Survey on Energy Consumption in Households 2011-2012 Press Release. Available online: https://www.statistics.gr/documents/20181/985219/Energy+consumption+in+households/ (accessed on 10 January 2020).

84. Braun, M.R.; Walton, P.; Beck, S.B.M. Illustrating the relationship between the coefficient of performance and the coefficient of system performance by means an R404 supermarket refrigeration system. *Int. J. Refrig.* **2016**, *70*, 225–234. [CrossRef]

85. Kyritsis, A.; Papanikolaou, N.; Christodoulou, C.; Gkonos, G.; Tselepis, S. Installation guidelines: Electrical. In *McEvoy's Handbook of Photovoltaics*, 3rd ed.; Kalogirou, S.A., Ed.; Academic Press: Cambridge, MA, USA, 2018; pp. 891–914. [CrossRef]

86. Odyssee- Mure. Available online: https://www.odyssee-mure.eu/project.html (accessed on 7 February 2020).

87. Lund, J.W.; Freeston, D.H.; Boyd, T.L. Direct utilization of geothermal energy 2010 worldwide review. *Geothermics* **2011**, *40*, 159–180. [CrossRef]

88. The European Commission's Science and Knowledge Service. Available online: https://ec.europa.eu/jrc/en/pvgis (accessed on 10 January 2020).

89. Kyritsis, A.; Voglitsis, D.; Papanikolaou, N.; Tselepis, S.; Christodoulou, C.; Gonos, I.; Kalogirou, S.A. Evolution of PV systems in Greece and review of applicable solutions for higher penetration levels. *Renew. Energy* **2017**, *109*, 487–499. [CrossRef]

90. Ministry of Economy and Development/General Secretariat of Commerce and Consumer Protection/Department of Prices Observatories. Available online: http://oil.gge.gov.gr/ (accessed on 10 January 2020).

91. Public Power Corporation, S.A. Hellas. Residential Tariff. Available online: https://www.dei.gr/en/oikiakoi-pelates/timologia/oikiako-timologio-xwris-xronoxrewsi-g1 (accessed on 10 January 2020).

92. Bloomberg NEF. Available online: https://about.bnef.com/ (accessed on 10 January 2020).

93. EASE/EERA, European Energy Storage Technology Development Roadmap. 2017. Available online: https://eera-es.eu/wp-content/uploads/2016/03/EASE-EERA-Storage-Technology-Development-Roadmap-2017-HR.pdf (accessed on 19 February 2020).

94. Tsiropoulos, I.; Tarvydas, D.; Lebedeva, N. Li-ion batteries for mobility and stationary storage applications–Scenarios for costs and market growth. 2018. Available online: https://publications.jrc.ec.europa.eu/repository/bitstream/JRC113360/kjna29440enn.pdf (accessed on 19 February 2020).

95. IRENA, Electricity Storage and Renewables: Costs and Markets to 2030. International Renewable Energy Agency, Abu Dhabi 2017. Available online: https://www.irena.org/publications/2017/Oct/Electricity-storage-and-renewables-costs-and-markets (accessed on 19 February 2020).

96. Heldeweg, M.A.; Saintier, S. Renewable energy communities as 'socio-legal institutions': A normative frame for energy decentralization? *Renew. Sustain. Energy Rev.* **2020**, *119*, 109518. [CrossRef]

97. Ceglia, F.; Esposito, P.; Marrasso, E.; Sasso, M. From smart energy community to smart energy municipalities: Literature review, agendas and pathways. *J. Clean. Prod.* **2020**, *254*, 120118. [CrossRef]

Article

The Cost-Optimal Analysis of a Multistory Building in the Mediterranean Area: Financial and Macroeconomic Projections

Roberto Bruno *, Piero Bevilacqua, Cristina Carpino and Natale Arcuri

Mechanical, Energetic and Management Engineering Department, University of Calabria, 87036 Rende, Italy; piero.bevilacqua@unical.it (P.B.); cristina.carpino@unical.it (C.C.); natale.arcuri@unical.it (N.A.)
* Correspondence: roberto.bruno@unical.it

Received: 6 February 2020; Accepted: 4 March 2020; Published: 7 March 2020

Abstract: Cost-optimal analysis was pointed out in the 2010/31 European Directive as a tool to evaluate the achievable building energy performance levels as a function of the corresponding costs. These analyses can be carried out by a financial projection for private investors and a macroeconomic approach to establish the minimal energy performance levels. Consequently, the financial projection provides different results that could stimulate private investors toward other cost-optimal solutions that do not match the minimal energy performance levels. For this purpose, both the projections were analyzed in the BEopt environment, developed by NREL, on a multistory building located in two contrasting climatic zones of the Mediterranean area, one cold and the other warm, highlighting the differences. The cost-optimal solutions were identified by a parametric study involving measures that affect thermal losses and solar gains, whereas the air-conditioning plant was left unchanged in order to include a fraction of renewable energy in the coverage of the building demands. Results showed that both the projections produced the same cost-optimal solutions, however, the latter matches the building designed to fulfill the minimal energy performance levels only in the cold climate. Conversely, noticeable deviations were detected in the warm location, therefore minimal energy performance levels should be revised, with preference for less insulated opaque surfaces and better performing glazing systems. Moreover, the macroeconomic scenario returns a more limited distance between the minimal energy performance levels and the cost-optimal solutions, therefore, it is far from the real economic frame sustained by private investors.

Keywords: building design; cost-optimal analysis; BEopt; economic projections

---

1. Introduction

Modern economy is based on the energy availability to guarantee development and benefit for society, the improvement of life quality and to satisfy human needs. In the EU, energy demand is strongly affected by the building sector that is responsible for 40% of the global energy consumption for final uses and for 36% of CO_2 emissions, with a residential sector that contributes by itself for 25.4% of the total demand [1–3]. Consequently, the spread of residential nZEB (nearly Zero Energy Buildings) [4] constitutes an appropriate strategy aimed at reducing the EU energy dependence and for limiting pollutant emissions. Clearly, suitable legislative plans have been developed in order to improve energy efficiency politics, as well as energy savings interventions and integration of renewable systems in the building sector. However, all these targets have to be attained in regard to a sustainable economic frame, as stated by article five of the 2010/31 European Directive [5,6]. The idea to design new buildings as nZEB with reduced investment costs, in fact, is still too far from an actual application. Other building configurations, instead, appear more attractive due to cheaper initial investments that result in more favorable outcomes despite a slight increase in running (operating) costs. Consequently, the building–plant system with the lowest energy demand represents the cost-optimal solution; conversely the latter can be identified as a favorable balance point between energy consumption, investment and operational costs [7]. The 2010/31 EU directive represents the first attempt for the definition of a comparative methodological framework for the calculation of the optimal energy performance levels as a function of the costs, with reference to new and existing buildings [8]. The procedure cannot be generalized across Europe, therefore Member States have to adapt it to the climatic context, the accessibility to the energy infrastructures and the local market. In order to overcome these drawbacks, appropriate guidelines (Regulation 244) [9] were formulated to support the implementation of the cost-optimal procedure in every country by different steps:

1. Definition of a reference building (RB) relative to its functionality and climatic context;
2. Identification of the energy efficiency measures (EEM) that apply to the RB at a global level or on the single component or on their combination;
3. Evaluation of the energy performance levels before and after the interventions, in accordance with the European technical standards;
4. Global cost calculation in accordance with the net present value (NPV) concept and the international standard EN 15459:2008 [10], taking into account initial, operative, maintenance and disposal costs;
5. Sensibility analysis on the costs as a function of different energy carriers;
6. Identification of the optimal solutions as a function of the cost and the corresponding energy performance levels.

The economic analysis can be carried out in two different ways. The financial projection, in which actual economic indexes are used and all the charges are included in the formation of the costs, is suitable for private investors. Alternatively, the macroeconomic projection, where the costs have to be evaluated by excluding every typology of taxes, but including the CO_2 emissions cost and applying a lower discount rate index than the prior case, can be used to emanate building regulations [11]. Indeed, at a national level, Member States use the results obtained with the macroeconomic projection to define the minimal energy performance requirements in buildings. Nevertheless, they have to verify periodically that the minimal energy performance requirements do not produce a worsened scenario than the energy performances corresponding to the cost-optimal solution. For instance, the latter could be subjected to variations due to the oscillations of the energy carrier costs. If the deviances are greater than 15%, appropriate modifications to the national legislation should be produced.

The general purpose of this paper is to perform the cost-optimal analysis by using the two economic projections, which will quantify the deviances in terms of corresponding costs detected on a reference building. The cost-optimal approach was widely considered in recent literature to identify suitable EEMs for the design of new buildings or for refurbishment planning. However, the majority of these investigations regard dominant heating climates. For instance, in [12] a cost-optimal approach for the identification of suitable insulation materials to adopt in the refurbishment of historical buildings was carried out, but the analysis was limited to heating applications because the RB was located in the Prealps with a prevalent continental climate. A cost-optimal analysis aimed for the attainment of nZEB schools located in Northeast Italy was carried out in [13], considering only the heating requirements due to the particular intended use. A school building was also investigated in [14] to define the optimum insulation thickness and the best glazing system by locating the edifice in a mountain city of South Italy. In [15], specific EEMs were contemplated for a Czech building stock, creating the basis for calculating the cost-effective solution at a national level. Finally, a study concerning the renovation of Swedish multistory buildings is reported in [16] by considering only heating and domestic hot water production, however, no information was provided for the design of new structures. Other investigations can be found for buildings located in warm climates, such as the Mediterranean area, where the role of cooling demands cannot be neglected. The involvement of the cooling requirements makes the cost-optimal analysis and the identification of suitable EEMs difficult, often because solutions employed to reduce heating needs produce cooling demand growth, and vice versa. Cooling requirements were involved in [17], where appropriate EEMs were identified specifically for the refurbishment of the typical Italian social housing stock. Furthermore, in [18] a cost-optimal analysis was developed for reference buildings located in each of the three different climatic zones of Cyprus, limiting the investigation only to the main components of the building envelope, without considering the replacement of the technical plant. In [19], by means of the results provided by the European project RePublic_ZEB, a cost-optimal analysis was carried out to achieve public nZEB located in five different countries, considering requirements for both heating and cooling. The cost-optimal analysis was the approach employed for the design of a new prototype of residential buildings, as shown in [20] where cost-effective nZEBs were achieved for different localities across Europe, highlighting also the role of appliances and lighting on energy consumption. With reference to the European situation, other studies focusing on the role of cooling demands were carried out in [21] but specifically for non-residential buildings located in Serbia. In [22], a cost-optimal analysis was conducted for warm climates in order to carry out a comparison between standard and high-performance, single residential buildings in the design phase. A multistory building, representative of existing Italian building stock, was investigated in [23] by proving the feasibility of nZEB in warm climates, especially in presence of massive structures. However, only the macroeconomic projection was employed, highlighting that it is very difficult to reduce the gap between cost-optimal and lowest energy consumption solutions. Finally, Portuguese buildings were investigated in [24] by considering both heating and cooling, but the cost-optimal analysis was addressed exclusively to the renovation of existing residential edifices. In Table 1, a list of the literature noted is summarized by highlighting the economic scenario followed, the energy services considered and the building typology investigated.

Table 1. Synoptic framework of the main documents found in literature, providing authors, economic scenario (F = Financial, M = Macroeconomic), energy services (H = Heating, C = Cooling, DHW = Domestic Hot Water, A = Appliance) and building typology (R = Residential, NR = Nonresidential).

Authors	Economic Scenario	Energy Services	Building Typology
Lucchi et al. [12]	M	H	R (historical)
Mora et al. [13]	F	H	NR (school)
D'Agostino et al. [14]	F/M	H	NR (school)
Karásek et al. [15]	M	H/DHW	R/NR
Bonakdar et al. [16]	F	H/DHW	R (multistory)
Carpino et al. [17]	M	H/C/DHW	R (social housing)
Loukaidou et al [18]	F	H/C	R/NR (nZEB)
Aelenei et al. [19]	Not indicated	H/C/DHW	NR (office)
D'Agostino&Parker [20]	F	H/C/DHW/A	R (two-story)
Stevanović [21]	Not indicated	H/C	NR (office)
Baglivo et al. [22]	F/M	H/C/DHW/A	R (single-story)
Zacà et al. [23]	M	H/C/DHW/A	R (multistory)
Brandão De Vasconcelos [24]	F/M	H/C/DHW	R (existing)

Clearly, a comparison between the cost-effective solutions provided by financial and macroeconomic projections for a large building located in the Mediterranean area is missing. For this reason, this document describes a parametric study carried out on a multistory building located in contrasting climatic zones to develop an optimized building envelope by means of the cost-optimal approach. Indeed, the main goal of this paper is the comparison between the solutions provided by the two economic scenarios and, eventually, to evaluate if private investors could be attracted by more favorable measures than those imposed by current regulations. Moreover, regarding the more influential features of the building envelope, the same analyses allow verification if the macroeconomic projection produces an optimal solution that matches the constraints imposed by regulations, in order to validate the minimal energy performance levels currently adopted in both the climatic zones. Finally, from the comparison between the two projections, the role of CO_2 emissions on the results obtained can be quantified.

2. Description of the Multistory Residential Building Considered

In the cost-optimal analyses, a building benchmark is required and this configuration is represented by the building–plant system that complies with all the measures in terms of minimal energy performance levels, as set by Italian legislation for the climatic zone considered, including a fraction over 50% of primary energy requirements provided by renewable sources [25]. The envelope morphologically consists of a single structure about 60 m long, with the ending modules (each about 20 m long) shifted back 5 m, configuring three well-identified blocks with a depth of about 10 m, see Figure 1. It was developed with five stories above the ground and a floor height of 2.85 m, with a neither habitable nor air conditioned attic. The large structure size was chosen in order to detect a greater impact of the CO_2 emissions for the provision of heating, cooling and domestic hot water.

Figure 1. Perspective view of the investigated multistory residential building.

Three different types of apartments were developed, as depicted in Figure 2, where all the rooms, including the corridor, were designed as conditioned indoor environments. The apartments were classified as a function of the net conditioned area as:

1. Small, about 60 m^2;
2. Medium, about 80 m^2;
3. Large, slightly over 120 m^2.

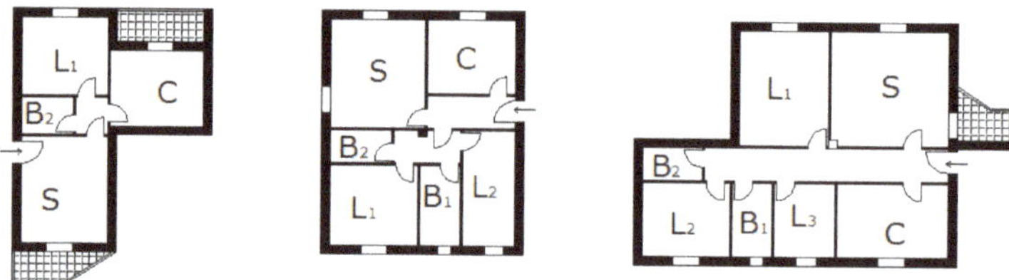

Figure 2. Typologies of apartments in the analyzed building (S: living room, B: bathroom, L: bedroom; C: kitchen).

In every floor (excluding the attic), two small, two large apartments and six medium flats were distributed around five external stairwells. Globally, the building included a total of 50 apartments for a total conditioned surface of about 4200 m^2.

For the energy evaluations, the building was located in a warm location (zone B) characterized by 899 winter degree-day (WDD) and 977 summer degree-day (SDD). The second cold location (zone F) was characterized by 3959 WDD and 74 SDD [26] with dominant heating needs. The structure was constructed of framed reinforced concrete. The external walls were developed with an internal plaster 2 cm thick with a thermal conductivity of 1 W/m·K, a hollow brick of 30 cm with thermal resistance of 0.79 m^2·K/W and a commercial ETICS panel (External Thermal Insulation Coating System) with thermal conductivity of, 0.0595 W/m·K), whose thickness varied as a function of the climatic zone to provide the thermal transmittance values listed in Table 2. Corresponding insulation thickness are indicated in centimeters (Table 2, values in round brackets). For the dynamic performances of opaque walls subjected to the solar radiation, in every building configuration surface mass greater than 250 kg/m^3 and periodic thermal transmittance lower than 0.06 W/m^{2}K were attained [25].

Table 2. Thermal transmittance of opaque walls (with correspondent insulation thickness in cm in round brackets) and windows (with normal solar factor of glazing in round brackets) for the building benchmark in climatic zones B and F.

Climatic Zone	Thermal Transmittance (W/m^{2}K), Insulation Thickness (cm) and Normal Solar Factor (-)			
	Vertical Wall	**Basement Floor**	**Second-Last Floor**	**Windows**
B	0.43 (8)	0.44 (8)	0.35 (10)	3.00 (0.75)
F	0.24 (15)	0.24 (15)	0.20 (18)	1.10 (0.60)

Regarding the non-dispersing surfaces, such as the horizontal inter-floor and vertical walls separating different apartments, a thermal transmittance of 0.8 W/m^{2}K was imposed, as well as for the pitched roof. Every room was equipped with a transparent surface whose area was set to one-eighth of the floor surface to exploit daylight adequately. Table 2 also reports window thermal transmittance and the corresponding normal solar factors (round brackets). In accordance with the minimal energy performance levels, a wooden frame mounting two clear glasses 4 mm thick with 12 mm of air-gap was required for the warm location. For zone F, instead, three panes 4 mm thick with the two external glasses subjected to low emission treatment and argon-filled in two 8 mm spaces were necessary [27]. The control of solar gains, especially crucial in the warm location, was modeled by using rolling shutters, controlled as a function of the exposure and time of the year, and by setting the reduction factors listed in Table 3 when incident solar radiation exceeded 300 W/m^2 [28]. In winter, instead, the same devices were considered fully closed during the night to reduce thermal losses. Internal gains of 352, 413 and 450 W for the apartments of 60, 80 and 120 m^2, respectively, and a natural ventilation with 0.5 air-change per hour, in accordance with Italian standards, were set [29]. Finally, energy requirements for heating and cooling have been determined with indoor set-point temperatures of 20 °C in winter and 26 °C in summer [29].

The air-conditioning plant was modeled as a split system (air–air heat pump) in every room, opportunely sized as a function of the cooling loads for Zone B and for heating in Zone F. No power integration was considered in winter because of the cut-off temperature of −10 °C and the bivalent temperature of −7.7 °C for Zone F. Independent heat pump water heaters (HPWH) were located in every bathroom to produce domestic hot water (DHW), satisfying the requirements calculated by setting a daily volume of hot water equal to 100, 122 and 165 L for the small, medium and large apartments, respectively [30].

Beyond the aerothermal source, the renewable fraction in the building–plant system benefits from a photovoltaic (PV) generator installed on the roof tilted southward and with a peak power of 16.1 kW$_p$ [31]. The renewable electricity was considered totally absorbed by heat pumps (if their operation is required) when energy surplus was not detected. When the electricity demand was greater than the PV production, the remaining part was absorbed from the external grid, conversely in presence of surplus, the renewable electricity was delivered outward.

Table 3. Solar gain reduction factors set for the investigated building, as a function of the exposure and for the month, used to manage solar shading devices during simulations (diurnal hours).

	North	East	South	West
Jan	1.00	0.72	0.57	0.79
Feb	1.00	0.74	0.56	0.71
Mar	1.00	0.65	0.57	0.66
Apr	1.00	0.62	0.61	0.67
May	1.00	0.62	0.67	0.66
Jun	1.00	0.60	0.70	0.64
Jul	1.00	0.61	0.67	0.61
Aug	1.00	0.60	0.59	0.62
Sep	1.00	0.61	0.56	0.64
Oct	1.00	0.62	0.54	0.68
Nov	1.00	0.67	0.55	0.84
Dec	1.00	0.73	0.54	0.78

3. Methodology

The cost-optimal analyses found in literature and previously mentioned were carried out by employing different tools for the energy performance analysis, for the optimization criteria or both [32]. Suitable software, in fact, was necessary to determine energy requirements as a function of a large number of possible combinations of measures, by considering different building envelope configurations, several energy provision systems, as well as the presence of renewable sources, thus requiring a large number of simulations [33]. Successively, an economic analysis with actualized costs had to be carried out in order to analyze the same EEMs also in monetary terms. Both these features can be attained simultaneously by using BEopt v. 2.8 (Building Energy Optimization tool), developed as freeware tool by NREL, the national renewable energy laboratory of the U.S. Department of energy, that can be considered as an Energy Plus front-end, aimed at the cost-optimal analysis of buildings [34]. The EnergyPlus engine was used for the calculation of the primary energy requirements for heating, cooling, domestic hot water and also lighting and appliances if needed, by separating the fossil and the renewable contributions and quantifying the CO_2 emissions required for the macroeconomic projection. In presence of renewable sources, the annual net primary energy, defined as the difference between the actual energy employed for the provision of the energy services and that exported outwards, was also evaluated. Since software and database were developed mainly for the North America context, appropriate adjustments were necessary. However, BEopt is highly flexible, allowing for the analysis of several building typologies in specific climatic contexts by introducing personalized components, upgrading the cost modules and including the economic indexes, as well as the use of an apposite weather database. In this regard, BEopt was used in [35] to investigate an optimal nZEB design characterized by the lowest costs in fourteen different locations across Europe, with particular reference to a single-story structure. Moreover, BEopt was used in [36] to investigate the cost-effectiveness of residential building stock retrofits in Italy and Denmark. BEopt was also used in [37] for building optimization in different Japanese cities.

The investigated building was simulated with different insulation thicknesses in the opaque dispersing walls and by equipping the envelope with two different glazing systems. This was done to investigate other configurations and to compare energy performances with the RB. These parameters affect thermal losses and solar gains and an optimal compromise had to be found to minimize the annual energy demand. Conversely, the air-conditioning plant for the provision of heating and cooling was maintained unvaried by always considering the use of electric air–air heat pumps. The choice to analyze these devices as a generation system exclusively was due to different reasons. Firstly, the same air conditioning system can be used for the provision of heating and cooling, resulting as less invasive from the installation point of view and consequently economically more attractive. Secondly, heat pumps assure the coverage of a noticeable fraction of the building requirements by means of the aerothermal renewable source [31]. Finally, these devices are able also to exploit the renewable electricity, rationalizing the electricity produced by the PV generator installed to meet the Italian regulation in terms of integrating renewable systems in buildings [31]. For the same reasons, the HPWHs equipped with electrical boosters were considered as an unchangeable system for the production of DHW. Despite that it affects the results noticeably [38], the influence of occupant behavior was not considered in the energy analysis because the main goal of this paper was the optimization of the building–plant system during the design process

For the financial and the macroeconomic projections, the cost-optimal analysis results were displayed in terms of annualized energy-related costs against the fossil contribution in the formation of the annual net primary energy absorbed from the grid. The first were defined by considering the electricity expenses sustained in the lifespan, annualized by a capital recovery factor and incremented by the EEM costs (and CO_2 emission cost in macroeconomic evaluations), obtaining more reduced values when compared with the items provided by the life cycle cost (LCC) analysis. For the purpose of this work, the CO_2 emissions were determined as a function of absorbed electric energy from the grid, whose production also requires the employment of fossil primary sources. As a precaution, a conversion factor of 1.95 to transform electricity into fossil primary energy was set constant for the whole building–plant lifespan, because of the high production percentage already obtained by means of renewable systems (especially photovoltaic and hydroelectric) and the massive use of natural gas in fired power plants [25]. Successively, appropriate emission factors related to the current Italian power generation system were employed [39].

Regarding the EEMs considered, the reference building was modified by including two improving and one worsening intervention, the latter to verify also if the cost-optimal solution could be detected in the presence of a pejorative energy scenario. The modified configurations present the same characteristics in terms of shape and orientation, whereas the thermal and optical properties of the envelope were varied. The potential optimization actions attainable by specific EEMs involve the reduction of the transmission losses through the dispersing surfaces and the optimization of solar gains through the glazing systems, the latter crucial for the reduction of the cooling requirements and avoiding the worsening of heating needs. These analyses were aimed at the identification of the optimal insulation thicknesses and the more appropriate windowed system to adopt in the building envelope, for the specific climatic context. In order to limit the simulation times, four different insulation thicknesses were considered for each of three dispersing surfaces, whereas two different typologies of glazing systems were investigated, producing 128 combinations of different measures. In detail, Table 4 lists the values of the insulation thicknesses and the corresponding thermal transmittances, the thermal transmittances and the normal solar factors of windows in the simulated building-plant configurations.

For the cost analysis, the approach adopted was based on the evaluation of the annualized energy-related costs, increased by the EEMs cost and determined in accordance with EU Regulation 244/2012 [9,10]. In particular, the economic analysis involved construction, management and maintenance costs of the building–plant system, beyond the operative (or running) costs derived from energy consumption during the lifespan, which the same regulation sets in 30 years for residential buildings. Since running costs are extremely volatile during the mean-long periods, their actualization to the initial years (when the initial investment is sustained) by means of appropriate inflation and discount rates, was required. For financial evaluations, the economic indexes employed in the cost-optimal analysis were set to 4% for the discount rate that represents the maximum interest that investors could obtain by an alternative safe investment. Currently, the latter are represented by a deposit account with restricted funds (generally greater than one year) that have a better yield than thirty-year government bonds. Greater discount rates were not considered because the latter represent the more favorable situation for a private investor, whereas a lower discount rate was contemplated in the macroeconomic projection. Regarding the other economic indexes, 0.3% was set for the general inflation rate [40] and 3.2% for the energy carrier inflation rate (electricity) [41]. Table 5 lists the economic indexes employed in both economic scenarios.

Table 4. Insulation thicknesses and thermal transmittances of opaque walls (VW, vertical wall; GS, ground slab; SF, second to last floor), thermal transmittances and normal solar factors of the windowed systems in climatic zones B and F.

CLIMATIC ZONE	Opaque Walls				Windowed Systems
	Insulation Thickness (cm), Thermal Transmittance (W/m^{2}K)				Thermal Transmittance (W/m^{2}K) Normal Solar Factor (-)
		VW	GS	SF	
B	4	0.60	0.62	0.46	3.00 75%
	8	0.43	0.44	0.40	
	10	0.38	0.38	0.35	1.10 60%
	12	0.33	0.24	0.31	
F	10	0.30	0.31	0.27	3.00 75%
	15	0.24	0.24	0.22	
	18	0.21	0.20	0.20	1.10 60%
	20	0.20	0.20	0.19	

Table 5. Economic indexes employed for the economic projections.

Economic Scenario	Discount Rate	General Inflation Rate	Energy Inflation Rate
Financial	4%	0.3%	3.2%
Macroeconomic	2%	0.3%	3.2%

For macroeconomic projections, only the discount rate was changed to 2% in order to consider a discount rate lower than that available on the market, as prescribed by regulation. The CO_2 emission cost was determined at annual level as an additional term in the evaluation of the global costs of the building-plant system. Nevertheless, the same regulation set a constant flat rate of 20 €/ton per year (valid until 2025) as stated by the current commission forecast of the carbon prices of the ETS (Emission Trading System) of the European Commission. It is worth mentioning that the CO_2 emission cost is very dissimilar from that found in literature; for instance, an average value of 375 € per ton of emitted CO_2 per year was determined in [42]. However, beyond the environmental costs, this is comprehensive of other items (for instance the social costs connected with human health effects).

For each package of measures, BEopt determined internally the corresponding NPV (Net Present Value) starting from the corresponding cost, also including installation costs. The NPV is usually employed in capital budgeting and investment planning to analyze the profitability of a projected investment. It is defined as the difference between the present value of cash inflows and outflows over a period of time. The construction costs that do not produce deviances in the comparison of the different building configurations (for instance structural frame, foundations, internal partitions or wall finish) were not considered. The different constructive solutions that minimize the global costs as a function of the different levels of energy savings were identified by adopting a sequential search optimization technique on discrete packages of measures to consider realistic design options. Because EU regulation establishes that measured costs have to be determined as a function of market investigations, a local regional data base [43], coherent with the locations and the construction times, was adopted. These lists include the installation costs, the transportation and the costs concerning the rental of machinery and equipment, provided per unit of surface or per component. The regional price lists guarantee cost uniformity across the involved territory, as well as adequacy to the market value, defining average costs for every component and material. In Table 6, the investment costs, as well as the main specific costs concerning the envelope and technical plants for the building benchmark located in the two climatic zones, are listed. It is worth mentioning that in the financial projection, a VAT of 4% was applied to the EEM components that were involved. Clearly, in zone F, the greater insulation thicknesses and the windows equipped with triple pane were major initial investments when compared to the RB in zone B (about 17%), as well as the air-conditioning plant due to the greater required heating power and the correspondingly larger size heat pumps. Taxes, excises and other charges were not considered in the macroeconomic projections. At annual level, the sustained costs were considered to involve building–plant management (running costs) and ordinary maintenance. In the formation of the running costs, by considering that the adopted energy carrier was represented exclusively by electricity, an average energy cost was evaluated at national level by considering the price determined for the quarter January–March 2019 [41]. In this regard, in BEopt, a user-specified electricity price as a function of the time of intended use, was implemented. In particular, three hourly intervals were set and distinguished with F1, F2 and F3 [41]. The corresponding electricity prices are listed in Table 7 including, in the financial projection, excises of 0.0227 €/kWh$_e$ and VAT of 10%. The same costs were reported without charges for the implementation of the macroeconomic projection. The PV electricity surplus delivered to the grid, due to the very limited average sales price (about 0.04 €/kW) [44] and the different production achievable among the climatic zones considered, was neglected and not included as positive cash flow to reduce the annual running costs. Finally, periodic costs were considered for the substitution of components due to obsolescence and wear, assuming the replacement times listed in Table 8 [9]. Building-envelope components were not replaced during the lifespan considered, whereas HVAC systems, HPWH and PV panels were changed at least one time during the economic analysis. These features allow for the depiction of the annualized energy-related costs corresponding to the different packages as a function of the net primary energy from fossil fuels. The annual CO_2 emissions costs for every building configuration were quantified only in the macroeconomic projection.

Table 6. Initial investment costs required for the reference building (RB) in climatic zones B and F, classified as a function of the main expense components.

	Total Initial Investment			
	Financial Projection Zone F		**Macroeconomic Projection**	
Components	**Zone B**	**Zone F**	**Zone B**	**Zone F**
	Costs (€)	**Costs (€)**	**Costs (€)**	**Costs (€)**
External Walls	849,937	1,148,000	798,197	1,078,116
Top Floor + Roof	79,666	103,389	73,300	95,128
Foundations + Basement	25,157	35,717	23,303	33,084
Internal Partition + Floor decks	300,327	300,327	283,074	283,074
Windows & Doors	127,384	146,883	103,667	119,536
Rolling shutters	25,154	25,154	20,770	20,770
Air Conditioning	257,265	266,252	205,860	213,051
Domestic Hot Water	115,832	115,832	91,230	91,230
PV Generator	25,766	25,766	21,262	21,262
INITIAL INVESTMENT	1,806,488	2,167,320	1,620,663	1,955,250

Table 7. Average electricity price determined for the quarter January–March 2019 concerning the Italian market, as a function of the three intervals defined by the time for intended use.

	Italian Electricity Price (€/kWh$_e$)		
	F1: from Monday to Friday 8:00–19:00	**F2: from Monday to Friday 7:00–8:00 and 19:00–23:00 Saturday 7:00–23:00**	**F3: from Monday to Saturday 23:00–7:00 Sunday and Holidays**
Financial	0.19615	0.18613	0.18613
Macroeconomic	0.15384	0.14482	0.14482

Table 8. Replacement times assumed for the main building–plant system components in a lifespan of 30 years.

Considered Duration (years)		
Typology	**Components/Systems**	**Years**
	Concrete masonry unit	Not replaced
	External insulation system	Not replaced
Building envelope	External plaster	30
	Windowed system	30
	Roof slab	Not replaced
	Roof covering	30
	Foundations	Not replaced
HVAC	Heat pumps	15
HPWH	Heat pump water heater	12
Renewable system	PV generator	25

4. Results and Discussion

Every package of measures was characterized by a precise value of the annual primary fossil energy demand and the corresponding energy-related annualized costs, opportunely plotted. Thus, the set of the packages considered determined a graph where the frontier points in the lowest part denote the Pareto front, representing the geometrical place that joins the optimal outcomes with a broken line. No points can be detected beyond the Pareto front, with the EEMs close to the y-axis that indicate interventions characterized by lower primary energy demands but higher costs, whereas the points close to the x-axis represents interventions that lead to lower costs but higher primary energy requirements and CO_2 emissions. Usually, the solution with the lowest annualized energy-related cost identifies the cost-optimal solution. In the graphs, only some representative points with remarkable results were highlighted. In particular, in every graph the points listed in Table 9 were emphasized.

Moreover, the other three combinations of interventions belonging to the Pareto front were considered, because these were representative of other alternative cost-effective solutions, and are indicated as OP1, OP2 and OP3. For every point considered, the deviances with COS (cost-optimal solution with the lowest annualized energy-related cost) were determined both in terms of costs and energy demands. The same procedure was carried out for both scenarios, in order to evaluate the differences between the building configurations considered when two different approaches concerning the economic analyses were carried out. Furthermore, in the macroeconomic projection, COS and RB (building benchmark designed as a function of the minimal energy performance levels) should be very close in order to verify if minimal energy performance levels were formulated as a function of the cost-optimal solution, as stated by the European regulations.

Table 9. Points indicating of the more representative building configurations.

Point	Description
COS	Cost-optimal solution with the lowest annualized energy-related cost
RB	Building benchmark designed as a function of the minimal energy performance levels
EEB	Building configuration with the lowest primary fossil energy demand
PEJ	Pejorative solution than minimal energy performance levels
MEC	Building configuration with the highest energy consumption
MC	Building configuration with the highest cost
OP1	Other building configuration belonging to the Pareto front
OP2	Further building configuration belonging to the Pareto front
OP3	Further building configuration belonging to the Pareto front

4.1. Cost-Optimal Analysis with Financial Projection

In accordance with the standard EN 15459, the annualized energy-related costs against the fossil primary energy demand provided by BEopt for climatic Zone B are shown in Figure 3. In Table 10, the results corresponding to the constructive solutions for each point considered are specified.

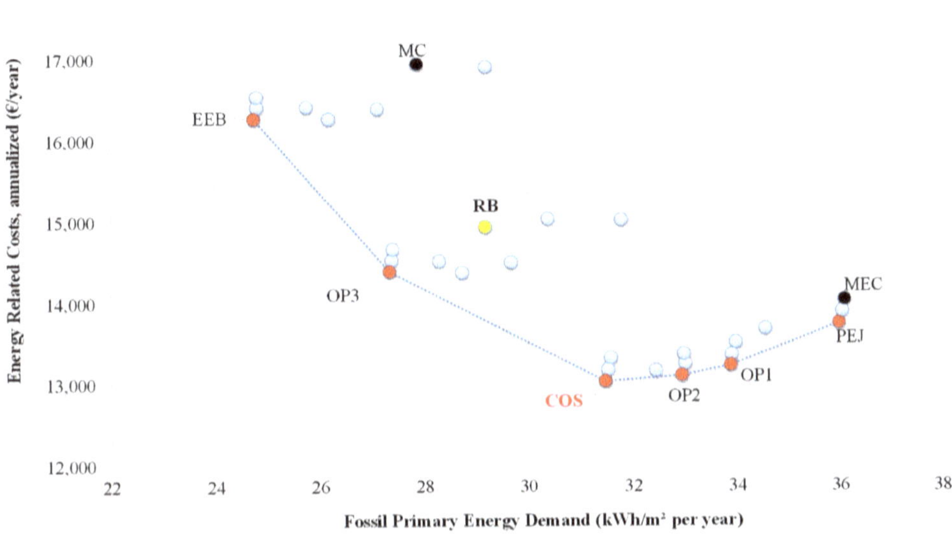

Figure 3. Annualized energy-related costs plotted against primary energy demands for the more representative EEMs analyzed for the warmer climatic zone B, including a financial projection.

Clearly, for the climatic context considered, the deviances in terms of cost between the COS and the RB point are evident; the latter produces an annualized extra cost of 1900 €, against a primary energy saving of 2.4 kWh/m². It means that, for the considered location and from the financial point of view, investors should prefer a solution energetically less efficient than that suggested by Italian legislation, because the reduction of the costs related to the attainment of the COS measures prevails on a slight annual running cost growth due to the higher electricity consumption. Moreover, the cost-optimal solution is placed on the right side of the graph, meaning that in the warm location the point economically looks more favorable toward greater energy consumption. The EEB solution allowed for reducing the primary energy demand to 24.9 kWh/m^2, with an energy saving of 17% and 26.5% when compared with RB and COS, respectively. However, the annualized extra cost increase of 3250 € (+25%) and 1300 € (+8.7%) make the result unattractive. The solutions OP1 and OP2 offer similar costs to the COS solution; however, the primary energy demands are slightly higher, therefore these packages of interventions are not recommended. Conversely, OP3 determined an augmented annualized global cost of about 1300 € (+10%); however, the primary energy demand was reduced significantly to 27.3 kWh/m^2, resulting in OP3 being more favorable than EEB. Finally, the PEJ point produced similar results to the MEC point and both were more expensive than COS. Therefore, the running cost growth due to the limited insulation thicknesses and the cheapest installed window system prevailed on the reduction of investment expenses. The analysis of the values listed in Table 10 suggests that in the warm location, the solution of a scarcely insulated ground slab is always preferred, whereas reduced insulation thicknesses in the dispersing vertical walls are recommended when triple pane systems are installed. Considering the cooling-dominant location, in fact, in summer the first allows for dissipation of thermal power toward the soil, and the second one permits the attainment of the best compromise between the augmentation of thermal losses and the limitation of solar gains. It is worth highlighting that, conversely to the cost-optimal solution, the EEB requires an envelope that is highly insulated, excluding the ground floor and glazing with reduced normal solar gains. The COS point, instead, requires only a large insulation thickness on the second last floor, whereas the thicknesses inside the vertical walls and ground slab are halved compared to those imposed by minimal energy performance levels. This is in agreement with [45], where excessive insulation thicknesses are not recommended for an office building in a warm climate, with thicknesses that have to be further reduced in the presence of high internal gains. In another investigation [11], instead, a reference office building in a warm climate met the nZEB target in conjunction with a favorable economic frame only when the building fabric is characterized by high thermal inertia.

Table 10. Detected optimal constructive solutions concerning different energy efficiency measures (EEMs) for climatic zone B (VWT, insulation thickness in vertical wall; GST, insulation thickness in the ground slab; SFT, insulation thickness in the second last floor; GS, glazing system).

EEMs	VWT	GST	SFT	GS
PEJ	4 cm	4 cm	4 cm	Double pane
OP1	4 cm	4 cm	8 cm	Double pane
OP2	4 cm	4 cm	12 cm	Double pane
OP3	8 cm	4 cm	12 cm	Triple pane
RB	8 cm	8 cm	10 cm	Double pane
EEB	12 cm	4 cm	12 cm	Triple pane
MEC	4 cm	12 cm	4 cm	Double pane
MC	12 cm	4 cm	4 cm	Triple pane
COS	4 cm	4 cm	12 cm	Triple pane

Moving to the climatic zone F, as shown in Figure 4, a different trend of the points depicting the several EEMs was detected. It seems that a major number of cases were analyzed when compared to the warm location, actually for the latter, some points are very close and tend to overlap due to the lower primary energy demands. The colder climatic zone involves greater heating requirements, consequently a noticeable increment of the net primary energy demand determined higher annualized energy-related costs. Consequently, for this case, the cost-optimal solution looks toward measures with reduced energy requirements. This effect is further amplified due to the scarce employment of the renewable sources related to the limited solar irradiance availability, and scarce PV production as a consequence, and the outdoor air temperatures that affect negatively the heat pump's performance in the winter. Comparison with Figure 3 shows a consistent limitation of the gap between the COS and the RB points. Nevertheless, in comparison with the warm location, the minimal energy performance levels determined a worsening of both the energy and economic scenarios, detecting a greater primary energy demand of 2.1 kWh/m^2 and an extra cost of 829 € per year compared to the cost-optimal solution, meaning the minimal energy performance levels should consider insulation thicknesses slightly greater than those currently imposed. In addition, the distance between the EEB and COS points is more limited than that determined for the warm location, with the first producing a reduction of the primary energy demand of 6.2 kWh/m^2 with an extra cost of 1356 € per year, confirming that the employment of insulated envelopes is suggested to match more closely the cost-effective solutions. The other considered EEMs provided, in every case, worsening results and therefore should not be taken into account. In particular, the worst package of measures is represented by the MEC point, representative of a medium-insulated envelope, with annualized energy-related costs similar to the MC point. Evidently, these building configurations determined a noticeable increment of the running costs that prevail on the reduction of expenses required for these packages. In Table 11, the results corresponding to the constructive solutions for each point considered were specified for the respective climatic zone. The only differences between COS and EEB concern the employment of a lower insulation thickness in vertical walls. Window systems equipped with triple pane are required to prevent, this time, excessive thermal losses during winter rather than to limit solar gains in summer. It is worth noticing that OP1 and OP2 allowed for the attainment of annualized energy-related costs limited to values lower than 1000 € per year when compared to COS, despite the employment of double pane windows. In addition, OP3 considered the same insulation thicknesses as OP2 but involving triple pane windows; however, the economic and energy scenarios worsened further, meaning that in this location the best compromise between thermal losses and solar gains also has to be identified.

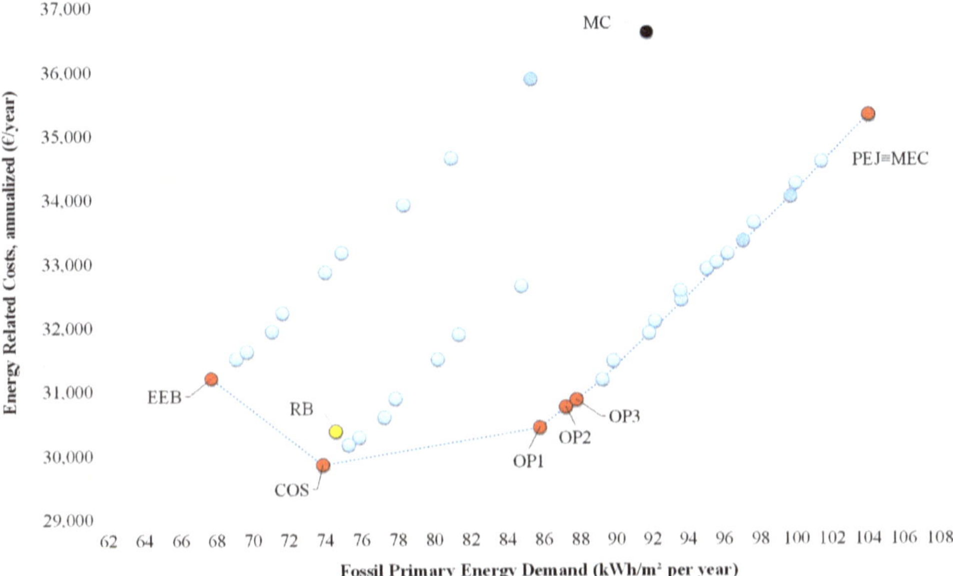

Figure 4. Annual energy-related costs against primary energy demands for different EEMs analyzed for the colder climatic zone F, including a financial projection.

Table 11. Detected optimal constructive solutions concerning different EEMs for climatic zone F (VWT, insulation thickness in vertical wall; GST, insulation thickness in the ground slab; SFT, insulation thickness in the second last floor; GS, glazing system).

EEMs	VWT	GST	SFT	GS
PEJ	10 cm	10 cm	10 cm	Double pane
OP1	15 cm	10 cm	15 cm	Double pane
OP2	15 cm	15 cm	15 cm	Double pane
OP3	15 cm	15 cm	15 cm	Triple pane
RB	15 cm	15 cm	18 cm	Triple pane
EEB	20 cm	20 cm	20 cm	Triple pane
MEC	10 cm	10 cm	10 cm	Double pane
MC	20 cm	10 cm	10 cm	Double pane
COS	15 cm	20 cm	20 cm	Triple pane

4.2. Cost-Optimal Analysis with Macroeconomic Projection

The same calculations were repeated for the localities by adopting the macroeconomic projection. It is clear that the point distribution is similar to the prior cases, however, a noticeable limitation of the annualized energy-related cost gap was detected. This limitation is due mainly to two factors:

1. The more limited EEMs cost due to the exclusion of taxes,
2. The lower annualized energy-related costs according to electricity prices considered without VAT and excises.

These effects prevail on the reduction of the discount rate that, conversely, produced greater annualized energy-related costs compared to the financial scenario when projected to the initial year. The combination of these items determined a downshift more or less constant for all of the EEMs, however, few differences among the points' position can be observed due to the different CO_2 emission costs among the investigated interventions. In Figure 5, the results of the macroeconomic projection are shown for the warm location, whereas Table 12 lists for the more representative packages the annual CO_2 emission, their costs, energy-related costs of financing and macroeconomic projections highlighting the differences. The deviance of the annualized energy-related costs between EEB and MEC, representing respectively, the better and the worst EEM from the point of view of energy consumption and CO_2 emission, is only 210 € per year. Therefore, the inclusion of the CO_2 emission costs in this location did not affect the results noticeably. Indeed, the CO_2 emissions in the formation of the annualized energy-related cost contribute from a minimum of 4.5% to a maximum of 7.5%; however, they did not modify the optimal building configuration and the point distribution. Nevertheless, expensive EEMs producing limited running costs, for instance MC, EEB and RB, provided higher differences between the two projections, meaning a major role of interventions cost compared to the running cost. Consequently, the results of the macroeconomic scenario are more precautionary in presence of buildings with high-energy performances that do not match the cost-optimal solution. Conversely, the differences decrease compared to the two cost-effective interventions and the minimum gap that was detected with OP2. Clearly, the macroeconomic scenario highlighted a consistent reduction of the cost gap between RB and COS, passing to 1000 € when compared with the financial projection (−45%). Nevertheless, in the warm location, the cost-optimal solution provided with the macroeconomic projection, also in this case, did not match the minimal energy performance levels, therefore the latter should be revised for the considered building typology.

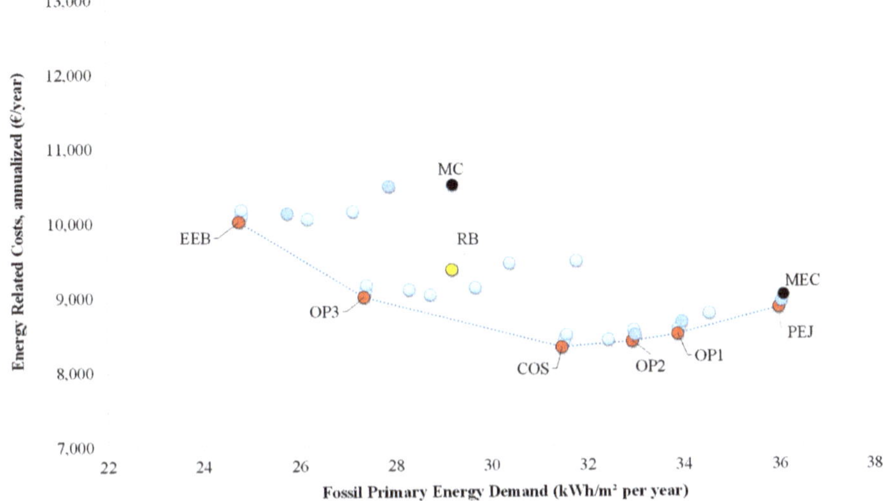

Figure 5. Annualized energy-related costs against primary energy demands for the more representative EEMs analyzed for the warmer climatic zone B, including a macroeconomic projection.

Table 12. CO_2 emissions and corresponding cost for the EEMs considered and comparison between the annualized energy-related costs (AERC) determined for the financial and macroeconomic projections in the warm location.

	CO_2 Emissions (Tonnes/Year)	CO_2 Emission Cost (€/Year)	Macroeconomic AERC (€/Year)	Financial AERC (€/Year)	Difference Macro/Financial (€/Year)
PEJ	33.3	666	8841	13,745	−4904
OP1	31.4	627	8490	13,229	−4739
OP2	30.5	610	8396	13,105	−4709
OP3	25.3	507	9001	14,390	−5389
RB	27.0	540	9360	14,933	−5573
EEB	22.9	458	10,024	16,272	−6248
MEC	33.4	668	9009	14,030	−5021
MC	25.8	516	10,500	16,940	−6440
COS	29.1	583	8319	13,034	−4715

The same evaluations were carried out for the building located in the cold location, as depicted in Figure 6; it is clear that the macroeconomic scenario produced a more evident limitation in the cost gap between the RB and the COS points, now separated by 300 € per year with a percentage deviation of 62%, when compared with the financial projection. Effectively, the latter confirms that minimal energy performance levels seem to be developed not far from the cost-optimal solution but exclusively for cold climatic context. In Table 13, the emitted CO_2 and corresponding costs are listed, as well as a comparison between the annualized energy-related costs determined for the two economic scenarios. The colder climatic context, due to the major heating requirements, made the role of CO_2 emission cost more influential, with an average percentage of 8% of the annualized energy-related cost. In this case, the running cost is augmented and significant for the quantification of the deviances between the financial and macroeconomic projection, as demonstrated by the EEMs with the greatest energy consumptions (MEC and PEJ). OP1 becomes the solution that allows for minimizing the gap between financial and macroeconomic projections.

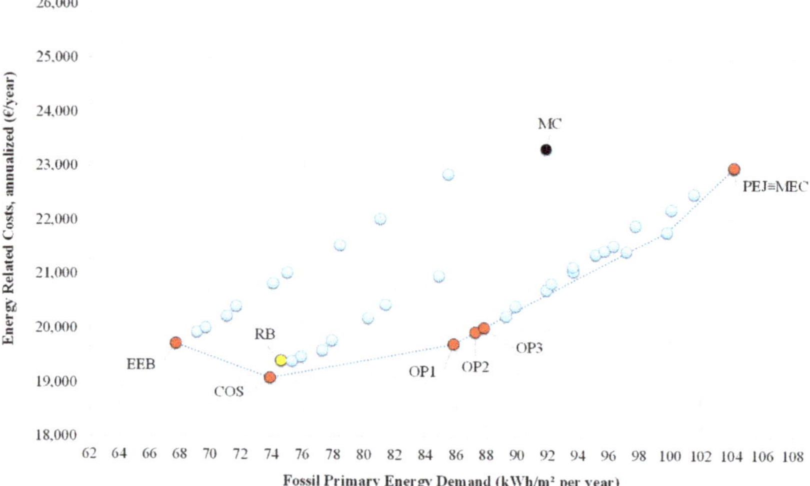

Figure 6. Annualized energy-related costs against primary energy demands for more representative EEMs analyzed for the colder climatic zone F, including a macroeconomic projection.

Table 13. CO_2 emissions and corresponding cost for the EEMs considered and comparison between the annualized energy-related costs (AERC) determined for the financial and macroeconomic projections in the cold location.

	CO_2 Emissions (Tonnes/Year)	CO_2 Emission cost (€/Year)	Macroeconomic AERC (€/Year)	Financial AERC (€/Year)	Difference Macro/Financial (€/Year)
PEJ	96.3	1926	22,998	35,306	−12,308
OP1	79.4	1588	19,724	30,429	−10,705
OP2	80.7	1614	19,945	30,749	−10,804
OP3	81.3	1625	20,024	30,861	−10,837
RB	70.3	1407	19,413	30,686	−11,273
EEB	62.7	1253	19,729	31,213	−11,484
MEC	96.3	1926	22,998	35,306	−12,308
MC	79.0	1581	22,869	36,604	−13,735
COS	68.4	1368	19,098	29,857	−10,759

5. Conclusions

The cost-optimal analysis was used to design a new multistory building in two contrasting climatic zones of the Mediterranean area in order to identify the best building envelope configurations. The latter designed to verify the current regulations in terms of energy performance levels for the climatic zones considered were used as benchmarks (RB). The main investigated parameters concern the insulation thickness in the dispersing opaque walls and the windowed system, with an unvaried air-conditioning plant to meet regulation constraints on the integration of renewable systems in buildings. The analyses were carried out by a financial approach (for private investors) and a macroeconomic projection (to plan regulations) by highlighting the differences among the detected optimal measures. Mainly, financial and macroeconomic projections provided the same cost-optimal solutions. Nevertheless, with a macroeconomic projection, the RB matches the cost-optimal solution only in the cold location, whereas in the warm climatic zone a noticeable deviance was detected and a less insulated envelope equipped with triple pane systems is recommended. In particular, a scarcely insulated ground floor allows for the transfer of thermal power to the soil during summer, limiting the cooling requirements.

Furthermore, lower insulation thicknesses in vertical opaque walls equipped with triple pane windows represent at an annual level the best compromise between thermal losses and employment of solar gains. Consequently, in the warm location, the minimal energy performance levels set for the RB should be revised, because less insulated building configurations allow for a reduction of the initial investment that prevails on a slight augmentation of the running costs. Regarding the cold climate, the macroeconomic projection produced a noticeable reduction of the cost gap between RB and the cost-optimal solution, however, a combination of measures that further reduce the thermal losses through the envelope are suggested. Nevertheless, the results confirm the good agreement between the measures imposed by regulation and the cost-effective solution.

The cost-optimal analyses carried out by the financial projection highlighted a major gap between the RB and cost-optimal solutions, especially for the warm location. In particular, the RB showed a greater annualized energy-related cost increment. Therefore, from the perspective of private investors, a further distance growth between the two points, equal to 859 € per year, 82% higher than the macroeconomic scenario, was detected. The role of the CO_2 emission on the points' distribution is marginal, by considering that the annualized energy-related costs varied between 4.5% and 7.5%. This is due to the favorable application of renewable systems connected with the advantageous climatic conditions. In the cold location, a distance growth between RB and the cost-optimal solution was detected again, this time at 514 € per year (+ 163%). However, the role of the CO_2 is more significant with an average weight of 8% on the global costs, due to the greater energy consumptions for heating. When the distance of the points denoting the several combination of measures is compared between the two economic projections, the deviances were more evident in the presence of expensive interventions in the warm location, denoting a greater role of the investment costs than the running costs. Conversely, the greatest deviances were detected also in correspondence to interventions with large energy consumption in the cold location, due to the major role of the CO_2 emissions. Globally, the financial projection tends to provide results close to the macroeconomic projection in warm climatic zones or in the presence of a high percentage of primary energy covered by renewable sources. However, the macroeconomic scenario returns a more limited distance between the minimal energy performance levels and the cost-optimal solutions, therefore it is far from the real economic frame sustained by private investors.

Despite the regulations being susceptible to further improvements, especially in warm climates, the results produced by the two projections confirm the possibility to design highly efficient buildings in the context of a sustainable economic frame. Regarding the required costs, the role of the renewable systems is marginal when compared with the other items, producing beneficial effects on the environmental impact point of view. In this way, the cost-optimal analysis can support the diffusion of more sustainable edifices because it is confirmed that use of this tool can effectively limit energy consumption in buildings and increase benefits for the society. Therefore, this approach is highly recommended in countries where directives concerning building energy performances are not yet promulgated.

Author Contributions: Conceptualization, R.B.; methodology, R.B. and P.B.; software, R.B.; validation, R.B.; investigation, R.B., Piero and Bevilacqua; data curation, R.B., P.B. and C.C.; writing—original draft preparation, R.B.; writing—review and editing, R.B., P.B. and C.C.; supervision, N.A.; project administration, N.A. All authors have read and agreed to the published version of the manuscript.

Funding: This research received no external funding.

Conflicts of Interest: The authors declare no conflict of interest.

References

1. European Commission. *Communication from the Commission to the European Parliament, the Council, the European Economic and Social Committee and the Committee of the Regions on an EU Strategy for Heating and Cooling;* European Commission: Brussel, Belgium, 2016.

2. Berardi, U. A cross-country comparison of the building energy consumptions and their trends. *Resour. Conserv. Recycl.* **2017**, *123*, 230–241. [CrossRef]

3. Pérez-Lombard, L.; Ortiz, J.; Pout, C. A review on buildings energy consumption information. *Energy Build.* **2008**, *40*, 394–398. [CrossRef]

4. D'Agostino, D.; Mazzarella, L. What is a Nearly zero energy building? Overview, implementation and comparison of definitions. *J. Build. Eng.* **2019**, *21*, 200–212. [CrossRef]

5. EU. *Directive 2010/31/EU of the European Parliament and of the Council of 19 May 2010 on the Energy Performance of Buildings*; EU: Brussel, Belgium, 2010.

6. Bruno, R.; Bevilacqua, P.; Cuconati, T.; Arcuri, N. Energy evaluations of an innovative multi-storey wooden near Zero Energy Building designed for Mediterranean areas. *Appl. Energy* **2019**, *238*, 929–941. [CrossRef]

7. Corgnati, S.P.; Fabrizio, E.; Filippi, M.; Monetti, V. Reference buildings for cost optimal analysis: Method of definition and application. *Appl. Energy* **2013**, *102*, 983–993. [CrossRef]

8. Almeida, M.; Ferreira, M. Cost effective energy and carbon emissions optimization in building renovation (Annex 56). *Energy Build.* **2017**, *152*, 718–738. [CrossRef]

9. EU. European Commission Commission Delegated Regulation (EU) No 244/2012 of 16 January 2012. *Off. J. Eur. Union* **2012**, *55*, 18–36.

10. Comite Europen de Normalisation. *Energy Performance of Buildings—Economic Evaluation Procedure for Energy Systems in Buildings*; CEN: Brussel, Belgium, 2007.

11. Congedo, P.M.; Baglivo, C.; D'Agostino, D.; Zacà, I. Cost-optimal design for nearly zero energy office buildings located in warm climates. *Energy* **2015**, *91*, 967–982. [CrossRef]

12. Lucchi, E.; Tabak, M.; Troi, A. The 'cost optimality' Approach for the Internal Insulation of Historic Buildings. *Energy Procedia* **2017**, *133*, 412–423. [CrossRef]

13. Mora, T.D.; Righi, A.; Peron, F.; Romagnoni, P. Cost-Optimal measures for renovation of existing school buildings towards nZEB. *Energy Procedia* **2017**, *140*, 288–302. [CrossRef]

14. D'Agostino, D.; Daraio, L.; Marino, C.; Minichiello, F. Cost-optimal methodology and passive strategies for building energy efficiency: A case-study. *Archit. Sci. Rev.* **2018**, *61*, 400–409. [CrossRef]

15. Karásek, J.; Pojar, J.; Kaločai, L.; Heralová, R.S. Cost optimum calculation of energy efficiency measures in the Czech Republic. *Energy Policy* **2018**, *123*, 155–166. [CrossRef]

16. Bonakdar, F.; Dodoo, A.; Gustavsson, L. Cost-optimum analysis of building fabric renovation in a Swedish multi-story residential building. *Energy Build.* **2014**, *84*, 662–673. [CrossRef]

17. Carpino, C.; Bruno, R.; Arcuri, N. Social housing refurbishment in Mediterranean climate: Cost-optimal analysis towards the n-ZEB target. *Energy Build.* **2018**, *174*, 642–656. [CrossRef]

18. Loukaidou, K.; Michopoulos, A.; Zachariadis, T. Nearly-zero Energy Buildings: Cost-optimal Analysis of Building Envelope Characteristics. *Procedia Environ. Sci.* **2017**, *38*, 20–27. [CrossRef]

19. Aelenei, L.; Paduos, S.; Petran, H.; Tarrés, J.; Ferreira, A.; Corrado, V.; Camelo, S.; Polychroni, E.; Sfakianaki, K.; Gonçalves, H.; et al. Implementing cost-optimal methodology in existing public buildings. *Energy Procedia* **2015**, *78*, 2022–2027. [CrossRef]

20. D'Agostino, D.; Parker, D. Data on cost-optimal Nearly Zero Energy Buildings (NZEBs) across Europe. *Data Brief* **2018**, *17*, 1168. [CrossRef]

21. Stevanović, S. Parametric study of a cost-optimal, energy efficient office building in Serbia. *Energy* **2016**, *117*, 492–505. [CrossRef]

22. Baglivo, C.; Congedo, P.M.; D'Agostino, D.; Zacà, I. Cost-optimal analysis and technical comparison between standard and high efficient mono-residential buildings in a warm climate. *Energy* **2015**, *83*, 560–575. [CrossRef]

23. Zacà, I.; D'Agostino, D.; Congedo, P.M.; Baglivo, C. Assessment of cost-optimality and technical solutions in high performance multi-residential buildings in the Mediterranean area. *Energy Build.* **2015**, *102*, 250–265. [CrossRef]

24. Brandão De Vasconcelos, A.; Pinheiro, M.D.; Manso, A.; Cabaço, A. EPBD cost-optimal methodology: Application to the thermal rehabilitation of the building envelope of a Portuguese residential reference building. *Energy Build.* **2016**, *111*, 12–25. [CrossRef]

25. Italian Republic. *Italian Iterministerial Decree 26th June 2015: Application of Calculation Methods for Energy Performance and Definition of Minimum Building Requirements*; Official Gazzette of Italian Republic: Rome, Italy, 2015; pp. 1–8.

26. Italian Unification Institution. *UNI 10349-3: Heating and Cooling of Buildings—Climatic Data Part. 3—Accumulated Temperature Differences (Degree-Days) and other Indices*; Italian Unification Institution: Roman, Italy, 2016.

27. International Organization for Standardization. *EN ISO 10077-1:Thermal Performance of Windows, Doors and Shutters—Calculation of Thermal Transmittance—Part. 1: General*; International Organization for Standardization: Geneva, Switzerland, 2017.

28. International Organization for Standardization. *EN ISO 52016: Energy Performance of Buildings—Energy Needs for Heating and Cooling, Internal Temperature and Sensible and Latent Heat Loads—Part. 1: Calculation Procedures*; International Organization for Standardization: Geneva, Switzerland, 2017.

29. Italian Unification Institution. *UNI TS 11300-1: Energy Performance of Buildings—Part. 1: Evaluation of Energy Need Far Space Heating and Cooling*; Italian Unification Institution: Roman, Italy, 2014.

30. Italian Unification Institution. *UNI 11300-2: Energy Performances of Buildings—Part. 2—Evaluation of Primary Energy Need and of System Efficiency for Space Heating, Domestic Hot Water Production, Ventilation and Lighting for Non-Residential Buildings*; Italian Unification Institution: Roman, Italy, 2014; pp. 37–40.

31. Italian Republic. *Implementation of Directive 2009/28/EC on the Promotion of the Use of Energy from Renewable Sources, Amending and Subsequently Repealing the Directives*; Italian Official Gazette N°71: Rome, Italy, 2011.

32. Attia, S.G.; De Herde, A. Early design simulation tools for net zero energy buildings: A comparison of ten tools. In Proceedings of the Building Simulation 2011: 12th Conference of International Building Performance Simulation Association, Sydney, Australia, 14–16 November 2011.

33. Bruno, R.; Pizzuti, G.; Arcuri, N. The Prediction of Thermal Loads in Building by Means of the en ISO 13790 Dynamic Model: A Comparison with TRNSYS. *Energy Procedia* **2016**, *101*, 192–199. [CrossRef]

34. Christensen, C.; Anderson, R.; Horowitz, S.; Courtney, A.; Spencer, J. BEopt Software for Building Energy Optimization: Features and Capabilities. In *Building America US Department Energy*; National Renewable Energy Laboratory: Golden, CO, USA, 2006.

35. D'Agostino, D.; Parker, D. A framework for the cost-optimal design of nearly zero energy buildings (NZEBs) in representative climates across Europe. *Energy* **2018**, *149*, 814–829. [CrossRef]

36. Di Pilla, L. A Methodology for Evaluating the Cost-Effectiveness of Residential Building Stocks Retrofits in Italy and Denmark. Ph.D. Thesis, University of Cagliari, Cagliari, Italy, 2013.

37. Yuan, J.; Farnham, C.; Emura, K. Inter-building effect and its relation with highly reflective envelopes on building energy use: Case study for cities of Japan. *Atmosphere* **2017**, *8*, 211. [CrossRef]

38. Carpino, C.; Loukou, E.; Heiselberg, P.; Arcuri, N. Energy performance gap of a nearly Zero Energy Building (nZEB) in Denmark: The influence of occupancy modelling. *Build. Res. Inform.* **2020**, 1–23. [CrossRef]

39. ISPRA. *Quality Assurance/Quality Control Plan for The Italian Emission Inventory Year 2018*; ISPRA: Roma, Italy, 2018.

40. Italian National Institute for Statistics ISTAT. Italian National Institution for Statistics. Available online: https://www.istat.it/ (accessed on 5 January 2020).

41. Italian Electric Market Manager (GME). Available online: www.mercatoelettrico.org (accessed on 5 January 2020).

42. Ricke, K.; Drouet, L.; Caldeira, K.; Tavoni, M. Country-level social cost of carbon. *Nat. Clim. Chang.* **2018**, *8*, 895–900. [CrossRef]

43. VV.AA. Regional Price List Calabria. 2019. Available online: https://www.regione.calabria.it/website/portalmedia/2018-11/PREZZIARIO.pdf (accessed on 5 January 2020).

44. Italian Energy Service Manager (GSE). Available online: www.gse.it (accessed on 5 January 2020).

45. D'Agostino, D.; de' Rossi, F.; Marigliano, M.; Marino, C.; Minichiello, F. Evaluation of the optimal thermal insulation thickness for an office building in different climates by means of the basic and modified "cost-optimal" methodology. *J. Build. Eng.* **2019**, *24*, 100743. [CrossRef]

Article

Energetic Retrofit Strategies for Traditional Sicilian Wine Cellars: A Case Study

Francesco Nocera *, Rosa Caponetto, Giada Giuffrida and Maurizio Detommaso

Department of Civil Engineering and Architecture, University of Catania, Viale Andrea Doria 6, 95125 Catania, Italy; rosa.caponetto@unict.it (R.C.); giada.giuffrida@unict.it (G.G.); maurizio.detommaso@phd.unict.it (M.D.)
* Correspondence: fnocera@unict.it; Tel.: +39-095-738-2366

Received: 5 May 2020; Accepted: 19 June 2020; Published: 22 June 2020

Abstract: Sicily is characterized by rural buildings, Palmenti, destined to wine production, which are scattered along the countryside and part of the local historical heritage. There are different types of rural buildings, but all have in common the use of ancient and well-established bioclimatic techniques for wine conservation and aging. Most of them were built with the double function of living space for the owner and productive spaces for all the activities correlated to the cultivations. Indeed, many rural houses, destined to the wine production, are characterized by wineries and wine cellars (the first for the wine production, the second to store the wine for the aging process). The growing production of high-quality Sicilian wines, very appreciated all over the world, leads to upgrade the ancient Palmenti to seek optimal hygrothermal conditions and, therefore, to guarantee high performance of the produced and stored wines. The purpose of this study is to investigate how the retrofit measures taken to comply with the energy regulations could affect the thermal behavior of a wine cellar constructed with consolidated bioclimatic technics. The results show the importance of not insulating the solid ground floor for maintaining suitable temperatures for the fermentation and aging of wine. This study can be useful for future analysis when comparing the optimal hygrothermal conditions of wine cellars located in homogeneous viticultural areas (with same climate, geology, soil, physical features, and height) in other parts of the world.

Keywords: bioclimatic strategy; traditional constructive technology; wine vinification; energetic retrofit

---

1. Introduction

The European Commission provides for reducing building energy consumption through the Energy Performance Building Directive 2002/91/EC [1] and 2010/31/EU [2]. The first directive is focused on new buildings while the second is concerned with existing buildings, not only when subjected to major renovation, but also when the building's technical elements and/or technical systems are retrofitted or replaced [3–5]. In the Eastern Sicilian territory, on the slopes of Etna volcano, the grapevine has the most widespread cultivation. The millenary viticultural tradition of the Mount Etna area dates back to Greek and Roman domination. Since 1968, Etna wines have acquired the Registered Designations of Origin (R.D.O.) denomination, established by Decree of the President of the Italian Republic on 11 August 1969 (one of the oldest in Italy) and today, Sicilian wines are highly valued throughout the world [6].

The vineyard, following the orography of the ground, hosts a number of diverse rural constructions adapted on the height of the site [7]. Between 700 and 1000 m above sea level, there are two prevalent kinds of farm buildings: two-story rural houses with the winery and warehouse on the ground floor and the owner's residence on the top floor, or small-sized buildings with a single room for wine pressing and storage in small barrels (Palmenti). The rural buildings of the Eastern Sicilian territory have a historical importance due to their cultural and architectural value. These buildings present elements of historical, ethnic, and anthropological interest, and are often subject to protection by the competent authorities. As it will be shown, this Etnean rural and vernacular heritage is significantly homogeneous concerning the use of local building materials, constructive technologies, traditional passive design features, indoor planimetric organization, and morphological aspects [8]. The rural houses are characterized by generally squared volumetry and lie on the natural irregular ground. Thus, farm buildings are organized on different levels which ensure that pressed grapes (must) naturally flow downwards through the vats into the tank. Despite the rural houses not having been constructed based on geometric plans, they always complied with a few features of simplicity and common sense considering the exposure, the sunlight, the view and many other bioclimatic aspects [9].

The Mediterranean climate heavily influenced design choices of the rural houses, which are constructed with natural materials and bioclimatic solutions, with the aim of mitigating winter and summer climatic extremes. Rural buildings display structural and thermophysical characteristics suitable not only for hygrothermal comfort and wellbeing of the occupants, but also for aging and conservation of the wine. The rural house exposure takes advantage of the natural ventilation provided by the winds of the Mount Etna area, which ensures that the building is always cooled. The rural houses are built by lava stone masonry (60 cm thick) with a loadbearing and buffering function. Moreover, the great mass, typical of high-density materials that characterize traditional architecture of Mediterranean area, improves the energy performance of the building envelope and guarantees thermal and acoustic comfort [10]. The building envelope behaves as a thermal flywheel and modulator element of the heat flow between the outside and the inside. This leads to a reduction of thermal load, especially in summer periods, increasing the temperature gap between the interior and exterior to more than 10 °C, and therefore improving indoor thermal comfort [11]. The openings are small, especially on the south exposure, preventing direct radiation and excessive heat transmission in summer days. The solid ground floor insulation is guaranteed by the fact that often the buildings are separated from the foundation ground by a widespread "crawl space", which physically separates the building from the foundation ground, therefore diminishing both energetic dispersions and rising dampness. In addition, the building's ceiling is often tall enough to guarantee adequate and natural ventilation for the indoor spaces. Furthermore, the roof stratigraphy is usually composed by a loadbearing wooden structure (beams and rafters) covered by intertwined canes and roof tiles [12]. In addition, the rural houses are often characterized by other elements that help mitigate climatic conditions that are typical of traditional dwellings in the Mediterranean area, such as loggias, balconies, verandas, pergolas, openings [13].

State of Art on Farm Wineries Thermal Performance

All materials and low energy and passive solutions not only made the rural houses comfortable, but allows for the control of the internal microclimate; this aspect is of fundamental importance concerning winemaking and conservation of wine, whose production and conservation activities were always located along the north façades or, when it was possible, in the basement, in order to preserve the quality of wine [14,15]. Temperature control during the winemaking process and storage of wine is of paramount importance for the final quality of the wine. Indeed, temperature affects the activity of enzymes, which are involved at various stages in the whole wine production process. In fact, enzymes and bacteria are already present in the grape and may influence its degradation during the mush–maceration phase or affect the final aroma of wine through oxidation. As it is known, the activity of micro-organisms always depends on the temperature of the surrounding environment. From a metabolic point of view, a temperature range from 20 to 25 °C is very favorable for alcoholic fermentation [16]. But at that temperature the fermentation activity becomes too strong and some aromatic compounds can diminish. Thus, in general, alcoholic fermentation should be operated at temperatures from 15 to 18 °C to achieve a satisfactory result [17] and make the wine age slowly, without losing quality [18]. Moreover, in Sicily, wine is traditionally bred and stored in wooden barrels.

Authors in general agree that relative humidity is of secondary importance when it comes to wine aging: in general, a high relative humidity is good because it reduces losses due to evaporation, but it is important to note that those high values should not produce mold in the barrels; ventilation can help in reducing mold-growth risk [18]. At the stage of aging, the stability of air is very important, and it should range between 12–16 °C (lower temperatures during winter would make wine aging slower) with a relative humidity between 70–82% (to prevent from mold growth in poorly ventilated spaces). In general, the more stable the wine cellar is, the less contraction–expansion phenomena will affect the quality of the wine. That is the reason why historically, some types of building were claimed to be more suitable than others to host some food industry like wineries; in particular, underground or partially underground building solutions, with massive envelopes, were constructed to maximize their thermal inertia, thus preventing effects from outdoor climate swings [19,20].

A common feature between traditional wineries from all over the world is the attention given to the relation between the envelope and the foundation ground. In this sense, many authors [21] confirmed that is precisely the thermal stability given by the contact with the ground, which makes traditional buildings (underground and partially underground) so competitive when compared to contemporary wineries (usually built aboveground and often provided by HVAC systems). Some authors proposed an analytical model to predict the thermal interaction between the wine cellar and underground [22], but in the following paragraphs we will explain that in our case study, the contact with the ground is slightly different. As Tinti et al. [22] point out, from the second half of the Twentieth Century, underground buildings were progressively abandoned and substituted with single-story aboveground buildings where indoor temperature and humidity are controlled by air-conditioning systems. Use of HVAC systems to maintain appropriate indoor temperatures and humidity is correlated to surrounding environmental conditions and, in some areas, it was quantified that 50% of the total energy demand is for wine production [23]. It is thus understandable why so much effort in recent years was focused on decreasing energy consumption in the food industry, in general, and in farm wineries [24]. Currently, renewed attention is given to historic small and medium farm wineries, which play a key role in the Italian wine sector [19] for their overall production, for the close relations exchanged with surroundings economies and landscapes, and for their good thermophysical performance.

Our study moves from these considerations with the aim to confirm (1) the effectiveness and efficiency of traditional wine cellars to maintain adequate indoor thermal conditions, thanks to ancient passive design solutions, and, subsequently, (2) to investigate which possible retrofit energy solutions could simultaneously comply with the Italian energy-saving regulation, improve indoor thermal stability, and guarantee the aspect of correct temperatures for the aging and storage of wines. The improvement of the hygrothermal condition for traditional wine cellars and, therefore, the resulting energetic retrofitting, is more notably motivated by the high quality required by the DOC denomination of Sicilian wines. The results of the study underline how a "proper balance between traditional bioclimatic ideas and modern technologies" [25] must be pursued and prioritized, in justified cases, over the national and international energetic regulations.

The results of the study are not merely useful for an in-depth study of the appropriate retrofit strategies aimed at optimizing the conservation and production conditions of wine in local wineries, but above all they could represent a useful feedback for similar research conducted in other parts of the world that share the same characteristics in terms of geology, pedology, climate, and topography. Obviously, the previous technical considerations are strictly related to other factors that contribute to the quality of the final product, which are the so-called organic factors (vine, rootstock) and the anthropogenic ones (cultivation techniques and oenological practices) [26–28].

2. Materials and Methods

In many cases, the palmenti of Eastern Sicilian territory are used for vinification, aging, and conservation of the wine. The retrofitting and/or the functional reorganization of vernacular buildings involve a change in the internal environmental conditions (temperature, relative humidity, quality of the air, lighting, etc.). Thus, attention must be paid to the improvement of energy performance of the building envelope. Since the thermal performance of buildings and the influences of thermal mass and passive strategies can be evaluated only through dynamic thermal simulations [29–31], the authors investigated the thermal performance of a vernacular building in free-running conditions, i.e., without any energy systems for indoor air conditioning. The thermophysical properties of the building were deduced by archival data, endoscopic inspections, and in situ measurements. Starting from an instrumental diagnosis based on heat flow meter and temperature probe data, a calibrated building model was created and yearly dynamic simulations were performed to evaluate the influence of different retrofit measures on the internal microclimate that is of fundamental importance for wine making and conservation of wine. The heat flow meter allowed estimation of wall thermal transmittance; the temperature probes were employed to measure air temperature inside the construction. Indoor humidity measurements were also carried out. The measurements were conducted using a climate measuring instrument, Testo 445, and a Thermozig heat-flux meter, according to ISO 9869. The weather data recorded by a local meteorological station were used as input for building simulation. Detailed dynamic thermal behavior and compliance with the normative in force of the buildings were analyzed using the software tool Design Builder, which is based on the calculation engine of Energy Plus. The building simulation model was calibrated using the temperature measured inside the wine cellar. The model validation was carried out by calculation of some statistical indices. The Mean Absolute Error (MAE), Mean Bias Error (MBE), Coefficient of Variation of the Root Mean Squared Error (CVRMSE), coefficient of determination (R^2) [32], Pearson correlation coefficient (r), and index of agreement (d) were calculated with the aim of evaluating the reliability of the numerical model [33].

Starting from the results of the annual dynamic simulations, it was possible to assess the suitable energy retrofit solutions to obtain the optimal thermo-hygrometric values for winemaking and conservation of wine. For the sake of brevity, from the yearly analysis, the authors extrapolated and analyzed the thermal behavior of the palmento in two representative weeks: 27 January–5 February and 5–14 September. The two weeks selected had the most extreme temperature and humidity values, and are fundamental for the fermentation, aging, and storage of wine on the Etna Volcano area. The winter week chosen is the coldest representative week of winter season and, moreover, is the week in which the wine is already suitable for consumption, with further possibility of improving its qualities by means of aging and storage. The summer week chosen corresponds to the typical harvest week and to the beginning of the long and delicate process of wine aging in order to confer the distinctive typical flavor; additionally, it is also the ending-week of the summer season, which is characterized by the highest temperature values.

3. Case Study: The Wine Cellar

3.1. Description of the Building Features and Constructions

The building (Figure 1) is located in Santa Maria di Licodia (latitude 37.66°, longitude 14.92°, Southern Italy), built during the late Eighteenth Century and currently used for production and storage of wine. The palmento has an overall net floor surface as large as 726 m². The property is located outside the town center, in Contrada Cavaliere, in a large area occupied by various kinds of cultivations close to the wooded areas of Etna volcano.

Figure 1. (**a**) The north façade of the wine cellar; (**b**) internal room of the wine cellar.

This building is undoubtedly representative of the rural building stock of that area. The main material used for the opaque envelope consists of rough-hewn stones obtained from remote lava fields of the Etna volcano, arranged with a lime mortar. This construction technique was very widespread in the Etna volcano area up until the end of the 1950s, and it is still quite common due to the great availability of volcanic rock. The exterior side of the wall is covered with a thick layer of lime-based plaster and its aggregate a volcanic sand derived from Etna activity, which is a local construction tradition. Lava stone masonry provides high thermal inertia, which is particularly suitable for the climate in this area; indeed, the summer season is quite long and hot, with peak daily outdoor temperatures that may easily exceed 35 °C. The thickness of the walls ranges from 60 to 80 cm.

Thermal transmittance of the walls was measured by using a heat flux meter. The heat flow plate and the external temperature probes were applied on a north-facing wall to avoid direct solar radiation. In agreement with standard ISO 9869, measurement time was chosen equal to three days (from 13 December until 16 December 2017). A value of thermal transmittance equal to 1.780 W/m^2 K was measured.

As a result of the transmittance measurement of the walls and their stratigraphy known by endoscopic inspections, the thermal conductivity of basalt stone with mortar was acquired by means of an inverse analytical expression for the calculation of U-values according to ISO 6946. The values of density and specific heat used to characterize the layers of the walls were acquired by ISO 10351.

Table 1 reports the thermophysical properties of the external walls for the building in its current configuration.

Table 1. Thermophysical properties of the external walls for the building in its current configuration.

Layers (from Inside to Outside)	s (m)	λ (W/m·K)	ρ (kg/m^3)	C$_p$ (J/kg·K)	R (m^2·K/W)
Internal surface resistance	-	-	-	-	0.13
Basalt stone with mortar	0.78	2.47	2550	1000	0.32
Plaster (lime)	0.02	0.80	1600	1000	0.025
External surface resistance	-	-	-	-	0.04

The building is also characterized by a traditional wooden pitched roof constructed as follows (from bottom to up): a double system of beams (main beams and rafters), an intertwined canes mat, and the typical terracotta roof tiles. This type of roof shows a U-value around 1.477 W/m^2·K.

The windows are single-glazed and with wooden frames (U = 4.82 W/m^2·K). The solid ground floor of the wine storage room is unpaved, and it consists of a tamped earth floor (made of a mixture of clay, gravel, and sand), laid over the top of a subfloor composed of a crawl space.

Generally, for a traditional tamped earth floor, the topsoil (with organic matter) is removed and filled up with a well compacted inorganic soil (composed of a mixture of clay, sand, gravel). More specifically, a 10 cm-thick barrier layer of clay-rich soil is first applied to a plane level and compacted. Then, a 20–25 cm layer of coarse to medium gravel is added, which acts as a capillary break.

Finally, 20 cm of tamped earth is compacted in layers approximately 6–7 cm thick. The consistency of the tamped earth material at the time of installation is semisolid to rigid. Each layer is thoroughly compacted and left to dry before the next layer is added. After leveling and compaction, the earth screed, which is the last layer, is additionally tamped with a flat board to tighten the "grain-on-grain" structure. After that, pore water is released, and the surface gathers a shiny appearance with improved mechanical strength.

In the crushing and primary (alcoholic) fermentation room, the ground floor is composed of the crawl space and the earthen layer and finished with lava stone slabs.

The superficial mass (SM) of the building components was calculated using the density values drawn from ISO 10351 and literature. The thermal transmittance of the roof, ground floor, and windows were calculated according to ISO 6946, and values of thermal conductivity acquired by ISO 10351 were used for each layer.

In Table 2, the values of Superficial Mass (SM) and thermal transmittance of the envelope (U) components are reported together with the threshold thermal transmittance of the envelope (U_{lim}) necessary to ensure compliance with Italian regulation [34].

Table 2. Thermal transmittance and surface density of the building components.

Building Components	SM (kg/m^2)	U (W/m^2·K)	Ul$_{im}$ (W/m^2·K)
External masonry	2070	1.78	≥0.45
Pitched roof	43.8	1.47	≥0.38
Solid ground floor	190	2.47	≥0.46
Windows	-	4.82	≥3.20

SM: Superficial Mass (SM); Thermal transmittance of the envelope (U_{lim}).

3.2. Building Modeling

The dynamic software Design Builder (Version 6.0, Design Builder Software Ltd., Stroud, UK), a well-known tool based on the calculation engine of Energy Plus, was used to create a model of the real building. The simulations were performed in free-running conditions (without HVAC system), referring only to the building used for the wine cellar (Figure 2), while all other adjacent rooms were considered as adiabatic zones.

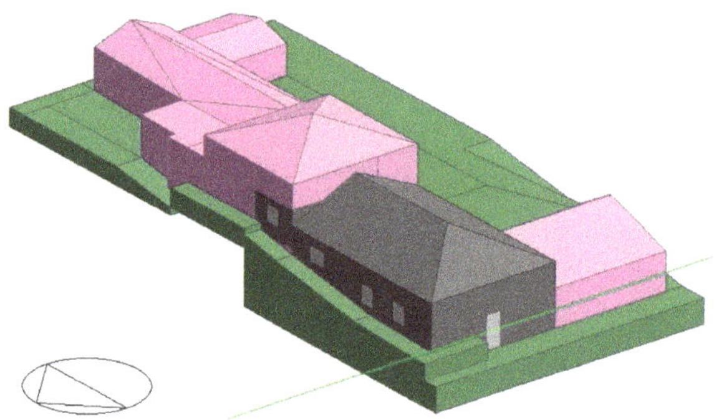

Figure 2. Design Builder Model of the wine cellar (gray).

According to the measured thermal transmittance value (U-value equal to 1.477 W/m^2 K) and to the endoscopic inspections, the wall's stratigraphy was built using the software. Moreover, according to the literature [31,35], the ventilation rate was set equal to 1.5 h^{-1} considering the air leakage within the building.

The model was simulated using the 2017 hourly weather data collected by a meteorological station located in Santa Maria di Licodia. Files acquired by weather stations of Sicilian Agrometeorological Information Service were used as input. The model was calibrated using the results of temperature measurements inside the wine cellar.

As a validation of the building model, the profile of indoor air temperature from the DesignBuilder simulation and that measured inside the wine cellar were compared. Figure 3 depicts the profiles of model simulation indoor air temperature plotted against those observed during the period 27 January–5 February 2017.

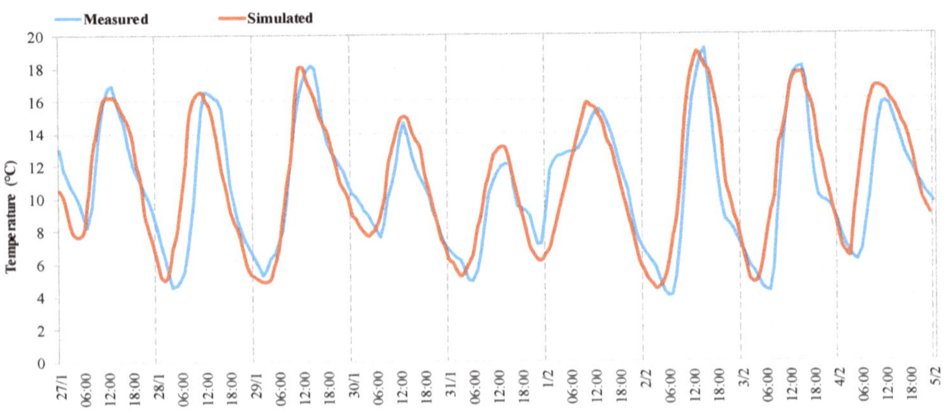

Figure 3. Trend of measured and the model-predicted indoor air temperatures.

The simulated air temperature values have the same trend as that measured. They fit well not only on the maximum values but also on the minimum values. The good reliability and agreement of the simulated values are confirmed by the calculation of statistical indexes (MAE, MBE, CVRSME). Indeed, the value of the coefficient of determination (R^2) is equal to 0.90, while the value of the Pearson coefficient (r) is equal to 0.86. In addition, an index of agreement (d) equal to 0.92 approaching unity reveals that the simulation model can be considered reliable enough for analysis of the thermal performance of retrofit strategies based on dynamic simulations. Table 3 summarizes the values of the statistical indexes calculated for the validation model.

Table 3. Statistical indexes used to validate the building simulation model.

Variables	MAE (°C)	MBE (%)	CVRSME (%)	r (-)	R^2 (-)	d (-)
Indoor temperature	1.53	4.8	6	0.86	0.90	0.92

The Mean Absolute Error (MAE); Mean Bias Error (MBE); Coefficient of Variation of the Root Mean Squared Error (CVRMSE).

3.3. Retrofit Strategies

In order to carry out the upgrading of the building envelope to comply with the Italian legislation, some possible interventions [36–38] are contemplated, not only considering the energetic retrofitting requirement of the rural house, but especially the thermal behavior of the unconditioned space like the wineries and wine cellars.

The retrofit strategies essentially consist of:

- application of thermal insulation on the inner surface of the external walls so that the façade is not disturbed by any intervention, with the aim of not interfering with the architectural attractiveness of this traditional building;
- replacement of existing wooden roof stratigraphy with an insulated roof with a new wooden loadbearing structure;
- realization of an insulated solid ground floor;
- replacement of windows components while keeping the chestnut wood as frame material.

The intervention on the external vertical closures is based on the installation of a 4 cm panel of Aeropan (λ = 0.015 W/m K, c_p = 1000 J/kg K, ρ = 230 kg/m^3) on the inner side of all masonries. Aeropan consists of a nanotechnological insulation in Aerogel coupled with a polypropylene breathing membrane reinforced with glass fiber. Concerning the roof upgrade, a 10 cm insulation layer of hemp fiber panel (λ = 0.038 W/m K, c_p = 1700 J/kg K, ρ = 30 kg/m^3) is placed on the wooden planks, upon a vapor barrier; the insulation is protected from the top by a waterproofing membrane. The upper part of the roof is completed by wooden laths, which define an unventilated air gap upwards of 8 cm.

The solid ground floor, simply consisting of a tamped earth layer on a crawl space subfloor, is replaced by a "modern" tamped earth floor. In order to ensure a better distribution of loads for buildings in seismic areas and to avoid cracks in the clay floor, a concrete screed with electrowelded mesh was inserted.

Therefore, starting from the original layer of sand and gravel, in the new solid ground floor there are, in order from bottom to top: a layer of expanded clay granules, a layer of granulated foamglass, a concrete screed reinforced with electrowelded mesh, and a tamped earth floor. Generally, the earth floor is finished with four to six coats of oil, often hemp or linseed, then topped with a combination of soft waxes. This treatment provides sheen and waterproofing to the floor. In our study, we propose not to complete this finishing to facilitate breathability.

This earthen floor allows regulation of the relative air humidity because clay absorbs water vapor when the air is too wet, whereas it is able to release water vapor when the air is drier. As some authors pointed out [39], the physical behavior of raw earth layers is typical of massive porous materials where coupled hygrothermal mechanisms coexist inside the micropore network geometry; in this case, the pavement experiences a heat transmission mechanism for materials with a high thermal mass, with consequent high thermal inertia and heat storage, evaporation, and condensation inside the pores, resulting from temperature changes produced by thermal waves. An absorption/release of moisture contained in the pores occurs because of the change in ambient humidity and the influence of rising dampness [40].

Due to their density and high thermal conductivity, dry earth floors are thermally efficient, capable of absorbing heat and releasing it gradually over time—the thermal "flywheel" effect. The English Heritage, in its guidance note titled, Energy Efficiency and Historic Buildings: Insulating Solid Ground Floors, observes that "the ground itself maintains a surprisingly stable temperature of around 10 °C" [41]. The intervention on the external windows consists in the replacement of the current single glazing with double glazing (s = 4 mm) separated by an air gap (s = 16 mm) with a chestnut wood frame. A low emissivity coating (ε = 0.10) on the inner glazing is proposed. Ventilation rate is set equal to 0.5 ac/h.

The thermophysical properties of each layer that composes the building construction after retrofit interventions are reported in Table 4.

Table 4. Thermophysical properties for each layer that composes opaque and glazing components after retrofit.

a) Wall				
Layer	**s (m)**	**λ (W/m·K)**	**ρ (kg/m3)**	**Cp (J/kg·K)**
Basalt stone with mortar	0.78	2.47	2550	1000
Lime mortar	0.01	0.900	1800	1000
Aeropan	0.04	0.015	230	1000
Lime mortar	0.01	0.419	1800	1000
Inner plaster (lime + gypsum)	0.02	0.80	1600	1000

b) Ground Floor				
Layer	**s (m)**	**λ (W/m·K)**	**ρ (kg/m3)**	**Cp (J/kg·K)**
Raw earth ground	0.03	0.80	2000	1000
Concrete screed	0.10	1.160	2000	1000
Granulated foamglass	0.04	0.045	125	900
Expanded clay granules	0.10	0.120	280	920
Sand and gravel	0.20	1.200	1700	1000

c) Roof				
Layer	**s (m)**	**λ (W/m·K)**	**ρ (kg/m3)**	**Cp (J/kg·K)**
Wooden Plank	0.03	0.15	450	1600
Vapor barrier	0.0002	0.130	950	1800
Hemp panel	0.09	0.038	30	1700
Waterproofing membrane	0.009	0.17	52.50	1800
Air gap	0.08		-	
Terracotta tiles	0.02	1.000	2000	840

The values of superficial mass and thermal transmittance of each opaque envelope component after RIs (retrofitting interventions) are reported in Table 5. It also shows the limit U-values (U_{lim}) that the same building components have to comply with the Italian regulation [34] for the climate zone of Catania. Tables 5 and 6 show the thermal transmittance and superficial mass of the building components after retrofit interventions (RIs). Moreover, the comparison between the U-values of the envelope components at the current state and the thermal transmittance of the same components after retrofit interventions (RIs) are reported.

Table 5. Thermal transmittance and surface density of the opaque components.

Building Components	**SM_{RI} (kg/m²)**	**U_{RI} (W/m²·K)**	**U_{Lim} (W/m²·K)**	**ΔU = U–U_{RI} (W/m²·K)**
External walls	2070	0.315	≤0.45	1.645
Pitched roof	59.0	0.309	≤0.38	1.168
Solid ground floor	562	0.460	≤0.46	2.051

Table 6. Main thermophysical properties of the windows.

Property	**Unit**	**Value**	
		Before	**After**
Emissivity of the glazed surface	-	0.84	0.10
Glass g-value	-	0.85	0.75
Thermal transmittance of the glazing	W/m²·K	5.88	2.00
Thermal transmittance of the frame	W/m²·K	3.50	3.50
Overall thermal transmittance of the window	W/m²·K	4.82	2.70

It can be observed that the proposed solutions for the building upgrading have lower thermal transmittance values than the ones imposed by current Italian regulations. It has to be highlighted that this choice results in a significant thickness of the insulation layer or in the use of very performative materials in the upgraded building components for satisfying the limit U-values. The results of the simulations will demonstrate the appropriateness of this choice.

4. Results

4.1. The Building in its Current State

The results of the free-running performance for the current building are shown in Figures 4 and 5. In Figure 4, the simulated hourly profile of the outdoor temperature (T_o), the measured indoor air temperature (T_{am}), and the simulated indoor air temperature (T_{as}) can be observed for a representative week in winter. There is good agreement between the measured and simulated values. The measured indoor air temperature ranges from a minimum of 10.4 °C to a maximum of 15.5 °C, and a mean daily oscillation of 4.1 °C is achieved during the period 27 January–5 February. The relative humidity presents a mean value of 71%, which is suitable for the storage of wine.

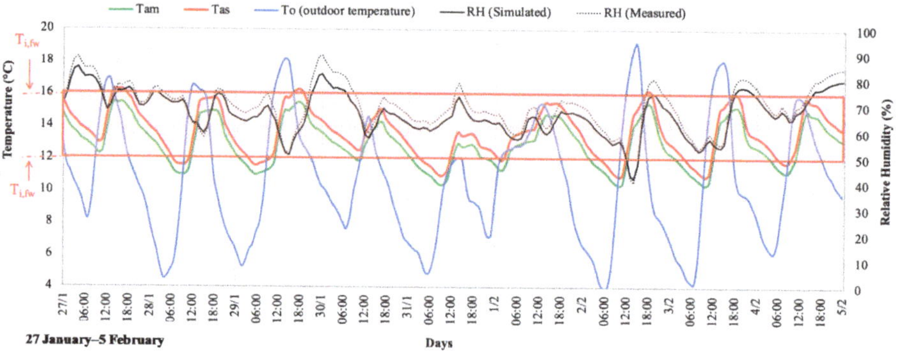

Figure 4. Daily profile of indoor temperatures at the current state of the building in free-running conditions during 27 January–5 February.

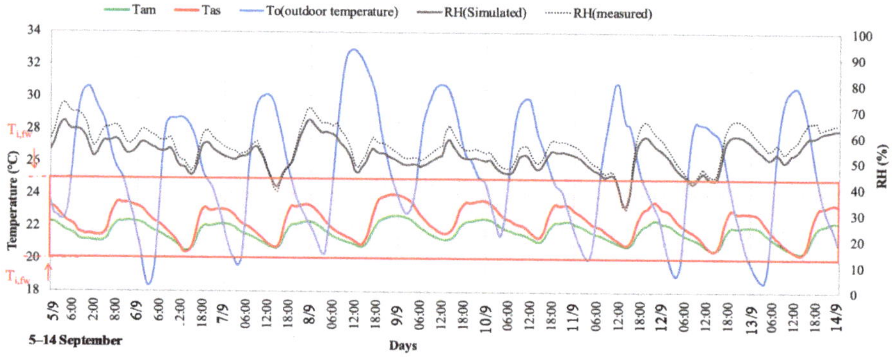

Figure 5. Daily profile of indoor temperatures at the current state of the building in free-running conditions during 5–14 September.

The trend of simulated temperature ranges from a minimum of 10.9 °C to a maximum of 16.3 °C and a mean daily oscillation of 5.3 °C, while the mean value of humidity is 68.8% and not congruent with the measured value.

The minimum values of T_{am} usually occur at 6:00 a.m. and its peak values are at around 3:00 p.m. The values of indoor temperature testify a good behavior of the building for aging and storage of the wine in wooden barrels during the winter period, considering that the range of optimal temperature is from 12 to 16 °C. Figure 5 depicts the simulated hourly trend of indoor thermal conditions during period (5–14 September).

The profile of measured indoor air temperature shows a minimum value of 20.4 °C and a maximum of 22.7 °C. Mean daily amplitude of air temperature wave of 3.0 °C is obtained.

In summer, the simulated air temperatures values are in good agreement with the measured ones, oscillating only between 20.3 and 24.0 °C. The mean value of humidity measured is 53.5% and is very close to the one simulated (57.0%).

The minimum values of T_{am} occur at around 5:00 a.m., whereas its maximum values are at 1:00 p.m. according to the profile of outdoor air temperature. It has to be highlighted that the values of air temperature are suitable for the alcoholic fermentation of wine because the variation of temperature of air is included in the range of 20–25 °C.

Some peculiarities regarding the constructive technology of our case study—like the lava stone masonry and the crawl space in the solid ground floor—make our results not immediately comparable with previous literature. The previous literature review evidence shows that there are few studies about the indoor thermal behavior of the wine cellars with a massive structure that, together with differences in terms of orientation, morphological characteristics, and climate conditions, make a useful comparison difficult.

4.2. Expected Results after Thermophysical Upgrading

The thermal behavior of the building after the upgrading is shown in order to evaluate the potentiality of the interventions on the performance of storage and fermentation of wine. In this scenario, all proposed abovementioned technological solutions for building retrofitting are implemented.

For this purpose, Figures 6 and 7 show the results of the simulations with the application of the retrofit solutions for two representative periods: 27 January–5 February and 5–14 September.

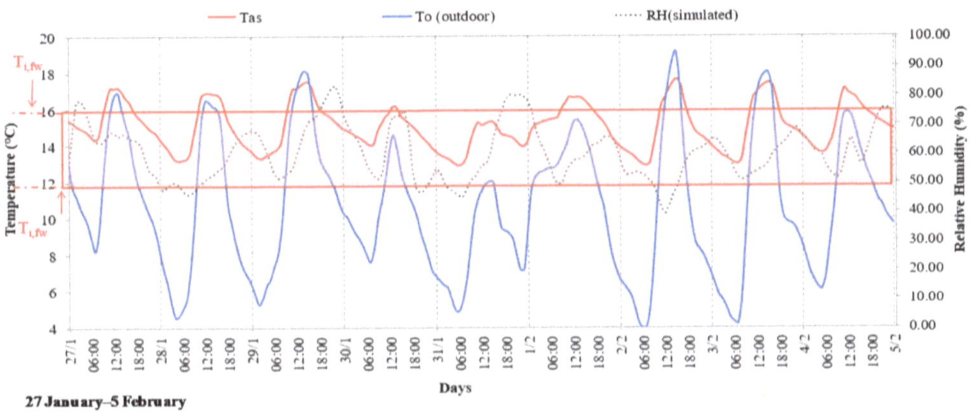

Figure 6. Daily profile of indoor air and operative temperature after retrofit of the building in free-running conditions during 27 January–5 February.

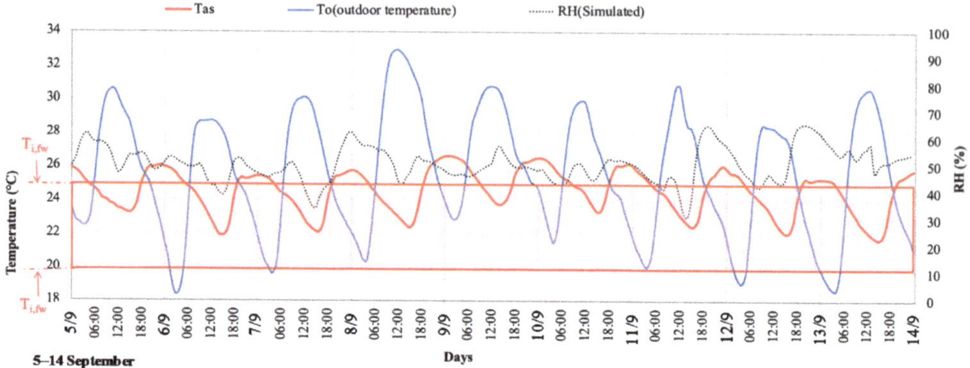

Figure 7. Daily profile of air and operative temperature after retrofit of the building in free-running conditions during 5–14 September.

Figure 6 shows that the hourly profile of air temperature ranges between 12.9–17.7 °C and maximum amplitude of temperature curve about 4.8 °C. As a consequence, the indoor thermal state continues to guarantee adequate conditions for the storage of the wine in the barrels for most of the investigated period, but the peak values of temperature are always above the upper optimal values.

During the summer period, the indoor temperature is up to the optimal limit due to the application of the insulation and the replacement of windows with an improvement of air tightness. Indeed, the profile of temperature of air ranges from 21.7 °C to 26.6 °C with a maximum daily amplitude of wave about 5.0 °C. As a result, it can be observed that the average indoor temperature (25.1 °C) is higher than the maximum value of temperature for a correct fermentation of wine. The relative humidity assumes a mean value of 53.8%.

According to these results, the indoor thermal behavior of the retrofit scenario (RI) is worse than the building's current state in summer conditions, if alcoholic fermentation and store process of wine are considered.

4.3. Potential Effects of Insulated Floor on Ground Surface

Comparing the thermal performance of the wine cellar at the current state (TR) and after its retrofit scenario (RI), it is evident that the implementation of standard energy efficient measures, applied to a bioclimatic structure, leads to a worst thermal behavior due, basically, to the insulation of the pavement that plays a fundamental role in heat and mass exchange for the maintenance of suitable temperature for wine storage and aging. This latter behavior is noticeable in Figures 8 and 9, where the temperature and humidity trend with (RI) are reported, and without thermal floor insulation (RI-IGF) and in current state (TR). In addition, in Table 7, the maximum and minimum values of air temperature, the maximum amplitude of temperature ($\Delta T_{a,max}$), and the average value of relative humidity (RH) related to the three scenarios mentioned during the period 27 January–5 February are reported. In Table 8, the maximum and minimum values of air temperature, the maximum amplitude of temperature, and the average value of humidity related to the three scenarios mentioned during period 5–14 September are reported.

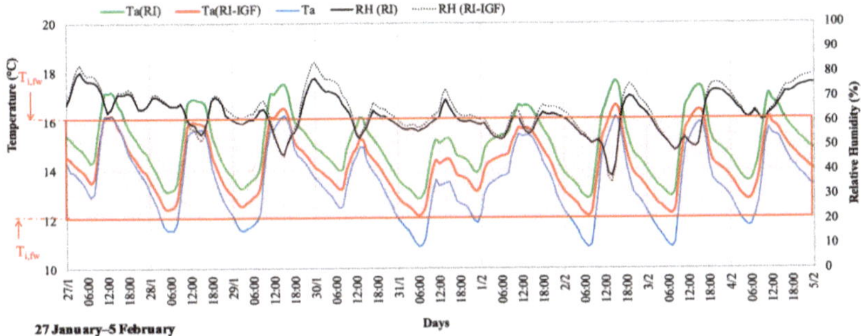

Figure 8. Daily profile of air temperature (T_a) in the current state, after retrofit Ta (RI), and retrofit scenario with uninsulated floor on ground surface (RI-IGF) during the period 27 January–5 February.

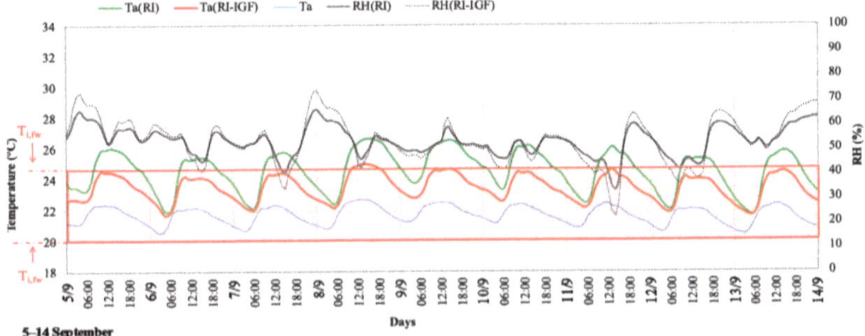

Figure 9. Daily profile of air temperature (T_a) in the current state, after retrofit Ta (RI), and retrofit scenario with no insulated floor on ground surface (RI-IGF) during the period 5–14 September.

Table 7. Maximum and minimum values of air temperature, maximum amplitude of temperature, and relative humidity of the three investigated scenarios during period 27 January–5 February.

Scenario		$T_{a,max}$ (°C)	$T_{a,min}$ (°C)	$\Delta T_{a,max}$ (°C)	RH %
Current state	TR	16.3	10.9	5.4	71.0
After retrofit	RI	17.7	12.9	4.8	63.7
After retrofit with no insulation on ground floor	RI-IGF	16.7	12.2	4.5	65.3

Table 8. Maximum and minimum values of air temperature, maximum amplitude of temperature, and relative humidity of the three investigated scenarios during the period 5–14 September.

Scenario		$T_{a,max}$ (°C)	$T_{a,min}$ (°C)	$\Delta T_{a,max}$ (°C)	RH %
Current state	TR	22.7	20.4	2.3	57.0
After retrofit	RI	26.6	21.7	4.9	53.8
After retrofit with no insulation on ground floor	RI-IGF	25.0	21.6	3.9	54.4

Indeed, looking at the results, it is worth noticing that, in winter, the minimum and maximum values of temperatures rise, but the optimal value of temperature for wine storage and aging is still guaranteed. Basically, only the humidity undergoes a decrement from ideal values. However, the peaks of temperature (RI) exceed the upper limit of 16 °C by about 1.7 °C. The latter value is lower in the

RI-IGF scenario configuration by around 0.7 °C. In any case, the insulation of the building leads to a reduction of oscillation between the peak values of temperature as reported in Table 7.

In summer, the thermal insulation applied leads to the overheating of the building and to the increment of indoor temperatures beyond the optimal values as reported in Figure 9. Indeed, the peak of maximum temperature is 26.6 °C in the RI scenario, on the other hand, although the increment of temperature occurs also in RI-IGF, the peaks lie within the ideal values (Table 8). On the contrary, the mean values of humidity do not undergo substantial variation compared to the current state.

The outcomes about the effectiveness of retrofitting scenarios on the indoor thermal conditions of the wine cellar are not comparable with those of other studies. In literature, there are a few studies [17] that deal only with the evaluation of indoor thermal conditions of traditional wine cellars in their current configurations. Poor attention is given to the thermal and physical properties of the materials used in the building components (such as the lava stone masonry and the crawl space solid ground floor) and in their retrofitted configurations. Finally, dynamic and energetic thermal simulations have rarely been used for the analysis of thermal performance of farm winery buildings.

5. Conclusions

The current research focused on the importance of traditional construction knowledge based on the use of locally sourced natural base materials and bioclimatic design choices, as those used for rural buildings in Etna volcano area. These dwellings are usually single or double-story buildings, resting on the slope of the site and using local building materials, such as lava stone. Lava stone is renowned for its use in masonry for the realization of thick walls, ranging from 60 to 80 cm, providing high thermal inertia, which is particularly important in Mediterranean climates. A good bioclimatic approach is always detectable in these buildings, as a result of the adaptation to natural phenomena that could contribute in creating better living spaces concerning sunlighting, exposure, and ventilation aspects.

This kind of traditional architecture was often used in farm wineries and wine cellars, productive spaces where typical temperature and humidity conditions must be maintained almost constant in order to allow an optimal wine fermentation, aging, and storing process.

This study therefore analyzed the effectiveness of a wine cellar located in the Mount Etna area, concerning its efficiency in maintaining adequate thermal and humidity indoor conditions for wine production. In this sense, the U-value measurements made onsite have allowed a certain level of accuracy for the Design Builder model, which was implemented in order to compare the current state of the building with different retrofit strategies. Temperature and humidity values of the current state simulation were calibrated with measurements made onsite, confirming the correctness of the model.

The observation of measured and simulated current state temperature and humidity showed an interesting adjustment within the limits considered to be optimal for wine production (including fermentation, aging, and storage processes). Nevertheless, the simulation highlighted that some usual retrofit interventions made to comply with the Italian energy saving regulation, such as the insulation of all elements of the envelope (roof, walls, and solid ground floor) and the replacement of the windows, cause an increase in average indoor temperatures during summer, affecting the process of alcoholic fermentation and storage of wine.

For this reason, a corrected retrofit scenario was investigated, where walls and roof are insulated whereas the ground insulation is not. This is because the existing earth floor, due to its density and high thermal conductivity, is particularly capable of absorbing heat and releasing it gradually over time (thermal "flywheel" effect). Therefore, the heat and mass exchange through the solid ground floor is not modified, with good consequences in the global thermal behavior of the building. In this scenario, indoor temperatures in summer conditions are higher than the current state ones, but they fall within the limits for the wine cellar activity, while humidity values remain closer to the optimal ones for aging of wine.

This paper showed the importance of contrasting usual upgrading and retrofit strategies for existing buildings, with the ones that are possible to apply to vernacular dwellings designed with a bioclimatic approach. In order to assess the best retrofitting options, several calibrated dynamic thermal simulations were run on Design Builder software, referring to the most extreme hygrothermal conditions during a year, and allowing a considerable reduction in data collection compared to previously used methodologies. The specificity of some traditional constructive practices (as the use of massive and uninsulated solid ground floors and walls), makes wine cellars an interesting case study. Because of this specificity, the environmental conditions (indoor temperature and humidity), which must be maintained for optimal fermentation and aging of wine, have to guide the upgrading and retrofit strategies on these buildings, even if this means not to comply with current national energetic regulations.

Author Contributions: Conceptualization, F.N., M.D.; methodology, G.G., R.C., M.D. and F.N.; formal analysis, M.D.; investigation, F.N. and M.D.; resources, G.G. and R.C.; data curation, G.G., R.C., M.D..; writing—original draft preparation, G.G., M.D., R.C. and F.N.; writing—review and editing, G.G., R.C., M.D. and F.N.; visualization, G.G.; supervision, R.C. and F.N.; project administration, G.G., R.C. and F.N.; funding acquisition, F.N. All authors have read and agreed to the published version of the manuscript.

Funding: This research received no external funding.

Conflicts of Interest: The authors declare no conflict of interest.

References

1. European Parliament. Directive 2002/91/EC of The European Parliament and of the Council of 16 December 2002 on the energy performance of buildings. *Official Journal of the European Union*. 1 April 2003. Available online: https://eur-lex.europa.eu/ (accessed on 20 June 2020).
2. European Parliament. Directive 2010/31/EU of The European Parliament and of the Council of 19 May 2010 on the energy performance of buildings (recast). *Official Journal of the European Union*. 18 June 2010. Available online: https://eur-lex.europa.eu/ (accessed on 20 June 2020).
3. Costanzo, V.; Evola, G.; Marletta, L.; Nocera, F. The effectiveness of phase change materials in relation to summer thermal comfort in air-conditioned office buildings. *Build. Simul.* **2018**, *11*, 1145–1161. [CrossRef]
4. Nocera, F.; Faro, A.L.; Costanzo, V.; Raciti, C. Daylight performance of classrooms in a Mediterranean school heritage building. *Sustainability* **2018**, *10*, 3705. [CrossRef]
5. Trovato, M.R.; Nocera, F.; Giuffrida, S. Life-cycle assessment and monetary measurements for the carbon footprint reduction of public buildings. *Sustainability* **2020**, *12*, 3460. [CrossRef]
6. Disciplinare di Produzione dei vini a Denominazione di Origine Controllata "Etna", Gazzetta Ufficiale della Repubblica Italiana, Serie Generale n. 243. 18 October 2011. Available online: https://www.gazzettaufficiale.it (accessed on 20 June 2020).
7. Caltabiano, I. The traditional architecture in Sicily: The rural area around the volcano Etna. In Proceedings of the PLEA 2006, the 23rd Conference on Passive and Low Energy Architecture, Geneva, Switzerland, 6–8 September 2006.
8. Palumbo, G.; Magnano di San Lio, E. *Le Residenze di Campagna nel Versante Orientale dell'Etna, Dipartimento di Architettura e Urbanistica*; Università degli Studi di Catania: Catania, Italy, 1991.
9. Zhai, Z.J.; Previtali, J.M. Ancient vernacular architecture: Characteristics categorization and energy performance evaluation. *Energy Build.* **2010**, *42*, 357–365. [CrossRef]
10. Nguyen, A.-T.; Tran, Q.B.; Tran, D.Q.; Reiter, S. An investigation on climate responsive design strategies of vernacular housing in Vietnam. *Build. Environ.* **2011**, *46*, 2088–2106. [CrossRef]
11. Zagorskas, J.; Zavadskas, E.K.; Turskis, Z.; Burinskiene, M.; Blumberga, A.; Blumberga, D. Thermal insulation alternatives of historic brick buildings in Baltic Sea Region. *Energy Build.* **2014**, *78*, 35–42. [CrossRef]
12. Cardinale, N.; Rospi, G.; Stazi, A. Energy and microclimatic performance of restored hypogeous buildings in south Italy: The "Sassi" district of Matera. *Build. Environ.* **2010**, *45*, 94–106. [CrossRef]
13. Ozay, N. A comparative study of climatically responsive house design at various periods of Northern Cyprus architecture. *Build. Environ.* **2005**, *40*, 841–852. [CrossRef]

14. Khalili, M.; Amindeldar, S. Traditional solutions in low energy buildings of hot-arid regions of Iran. *Sustain. Cities Soc.* **2014**, *13*, 171–181. [CrossRef]

15. Giucastro, F.G.S.; Giordano, D. Et(h)nic Architecture in Mediterranean Area. *Energy Procedia* **2016**, *96*, 868–880. [CrossRef]

16. Troost, G. *Technologie des Weines (Handbuch der Lebensmitteltechnologie)*, 6th ed.; Ulmer Verlag: Stuttgart, Germany, 1988.

17. Ribéreau-Gayon, P.; Dubourdieu, D.; Donèche, B.; Lonvaud, A. *Handbook of Enology*; John Wiley and Sons: Chichester, UK, 2006; Volume 1.

18. Mazarrón, F.R.; Cid-Falceto, J.; Cañas, I. Ground Thermal Inertia for Energy Efficient Building Design: A Case Study on Food Industry. *Energies* **2012**, *5*, 227–242.

19. Cañas Guerrero, I.; Martin Ocaña, S. Study of the thermal behaviour of traditional wine cellars: The case of the area of "Tierras Sorianas del Cid" (Spain). *Renew. Energy* **2005**, *30*, 43–55. [CrossRef]

20. Tassinari, P.; Barbaresi, A.; Benni, S.; Torreggiani, D. Farm wineries design: Preliminary indications for integrating energy efficiency in building modelling. In Proceedings of the International Conference of Agricultural Engineering, Valencia, Spain, 8–12 July 2012.

21. Barbaresi, A.; Dallacasa, F.; Torreggiani, D.; Tassinari, P. Retrofit interventions in non-conditioned rooms: Calibration of an assessment method on a farm winery. *J. Build. Perform. Simul.* **2017**, *10*, 91–104. [CrossRef]

22. Tinti, F.; Barbaresi, A.; Benni, S.; Torreggiani, D.; Bruno, R.; Tassinari, P. Experimental analysis of thermal interaction between wine cellar and underground. *Energy Build.* **2015**, *104*, 275–286. [CrossRef]

23. Tassinari, P.; Galassi, S.; Benni, S.; Torreggiani, D. The built environment of farm wineries: An analysis methodology for defining meta-design requirements. *J. Agric. Eng.* **2011**, *42*, 25–31.

24. Giuffrida, S.; Gagliano, F.; Nocera, F.; Trovato, M.R. Landscape assessment and economic accounting in wind farm programming: Two cases in sicily. *Land* **2018**, *7*, 120. [CrossRef]

25. Widera, B. Bioclimatic Architecture as an Opportunity for developing countries. In Proceedings of the 30th International PLEA Conference, CEPT University, Ahmedabad, India, 16–18 December 2014.

26. Johnson, H.; Robinson, J. *Atlante Mondiale Dei Vini*; Mondadori Electa: Milan, Italy, 2014; ISBN 9788837099060.

27. Ribéreau-Gayon, P.; Dubourdieu, D.; Donèche, B. *Trattato di Enologia Volume 1. Microbiologia del vino e Vinificazioni, Edagricole*, 4th ed.; New Business Media: Milano, Italy, 2017; ISBN 885065507X.

28. Bonfante, A.; Terribile, F. Zonazione viticola. Un approccio su base fisica. In Proceedings of the Suolo e Vino Conference, Imola, Italy, 11–12 October 2006; pp. 49–57.

29. Ascione, F.; De Rossi, F.; Vanoli, G.P. Energy retrofit of historical buildings: Theoretical and experimental investigations for the modelling of reliable performance scenarios. *Energy Build.* **2011**, *43*, 1925–1936. [CrossRef]

30. Cornaro, C.; Puggioni, V.A.; Strollo, R.M. Dynamic simulation and on-site measurements for energy retrofit of complex historic buildings: Villa Mondragone case study. *J. Build. Eng.* **2016**, *6*, 17–28. [CrossRef]

31. Gagliano, A.; Nocera, F.; Patania, F.; Detommaso, M.; Sapienza, V. Deploy energy-efficient technologies in the restoration of a traditional building in the historical center of Catania (Italy). *Energy Procedia* **2014**, *62*, 62–71. [CrossRef]

32. Ramos Ruiz, G.; Fernández Bandera, C. Validation of Calibrated Energy Models: Common Errors. *Energies* **2017**, *10*, 1587. [CrossRef]

33. Willmott Cort, J.; Robeson Scott, M.; Matsuura, K. Short Communication A refined index of model performance. *Int. J. Climatol.* **2012**, *32*, 2088–2094. [CrossRef]

34. Italian Republic. *Ministerial Decree 26 June 2015. Appx. A, Att. 1. Adeguamento del Decreto del Ministro Dello Sviluppo Economico, 26 Giugno 2009—Linee Guida Nazionali per la Certificazione Energetica Degli Edifici*; OJ of the Italian Republic: Rome, Italy, 2015. Available online: https: //www.mise.gov.it/index.php/it/normativa/decreti-interministeriali/2032968-decreto-interministeriale-26-giugno-2015-adeguamento-linee-guida-nazionali-per-la-certificazione-energetica-degli-edifici (accessed on 20 June 2020).

35. Lagüela, S.; Martínez, J.; Armesto, J.; Arias, P. Energy efficiency studies through 3D laser scanning and thermographic technologies. *Energy Build.* **2011**, *43*, 1216–1221. [CrossRef]

36. Bevilacqua, P.; Benevento, F.; Bruno, R.; Arcuri, N. Are Trombe walls suitable passive systems for the reduction of the yearly building energy requirements? *Energy* **2019**, *185*, 554–566. [CrossRef]

37. Bevilacqua, P.; Bruno, R.; Arcuri, R. Green roofs in a Mediterranean climate: Energy performances based on in-situ experimental data. *Renew. Energy* **2020**, *152*, 1414–1430. [CrossRef]

38. Bruno, R.; Bevilacqua, P.; Cuconati, T.; Arcuri, N. Energy evaluations of an innovative multi-storey wooden near Zero Energy Building designed for Mediterranean areas. *Appl. Energy* **2019**, *238*, 929–941. [CrossRef]

39. Fabbri, A.; Morel, J.C. *Earthen materials and constructions. Nonconventional and Vernacular Construction Materials*; Woodhead Publishing Series in Civil and Structural Engineering; Woodhead Publishing: Cambridge, UK, 2013.

40. Giuffrida, G.; Caponetto, R.; Nocera, F. Hygrothermal Properties of Raw Earth Materials: A Literature Review. *Sustainability* **2019**, *11*, 5342.

41. Ogley, P.; Pickles, D.; Wood, C.; Brocklebank, I. *Energy Efficiency and Historic Buildings: Insulating Solid Ground Floors*; English Heritage: London, UK, 2012.

 *energies*

Article

An Innovative Trombe Wall for Winter Use: The Thermo-Diode Trombe Wall

Jerzy Szyszka [1], Piero Bevilacqua [2,*] and Roberto Bruno [2]

[1]   Faculty of Civil and Environmental Engineering and Architecture, Rzeszow University of Technology, 35-959 Rzeszów, Poland; jszyszka@prz.edu.pl
[2]   Department of Mechanical, Energetic and Management Engineering, University of Calabria, Arcavacata di Rende, 87036 Cosenza, Italy; roberto.bruno@unical.it
*   Correspondence: piero.bevilacqua@unical.it

Received: 26 March 2020; Accepted: 28 April 2020; Published: 30 April 2020

Abstract: The use of passive solutions for building envelopes represents an important step toward the achievement of more efficient and zero-energy building targets. Trombe walls are an interesting and viable option for the reduction of building energy requirements for heating, especially in cold climates. This study presents the experimental analysis of an innovative Trombe wall configuration, named a thermo-diode Trombe wall, which was specifically designed to improve the energy efficiency by providing a proper level of insulation for the building envelope. Such a design is essential in cold climates to limit the thermal losses whilst increasing solar heat gains to the heated spaces. An experimental campaign was conducted from December to March that involved monitoring the external climatic conditions and the main thermal parameters to assess the thermal performance of the proposed solution. The results demonstrated that in the presence of solar radiation, the thermo-diode Trombe wall was able to generate significant natural convection inside the air cavity, with temperatures higher than 35 °C in the upper section, by providing consistent heat gains for the indoor environment, even on cold days and for hours after the end of the daylight. The efficiency, relative to the incident solar radiation, reached 15.3% during a well-insolated winter day.

Keywords: trombe wall; experimental analysis; solar gains; PCM thermal storage

1. Introduction

To achieve more sustainable and lower energy-consuming buildings, particular attention has been paid to the development of green standards for building construction [1]. The use of passive solutions in the building envelope appears to be more than an obligatory step. In this regard, several solutions have been proposed and studied by researchers, including green roofs [2], the Barra–Constantini system [3], the adoption of proper control strategies of venetian blinds [4], and the use of agricultural building insulating materials [5]. Trombe walls represent an interesting solution in this direction since they are capable of adequately exploiting solar energy and provide both heating and ventilation for indoor environments. In a Trombe wall, the solar radiation incident on the vertical surface is captured by an absorber with a high absorptivity coefficient. Thermal energy is stored in the Trombe wall structure thanks to the elevated thermal mass. Subsequently, thermal energy is released toward the indoor environment via air thermo-circulation, convection and long-wave radiation from the absorber inner surface. Airflow is usually activated and managed through proper vents, and it occurs in an air gap formed between an external glazed façade and the massive wall. The air in the air gap is heated and delivered to the adjacent rooms thanks to thermo-circulation, passively providing thermal energy. The massive wall is the most crucial element of a Trombe wall since it is responsible for the thermal storage [6]. Therefore, the choice of suitable materials for a massive wall is crucial as well as the choice of an appropriate glazing system according to the climatic location [7]. The capacity of a Trombe wall to achieve energy savings has been investigated in several climatic conditions: from subtropical locations [8], to semi-arid [9] and hot Mediterranean climates [10], to cold Polish loc tions [11]. A numerical model of a Trombe wall was developed in the TRNSYS environment and validated using a small-scale experimental prototype [12], producing a 77% reduction in heating energy demand for a 16 m^2, non-insulated simple Tunisian building. In a Mediterranean location, it was found that the Trombe wall did not reduce the maximum heating load; however, an annual heating energy saving of up to 32.1% can be reached with a 37% area ratio [13]. A ventilated low-cost Trombe wall using low-tech prefab components was designed and tested for passive heating and cooling of existing buildings in Chile [14]. The results showed predicted energy savings of 44.14% and 25.35%, respectively, for two cities in different winter microclimates. More sophisticated approaches involved computational fluid dynamics (CFD) [15] for evaluating the achievable thermal performance using a Trombe wall for either a single room [16] or an entire house [17]. A three-dimensional CFD model was developed and validated to investigate a Trombe wall equipped with a venetian blind [18]. CFD was also used to investigate a composite Trombe wall, which combined a water wall and a traditional Trombe wall [19]. A building energy simulation and CFD were coupled to analyze the thermal performance of a Trombe wall with a venetian blind with the aim of regulating shading and airflow in the cavity of the solar wall in the cooling season by observing the reduction of energy consumptions [20].

Indeed, an important issue with a conventional Trombe wall is the excessive solar heat gain during summer, which can lead to indoor environments overheating by producing a worsening of the cooling requirements. Stazi et al. [21] also stressed the importance of summer shading of the Trombe wall to prevent excessive overheating and a counterproductive heat transfer to the indoor environment. The possibility of achieving summer energy savings thanks to the adoption of proper ventilation strategies was analyzed in different climatic contexts [22]. The variation of the blind tilt angle of a Trombe wall was found to have a significant effect on the limitation of cooling loads [23].

The possibility of improving the performance of a Trombe wall with appropriate modifications has been addressed by some authors. An unvented Trombe wall with an extra window in the massive wall was simulated using Solidworks in the locality of Athens, Greece, to verify the possibility of producing greater indoor temperatures and improving the internal space lighting [24]. The addition of vertical thermal fins to the internal surface was considered to improve the efficiency of an unvented Trombe wall and to maximize the heat transfer [25]. The application of vertical thermal fins on the absorber of a Trombe wall was also experimentally addressed in the arid climate of Yazd (Iran), obtaining an increase in the energy stored by up to 3% [26]. A Trombe wall, together with a solar chimney and a water spraying system, was experimentally studied in a test room under a desert climate, which resulted in enhancing the thermal efficiency by approximately 30% [27]. A novel application proposed a Trombe wall with blinds for shading and water flowing channels, obtaining a higher overall thermal efficiency [28]. An interactive glass wall, whose working principle is analogous to a Trombe wall, was presented and experimentally analyzed [29]. A modified Trombe wall prototype called a collector–accumulation wall (CAW) was experimentally assessed using a laboratory simulator to evaluate the heat distribution efficiency [30].

Some studies proposed various configurations to improve the storage capacity of the massive wall. In an experimental study of a small-scale Trombe wall, phase-change material (PCM) was inserted in the form of a brick-shaped package [31]. The integration of PCM in lightweight building components has also been considered as an important strategy to compensate for the lack of thermal storage mass of such buildings [32]. A Trombe wall integrated with a double layer of PCM wallboard on a south façade was proposed and simulated using TRNSYS software. The results showed a reduction of the peak cooling and heating loads by 9% and 15%, respectively, compared with a reference Trombe building, showing that PCM can alleviate summer overheating and maintain indoor thermal comfort like a classical Trombe building in winter [33].

In this study, we analyzed an innovative design and configuration called a thermo-diode Trombe wall (TWTD). The main feature of the proposed systems lies in the capability of providing an adequate level of thermal insulation to the building envelope that would conversely produce a strong disadvantage in common Trombe wall configurations, limiting the thermal performance. The proposed wall is characterized by a highly insulated external wall and a limited glazed surface compared to a common Trombe wall, which allows for solar radiation to be caught on the absorber surface that does not extend for the entire wall height. The heat transfer takes place due to natural convection between the lower and upper sections of the wall air gap that are established because of the peculiar configuration and thermal insulation disposition. The thermal mass is further enhanced thanks to the presence of PCM, which has been properly placed in the air cavity to prolong the passive heat gain, even after the end of the daily solar radiation.

After presenting the operating principle and the experimental set-up, the results of a monitoring campaign from December to March are reported, which demonstrates the excellent thermal behavior of the proposed system.

2. Materials and Methods

2.1. Experimental Set-Up

The experimental set-up was located at the University of Rzeszów (Poland). The study was conducted in winter in the climatic conditions of the same city. A proper heating system was able to maintain the temperature of the inside air at the desired value. During the experiments, the internal air temperature was set to 20 °C.

The TDTW apparatus in the experimental field had a total height of 2.3 m and a southern orientation (Figure 1). The system consists of two main sections. The lower zone is responsible for the absorption of solar radiation and the upper zone is responsible for heat storage and distribution. An internal air gap, developed through the entire height of the wall, connects the two sections. The thermal insulation of both sections was made of 10 cm thick expanded polystyrene. The air gap of the upper section was separated from the inside of the chamber by a 1.25 cm thick plasterboard. The glazing system consisted of a single-pane glazed unit mounted on an insulated frame. The absorber was made of a perforated black matter painted stainless-steel sheet with an absorption coefficient of 0.95, as reported in the product data sheet. Thermal properties of the TDTW materials are reported in Table 1.

Table 1. Trombe wall material and thermal properties.

Material	Function	Thermal Conductivity (W/m·K)	Density (kg/m^3)	Specific Heat (J/kg·K)
Expanded polystyrene	Upper external insulation	0.036	20	1460
Plasterboard	Internal wall	0.17	900	1000
Expanded polystyrene	Lower internal insulation	0.036	20	1460
Stainless-steel sheet	Absorber	25	7900	460
	Type	Glass Thermal Transmittance (W/m^2·K)	Frame Thermal Transmittance (W/m^2·K)	g Value
Glazing	4/16/4	1.2	0.9	0.68

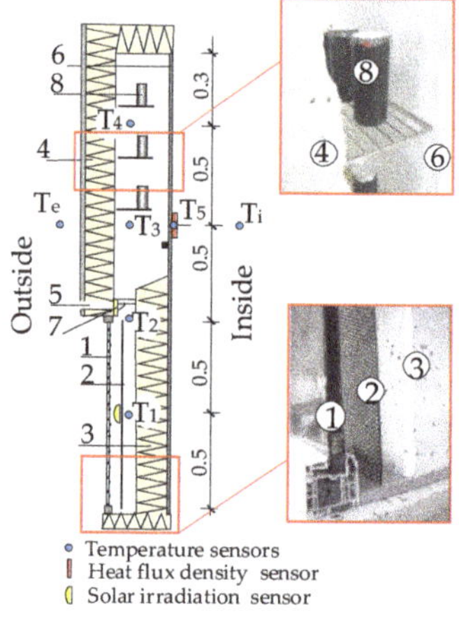

Figure 1. Scheme of the proposed thermo-diode Trombe wall (TDTW): (**a**) three-dimensional representation and (**b**) a cross-section with labels of the main components (1—glazing system, 2—black steel absorber, 3—lower insulation, 4—upper insulation, 5—vent, 6—internal wall, 7—summer mode separator, and 8—phase-change material (PCM) containers.

To increase the storage capacity of the upper section, appropriate materials with greater thermal capacities can be added. The heat can be stored during the hours of solar irradiation and then transferred to the interior of the building after sunset. Due to the high specific heat value or phase change heat, water or phase change materials (PCMs) are the most suitable materials for this function. When using a PCM, it is important to choose an appropriate melting temperature, which should correlate with the temperature range of the air heated inside the gap of the heat storage and distribution section, which accompanies the photothermal conversion of the solar radiation in the lower section of the TDTW. To increase the system's heat capacity, the storage and distribution section was equipped with five containers of 0.33 l each of PCM (RT 28, Rubitherm, Germany) placed in the middle of the height. The PCM has an enthalpy of fusion of 220–225 kJ/kg and declared melting and solidification temperatures in the range 27–29 °C.

The temperature in the air gap cavity was measured at four points at different heights. Sensors T1 to T4 were placed such that they were spaced 0.5 m from each other (Figure 1). Furthermore, the internal and external air temperature was monitored at a height of 1.5 m from the ground, while the heat flux was measured on the internal surface at approximately 1.5 m from the internal floor. The solar radiation was monitored using a sensor placed vertically in the air cavity of the lower section at a height of around 0.5 m from the ground to measure the net incident solar radiation on the absorber surface and by another sensor placed on the vertical external side of the wall at a height of 2 m. The accuracy and type of sensors used in the experiments are reported in Table 2.

Table 2. Sensors type and accuracy.

Measurement	Type	Accuracy
Temperature	ZA 9020-FS Thermo E4	±0.05 K, ±0.05% of the measured value
Heat flux density	ALMEMO FQ A020 C	<6% of the measured value
External air temperature	PT1000	±0.2 °C (−200 °C to +100 °C)
Solar irradiance	DeltaOhm LP Pyra12	<1% (first class)

The study was conducted between December and March 2019 to assess the performance of the TDTW in winter conditions.

The data were recorded every ten minutes using a 16 channels Data Acquisition System Comet MS6D. In the following analysis, the hourly averages of the data are reported and discussed for clarity of interpretation.

2.2. Winter Operating Principles

The proposed thermo-diode Trombe wall (TDTW) system exploits the thermal stratification of air and natural convection to provide thermal energy to the indoor environment. Solar radiation penetrating through the glazing located on the lower zone of the wall is absorbed on the black steel-plate absorber. The air heated by the absorber as a result of the photothermal conversion rises via natural convection, transporting heat to the top part of the wall. The heat is finally transferred to the interior room through the internal wall from the accumulation and distribution section (Figure 2).

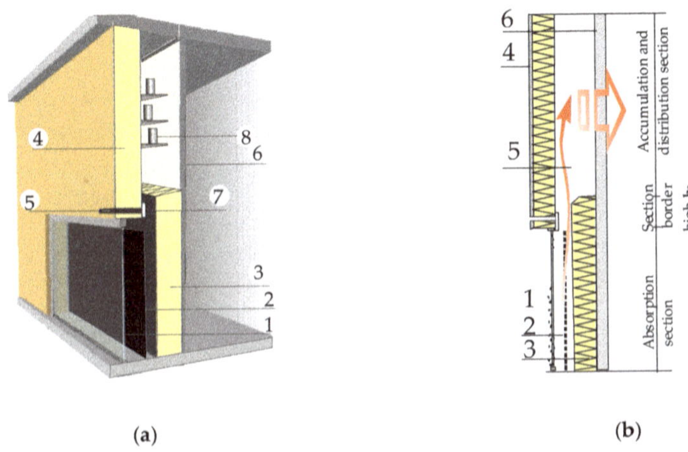

(a)   (b)

Figure 2. Scheme of the proposed thermo-diode Trombe wall: (**a**) three-dimensional representation and (**b**) a cross-section, both with labels of the main components (1—glazing system, 2—black steel absorber, 3—lower insulation, 4—upper insulation, 5—vent, 6—internal wall, 7—summer mode separator, and 8—thermal storage material (PCM containers)).

In periods of low solar exposure or at night, air movement stagnation was observed due to the thermal stratification phenomenon.

To limit the heat losses, the TDTW is protected by a thermal insulation layer installed over the entire wall height but in a different position in the two sections. In the lower part, the insulation is directly placed on the internal wall, a few centimeters away from the absorber. In the upper section, the thermal insulation is placed above the glazing and constitutes the physical separation from the outside. To avoid thermal bridge effects over the border of the two sections, the bottom thermal insulation should have a height of several centimeters above the upper edge of the glazing system. This height can be initially determined by imposing a rule stating that the thermal resistance in the zone where the two insulations overlap (R_{ha}) must be equal to or greater than the thermal resistance evaluated in a perpendicular direction in each of the upper (R_{us}) and lower (R_{ls}) sections, namely:

$$R_{ha} \geq R_{us}$$
$$R_{ha} \geq R_{ls}$$

(1)

R_{ha} is the thermal resistance of the air layer at the height of the overlap of insulation h_a (m), obtainable using:

$$R_{ha} = \frac{h_a}{\lambda_a},\tag{2}$$

where λ_a is the thermal conductivity of stagnant air (0.024 W/mK).

R_{ls} is the thermal resistance of the lower section, which is given by:

$$R_{ls} = \sum R_{si} + R_g + R_{agl} + R_{ild} + R_{wll} + R_{se}.\tag{3}$$

R_{us} is the thermal resistance of the upper section, which is given by:

$$R_{us} = \sum R_{si} + R_{agu} + R_{ilu} + R_{wlu} + R_{se}.\tag{4}$$

where:

- R_g is the thermal resistance of the glazing system;
- $R_{agl/agu}$ is the thermal resistance of the lower/upper section of the air gap;
- R_{ild}/R_{ilu} is the thermal resistance of the lower/upper section of the insulation layer;
- R_{wll}/R_{wlu} is the thermal resistance of the lower/upper section of the wall layer;
- R_{si} and R_{se} are the surface air-wall thermal resistances.

2.3. Summer Operating Principles

The TDTW can be additionally modified to provide building cooling by exploiting the ventilation and the stack effect due to buoyancy. For this purpose, the vent duct should be connected to a solar chimney, as shown in the example in Figure 3.

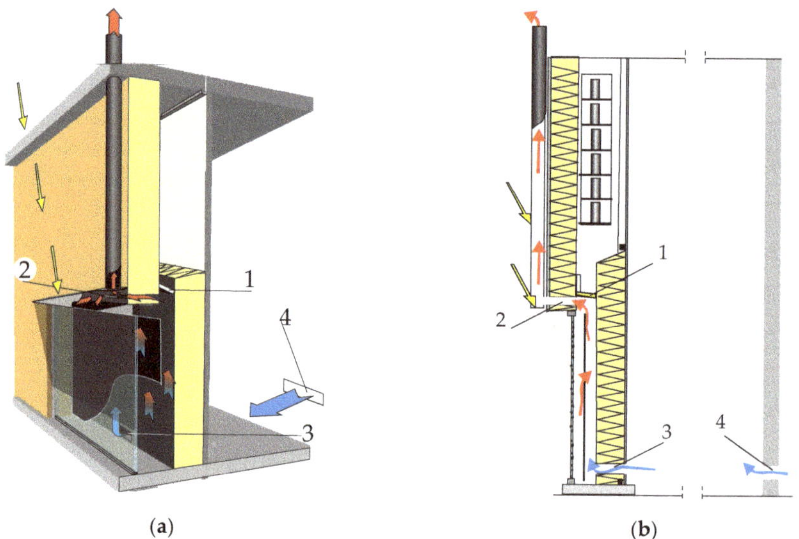

Figure 3. Diagram of the TDTW operation in summer conditions: (**a**) general view and (**b**) cross-section of the wall (1—summer flap separator, 2—vent channel, 3—lockable infiltration channel, and 4—ventilation hole located on the room of the building).

In summer, the high angle of incidences of solar radiation limit direct radiation of the glazing and absorber; however, despite this, the air in the lower absorption section can be subjected to intense heating and consequently rise to the upper distribution zone, providing an additional load for the indoor spaces from the upper distribution section of the wall. This undesirable effect can be counteracted by using shading elements (e.g., louvers, blinds) or by using a summer mode separator flap. The rising of the flap (Figure 3) blocks the free flow of warm air and unlocks the ventilation duct, which allows for the connection of the air gap to the solar chimney. In this operational configuration, the absorber provides an increase in the air temperature of the lower section of the wall that triggers the stack effect, forcing airflow toward the external solar chimney connected to the cavity, consequently drawing a colder airflow from indoor spaces through proper vents positioned at the bottom of the internal wall.

The effectiveness of the chimney in terms of creating draft-intensifying ventilation using the TDTW in the building depends on the level of air heating in the chimney and its height. The chimney draft can be estimated using a simple relationship:

$$p_d = h_{ch} \times g \times (\rho_{ae} - \rho_{ach}) \tag{5}$$

where:

- p_d is the chimney buoyancy force (Pa);
- h_{ch} is the chimney height (m);
- g is the gravitational acceleration (m/s^2);
- ρ_{ae} is the external air density (kg/m^3);
- ρ_{ach} is the air density inside the chimney (kg/m^3).

Since the proposed TDTW was developed mainly for continental climates, the article presents and discusses the results of initial experiments conducted to assess the winter operation mode performance.

2.4. Thermal Storage Tests

A first analysis was performed to evaluate the thermal behavior and the heat storage properties of the material used in the TDTW. The study aimed to assess the heat storage capacity of PCM material placed in an aluminum can.

The main advantage of using a phase change material lies in the capacity to store heat during the melting process, therefore exploiting the latent heat of fusion of the entire container volume. To verify the heat storage capacity, the PCM cans were tested in a Memmert Humidity Chamber HCP108, with an operation range +18–160 °C and a set-point temperature accuracy of 0.1 °C, where controlled air temperatures of 30 °C, 35 °C, and 40 °C were set in three respective experiments. The experiments aimed at reproducing the real operation conditions of the can in the TDTW; indeed, these temperatures expected in the Trombe wall air cavity when solar radiation is available. The temperature was measured and recorded at intervals of 1 min. Four temperature sensors distributed along the radius were used to measure the radial temperature profile (Figure 4).

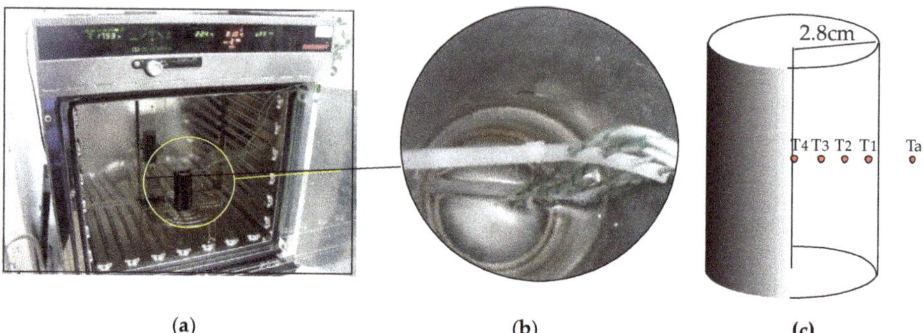

Figure 4. (**a**) Memmert climatic chamber, (**b**) top view of cans with liquefied PCM and temperature sensors, and (**c**) temperature sensor distribution in the container.

3. Results and Discussion

3.1. Thermal Storage Capacity of the PCM Containers

The temperatures found during the three tests are reported in Figure 5. Considering the test with a chamber temperature of 30 °C (Figure 5a), it was possible to observe the behavior of the PCM material. As soon as the chamber was activated, the air temperature jumped to the set-point value of 30 °C. As a consequence, the PCM started to warm rapidly with the most external sensor (T1) increasing before the others placed in the inner part of the container. When T1 reached a value of about 23 °C, a sudden variation of the curve slope was observed, denoting the beginning of the melting process, and when it finished, the temperature increase was much slower. When a value of 26 °C was reached, there was another sudden slope variation, where the temperature increased with the same rate as the initial phase until the achievement of the set-point value imposed by the chamber. It is interesting to observe that when the melting front reached the most external sensor (T4), all the curves overlapped, indicating a uniform temperature distribution due to the liquid phase. When the chamber was turned off, the air temperature swiftly fell to a value of around 20 °C. Consequently, the PCM followed the rapid decline just as quickly until all the sensors registered a value of around 24.5 °C. Then, an abrupt change of slope was observed in the curves, where they tended to be almost flat, denoting, in this case, the start of the solidification process. It is possible to observe that the most external part of the container, being directly subject to the cooling effect of the chamber air, was the first to complete the solidification, with its temperature that started to decrease while the other parts were still completing the phase change.

Similar trends were observed when setting the chamber temperature to 35 °C (Figure 5b) and 40 °C (Figure 5c).

In light of the obtained results, it is possible to define the effective contribution of PCM in thermal storage as the period between the time of the discontinuity in the temperature curve during the cooling process and the time when the most external sensor T4 deviates from the other curves (which denotes complete solidification of the PCM container). In the three tests, the discharge period was found to be in the range of 3.98 h to 4.22 h, confirming the positive contribution that the use of such a material can provide in the proposed TDTW.

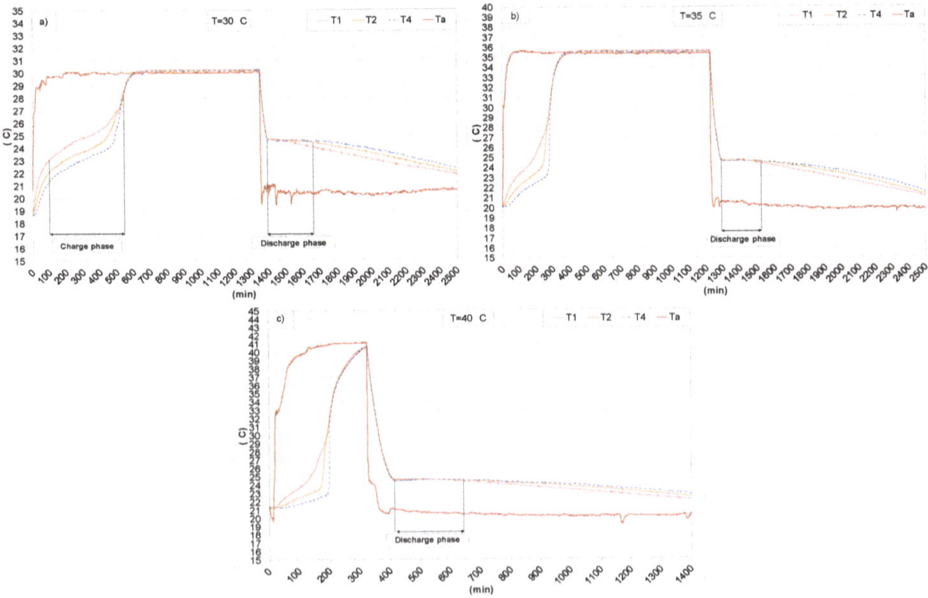

Figure 5. Temperature measurements along the radial direction of the PCM container during the tests on the climatic chamber with set-point temperatures of 30 °C (**a**), 35 °C (**b**), and 40 °C (**c**).

3.2. Analysis in Clear-Sky Conditions

To assess the performance of the proposed TDTW, Figure 6 represents the monitoring of variables over three days (17–19 February 2019), which were characterized by clear-sky conditions with the presence of solar radiation for almost the entire daylight period. In these conditions the registered external temperatures were low, varying from 1.1 °C to 16.4 °C, without falling below 0 °C. It must be highlighted that the solar radiation shown in the following graphs was recorded using the sensor placed in the air cavity of the lower section of the wall, and therefore represents the amount of radiation that struck the absorber after being transmitted by the glazed surface. The daily peak of solar radiation ranged between 335.4 W/m^2 and 375.5 W/m^2, with a distinct daily pattern that is typical of clear-sky conditions. The indoor environment was kept at 20 °C via the climatic chamber operation.

A similar daily pattern appeared for the monitored variables for all three days. For the first day, during the night, the air gap in the lower section of the Trombe wall (T1 and T2) showed low temperatures because of the great thermal exchanges toward the outdoor environment throughout the glazed surface. As soon as the solar radiation appeared, the temperature rose abruptly, with T1 growing from 6.5 °C to a peak of 45.2 °C at 13:00, and T2 followed a similar trend. The solar radiation transmitted through the glazed surface was absorbed by the black steel absorber that consistently heated the air in the cavity's lower section. The increase in air temperature produced an intensification of natural convection inside the air cavity by consequently heating the upper section of the wall. This can be seen by observing temperatures T3 and T4, which registered peaks of 38.9 °C and 38.4 °C, respectively, at 14:00, with one hour of time lag compared to the peak temperature of the bottom section.

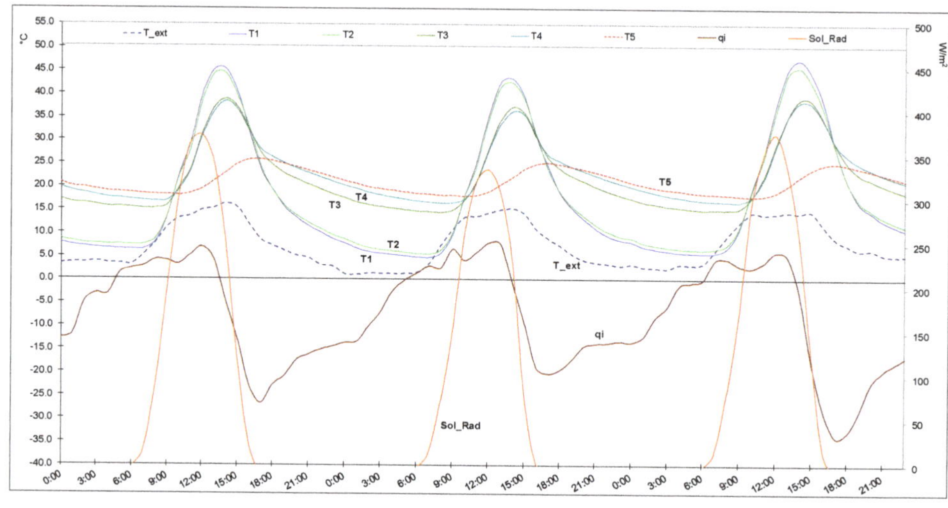

Figure 6. Solar radiation (Sol_Rad), external air temperature (T_ext), air gap temperature (T1, T2, T3, T4), indoor surface temperature (T5), and heat flux (qi) for 17–19 February 2019 (Sol_Rad and qi on the right *y*-axis, temperature on the left *y*-axis).

After sunset, because of the greater thermal losses toward the external environment, T1 and T2 dropped swiftly to reach a daily minimum value that was below 10 °C. The effect produced by the PCM containers was more interesting: it enabled the heat produced by natural convection to be stored in the upper cavity (T3 and T4) during the peak temperature hours and then release it for a prolonged time (up to 4 h, as demonstrated in Figure 5). It can be seen from Figure 6 that because of the heat released by the PCM containers during the solidification process, the upper section cavity temperatures (T3 and T4) decreased with a lower slope, with T4 remaining higher than T3, as a result of the presence of the PCM.

The heat flux was positive, indicating a net loss of thermal energy for the indoor environment, from early morning (04:30) until 13:30. After this time, an inversion of heat flux occurred, denoting net thermal energy entering the indoor spaces, until the late afternoon (17:30). Then, the heat flux reversed again in the early morning of the following day. The same daily cycle was repeated for the next two days.

The results clearly show how the proposed Trombe Wall generated heat gains that shifted in time after the end of the daylight hours, with a positive contribution until the first hours of the morning. The peak entering heat flux of 26.6 W/m^2, as an absolute value, was found at 17:00.

It must be highlighted that the greater temperature values in the cavity upper section were also attributable to the heat flux from the conditioned indoor space that provided heat to the cavity (when a positive heat flux was observed). Yet, the positive effect of the proposed Trombe wall, with staggered and alternating thermal insulation in the two sections, allowed the upper cavity temperature to not fall below 15 °C. The results also showed that the thermal energy lost through the internal wall toward the air cavity (36.0 Wh/m^2 for the first day) was considerably lower, as an absolute value, than the thermal energy provided to the indoor spaces (207.7 Wh/m^2 for the same day). The other two days repeated almost the same trends of the examined variables.

A further interesting aspect is the behavior of the wall in the presence of solar radiation but with a much colder external temperature for 22–23 February 2019, as reported in Figure 7. The solar radiation still reached a peak of 398.1 W/m^2 on 22 February and a peak of 332.1 W/m^2 on 23 February, but the air temperature dropped considerably, reaching minimum values of −5.1 °C and −7.0 °C, respectively, and daily peaks of 4.9 °C and 3.3 °C respectively.

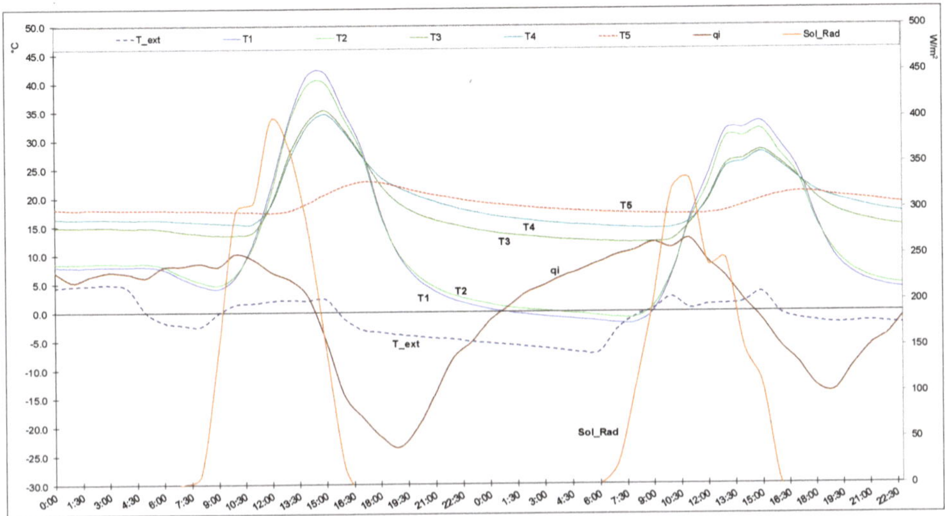

Figure 7. Solar radiation (Sol_Rad), external air temperature (T_ext), air gap temperature (T1, T2, T3, T4), indoor surface temperature (T5), and heat flux (qi) for 22–23 February 2019 (Sol_Rad and qi on the right *y*-axis, temperature on the left *y*-axis).

In this scenario, the TWTD was still able to generate high temperatures in the air cavity, even though the lower solar radiation recorded in the second monitored day produced markedly lower temperature levels. From Figure 7, it is evident how the Trombe wall was able to increase the air cavity temperature, even after the solar radiation peak. On 22 February, T1 started from a value of 8.0 °C and rapidly increased to reach a peak of 41.8 °C at 15:00, with T2 following the same trend with a slightly lower peak value. It then decreased as swiftly as it rose, falling to values close to 0 °C at midnight because of the consistent drop of the external temperature and the consequent thermal losses through the glazed surface. Nevertheless, the cold external air did not prevent the possibility of reaching high temperatures in the upper section of the cavity, where T3 (and T4 being close) reached a value of 35.3 °C due to the natural convection triggered by the lower cavity. As a result, the heat flux inverted at the time of the temperature peak, showing that thermal energy was provided to the indoor spaces until the end of the day. For the whole day, the heat gains amounted to 130 Wh/m^2 against thermal losses of 106.2 Wh/m^2. Furthermore, it is possible to appreciate how the presence of PCM material in the upper cavity of the wall produced a more stable temperature (both T3 and T4), which declined with a low gradient. The heat flux registered a peak of 23.6 W/m^2 at 19:00.

On 23 February, the lower available solar radiation induced a lower temperature in the air gap: in the lower section T1 reached 33.2 °C, whereas in the upper section, a maximum of 28.2 °C was attained by T5, again at 15:00. Despite the colder outdoor temperature and lower solar radiation, the TDTW was still able to generate positive heat gains with a heat flux that inverted at 15:00, similar to the previous day, showing that thermal energy was transferred toward the indoor space. The heat gains amounted to 63.7 Wh/m^2, lower than the 104.9 Wh/m^2 lost toward the air cavity.

3.3. Analysis in Overcast Sky Conditions

Since the main driving force of a Trombe wall is the solar radiation that is converted and transferred toward the indoor spaces, it is important to verify the thermal behavior of the proposed TDTW in overcast conditions, which can often occur in winter. The results in Figure 8 show the trend of the monitored variables for two overcast days in February (21/02/2019 and 24/02/2019).

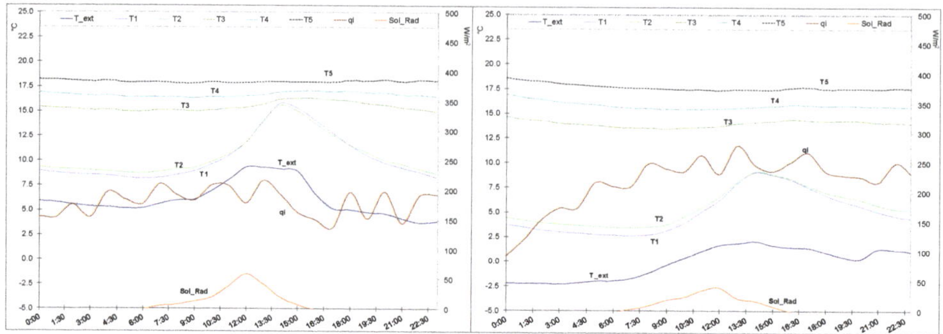

Figure 8. Solar radiation (Sol_Rad), external air temperature (T_ext), air gap temperature (T1, T2, T3, T4), indoor surface temperature (T5), and heat flux (qi) for 21 February 2019 (**left**) and 24 February 2019 (**right**) (Sol_Rad and qi on the right *y*-axis, temperature on the left *y*-axis).

The scarce amount of solar radiation, which reached a peak of 58.7 W/m^2 on 21 February and 39.1 W/m^2 on 24 February, was not sufficient to activate the passive heating of the internal space. On 21 February (Figure 8 on the left), the lower section of the cavity was not able to reach a proper temperature to adequately transfer heat in the upper section via natural convection, with T1 showing a modest peak of 15.8 °C. T3 and T4 showed a rather stable trend around their mean values of 15.5 °C and 16.7 °C, respectively, during the whole day. Even the indoor surface temperature was fairly stable around its mean value of 18.0 °C. In this condition, the upper section TDTW temperature was mainly produced by the heat flux coming from indoor spaces through the internal wall. A similar situation was observed on 24 February (Figure 8 on the right), where because of the more stringent temperatures, T1 in the lower section did not rise above 9.1 °C. T3 and T4 again showed a stable trend around their mean values of 14.0 °C and 15.8 °C, respectively, whereas T5 showed an almost flat curve around its mean of 17.7 °C.

As already observed for the clear-sky conditions, an evident correlation between thermal flux and external air temperature appeared. When the average external air temperature was 6.1 °C (on 21 February) the registered heat flux peak was 8.0 W/m^2, with a total daily thermal energy loss of 138.9 Wh/m^2; with an average external air temperature of −0.1 °C (on 24 February), the registered heat flux peak was 11.7 W/m^2, with a total daily thermal energy loss of 193.5 Wh/m^2.

3.4. Trombe Wall Efficiency

To assess the overall winter performance of the proposed TWTD system, Table 3 reports the daily summary of the main monitored variables for the whole period of the experimental campaign. The daily efficiency of the Trombe wall η_d was determined using:

$$\eta_d = \frac{rQ_{gain}}{Q_{sol}} = \frac{r \int |q_i^-| \, dt}{\int \varphi \, dt}, \tag{6}$$

where:

- φ is the solar irradiation available on the TWTD absorption surface (W/m^2);
- q_i^- is the heat flux entering the conditioned space through the internal wall (W/m^2);
- r is the ratio between the distribution area A_d that provides heat flux to the indoor spaces to the external glazing area A_g that receives the solar radiation. The integrals are performed over the entire day considered.

Table 3. Cumulative daily solar irradiation on the wall's surface, average external air temperature, thermal energy leaving the indoor room through the internal, and the thermal efficiency of the proposed TWTD.

Date	Rad_sol (Wh/m^2)	T_ext,ave (°C)	Thermal Loss (Wh/m^2)	Efficiency (-)
15 Feb	360.6	5.2	178.5	0.0%
16 Feb	4316.6	6.3	129.9	8.7%
17 Feb	4639.4	7.9	36.0	12.0%
18 Feb	3962.2	6.9	42.3	12.8%
19 Feb	4621.4	8.0	31.5	15.3%
20 Feb	1266.2	6.3	46.3	9.4%
21 Feb	516.2	6.1	138.2	0.0%
22 Feb	4643.8	−0.1	105.3	7.6%
23 Feb	3961.3	−2.0	107.0	4.5%
24 Feb	448.1	−0.1	193.8	0.4%
25 Feb	2896.7	5.5	148.2	5.8%
26 Feb	3163.0	7.2	69.1	7.4%
27 Feb	2315.1	6.0	58.3	5.9%
28 Feb	3722.4	10.4	51.8	7.4%
1 Mar	2499.2	4.0	49.8	9.7%

As seen in Table 3, the TDTW was only unable to provide heat gains in severe weather conditions with a low amount of daily solar irradiation, where the entering heat flux was not registered. Nevertheless, the amount of thermal energy leaving the indoor space toward the air cavity (maximum of 193.8 Wh/m^2 on 24 February) was not consistently greater than the amount registered in the days of normal operation of the Trombe wall. The efficiency reached a maximum of 15.3% on 19 February, a greatly insolated day. The higher availability of solar radiation produced higher efficiencies; however a correlation with the external air temperature also appeared. With the same solar irradiation, the efficiency grew with the daily average air temperature: considering 17, 19, and 22 February, with similar levels of daily solar irradiation (4639.4 Wh/m^2, 4621.4 Wh/m^2, and 4643.8 Wh/m^2, respectively), the efficiency grew from 7.6% to 12.0% when the average air temperature increased from −0.1 °C to 7.9 °C, and reached a maximum of 15.3% for a temperature of 8.0 °C. A significant slump in efficiency from 12.8% to 4.5% was observed when the air temperature dropped from 6.9 °C to −2.0 °C with a similar level of solar irradiation (3962.2 Wh/m^2 and 3961.3 Wh/m^2 on 18 and 23 February, respectively).

It must be noted that the performance of the wall was evaluated in cold winter conditions. It is possible to assume, based on the results obtained, that in intermediate seasons, where there is a non-negligible demand of energy for space heating in the particular climate condition, the TDTW can reach even greater efficiency and produce higher heat gains for indoor spaces.

4. Conclusions

To reach the target of sustainable and lower energy-consuming buildings, the adoption of proper technical solutions for the envelope appears to be a necessary step. The use of passive systems, such as Trombe walls, can generate important heat gains by converting the solar radiation striking the absorbing surface into thermal energy that is available for heating requirements. Especially in cold climates, the scarce availability of solar radiation can often lead to the adoption of highly insulated envelopes, limiting the exploitable solar gains. To combine the effects of a high-thermal-resistance wall and convert the available solar radiation into a positive heat gain, a new Trombe wall concept, called a thermo-diode Trombe wall, was proposed. An experimental campaign conducted in winter allowed for analysis of the thermal behavior of the system in different meteorological conditions. The use of PCM as a thermal storage material proved to be an interesting solution since preliminary tests conducted in a climatic chamber demonstrated the capacity of the containers to release heat during the solidification process for a period of 4 h when subject to typical temperatures reached in the Trombe wall.

The results of the analysis demonstrated the capability of the TDTW to trigger consistent natural convection, producing temperatures in the air cavity that exceeded 35 °C on winter days with available solar radiation. For a clear-sky day, the thermal energy provided to the indoor room was 207.7 Wh/m^2 compared with the 36.0 Wh/m^2 lost toward the air cavity. Even on cold days, the system was able to convert solar radiation and produce considerable benefits for the conditioned room, with a heat gain of 130 Wh/m^2 compared with a thermal loss of 106.2 Wh/m^2. There were only a few days where an almost negligible solar radiation was registered and the TDTW was unable to generate positive heat gains, but being highly insulated meant the thermal losses toward the external environment were constant.

When comparing days with similar solar radiation, the efficiency was found to increase with the daily average external air temperature. The maximum efficiency of 15.3% was produced on a day with solar irradiation of 4621.4 Wh/m^2 and an average external air temperature of 8.0 °C.

Finally, the analysis demonstrated that the proposed system represents an important solution to improve the thermal efficiency of the buildings, leading to the realization of the nZEB concept, especially in cold climates where a highly insulated envelope is important.

5. Patents

Szyszka, J. and Lichołai, L., Rzeszow University of Technology, A collector–accumulative barrier; patent application 21.02.2019 r. nr PL.428996 (in the Polish language).

Author Contributions: Conceptualization, J.S.; methodology, J.S.; formal analysis, P.B. and R.B.; investigation, P.B.; resources, J.S.; data curation, J.S.; writing—original draft preparation, P.B. and R.B.; writing—review and editing, P.B., R.B., and J.S. All authors have read and agreed to the published version of the manuscript.

Funding: This research received no external funding.

Conflicts of Interest: The authors declare no conflict of interest.

References

1. Telichenko, V.; Benuzh, A.; Eames, G.; Orenburova, E.; Shushunova, N. Development of Green Standards for Construction in Russia. *Procedia Eng.* **2016**, *153*, 726–730. [CrossRef]
2. Bevilacqua, P.; Bruno, R.; Arcuri, N. Green roofs in a Mediterranean climate: Energy performances based on in-situ experimental data. *Renew. Energy* **2020**, *152*, 1414–1430. [CrossRef]
3. Imessad, K.; Messaoudene, N.A.; Belhamel, M. Performances of the Barra-Costantini passive heating system under Algerian climate conditions. *Renew. Energy* **2004**, *29*, 357–367. [CrossRef]
4. Nicoletti, F.; Carpino, C.; Cucumo, M.A.; Arcuri, N. The Control of Venetian Blinds: A Solution for Visual Comfort. *Energies* **2020**, *13*, 1731. [CrossRef]
5. Cascone, S.M.; Cascone, S.; Vitale, M. Building insulating materials from agricultural by-products: A review. In *Sustainability in Energy and Buildings*; Springer: Berlin/Heidelberg, Germany, 2020.
6. Hami, K.; Draoui, B.; Hami, O. The thermal performances of a solar wall. *Energy* **2012**, *39*, 11–16. [CrossRef]
7. Koyunbaba, B.K.; Yilmaz, Z. The comparison of Trombe wall systems with single glass, double glass and PV panels. *Renew. Energy* **2012**, *45*, 111–118. [CrossRef]
8. Krüger, E.; Suzuki, E.; Matoski, A. Evaluation of a Trombe wall system in a subtropical location. *Energy Build.* **2013**, *66*, 364–372. [CrossRef]
9. Dabaieh, M.; Elbably, A. Ventilated Trombe wall as a passive solar heating and cooling retrofitting approach; a low-tech design for off-grid settlements in semi-arid climates. *Sol. Energy* **2015**, *122*, 820–833. [CrossRef]
10. Soussi, M.; Balghouthi, M.; Guizani, A. Energy performance analysis of a solar-cooled building in Tunisia: Passive strategies impact and improvement techniques. *Energy Build.* **2013**, *67*, 374–386. [CrossRef]
11. Błotny, J.; Nemś, M. Analysis of the Impact of the Construction of a Trombe Wall on the Thermal Comfort in a Building Located in Wrocław, Poland. *Atmosphere* **2019**, *10*, 761. [CrossRef]
12. Abbassi, F.; Dimassi, N.; Dehmani, L. Energetic study of a Trombe wall system under different Tunisian building configurations. *Energy Build.* **2014**, *80*, 302–308. [CrossRef]
13. Jaber, S.; Ajib, S. Optimum design of Trombe wall system in mediterranean region. *Sol. Energy* **2011**, *85*, 1891–1898. [CrossRef]

14. Agurto, L.; Allacker, K.; Fissore, A.; Agurto, C.; De Troyer, F. Design and experimental study of a low-cost prefab Trombe wall to improve indoor temperatures in social housing in the Biobío region in Chile. *Sol. Energy* **2020**, *198*, 704–721. [CrossRef]

15. Bajc, T.; Todorović, M.N.; Svorcan, J. CFD analyses for passive house with Trombe wall and impact to energy demand. *Energy Build.* **2015**, *98*, 39–44. [CrossRef]

16. Abdeen, A.; Serageldin, A.A.; Ibrahim, M.G.E.; El-Zafarany, A.; Ookawara, S.; Murata, R. Experimental, analytical, and numerical investigation into the feasibility of integrating a passive Trombe wall into a single room. *Appl. Therm. Eng.* **2019**, *154*, 751–768. [CrossRef]

17. Shashikant, K.N. Numerical analysis of house with trombe wall. *Int. Res. J. Eng. Technol.* **2016**, *3*, 966.

18. Hong, X.; He, W.; Hu, Z.; Wang, C.; Ji, J. Three-dimensional simulation on the thermal performance of a novel Trombe wall with venetian blind structure. *Energy Build.* **2015**, *89*, 32–38. [CrossRef]

19. Zhou, L.; Huo, J.; Zhou, T.; Jin, S. Investigation on the thermal performance of a composite Trombe wall under steady state condition. *Energy Build.* **2020**, *214*, 109815. [CrossRef]

20. Hong, X.; Leung, M.K.H.; He, W. Effective use of venetian blind in Trombe wall for solar space conditioning control. *Appl. Energy* **2019**, *250*, 452–460. [CrossRef]

21. Stazi, F.; Mastrucci, A.; di Perna, C. Trombe wall management in summer conditions: An experimental study. *Sol. Energy* **2012**, *86*, 2839–2851. [CrossRef]

22. Bevilacqua, P.; Benevento, F.; Bruno, R.; Arcuri, N. Are Trombe walls suitable passive systems for the reduction of the yearly building energy requirements? *Energy* **2019**, *185*, 554–566. [CrossRef]

23. Hu, Z.; He, W.; Hong, X.; Ji, J.; Shen, Z. Numerical analysis on the cooling performance of a ventilated Trombe wall combined with venetian blinds in an office building. *Energy Build.* **2016**, *126*, 14–27. [CrossRef]

24. Bellos, E.; Tzivanidis, C.; Zisopoulou, E.; Mitsopoulos, G.; Antonopoulos, K.A. An innovative Trombe wall as a passive heating system for a building in Athens—A comparison with the conventional Trombe wall and the insulated wall. *Energy Build.* **2016**, *133*, 754–769. [CrossRef]

25. Abbassi, F.; Dehmani, L. Experimental and numerical study on thermal performance of an unvented Trombe wall associated with internal thermal fins. *Energy Build.* **2015**, *105*, 119–128. [CrossRef]

26. Rabani, M.; Rabani, M. Heating performance enhancement of a new design trombe wall using rectangular thermal fin arrays: An experimental approach. *J. Energy Storage* **2019**, *24*, 100796. [CrossRef]

27. Rabani, M.; Kalantar, V.; Dehghan, A.A.; Faghih, A.K. Empirical investigation of the cooling performance of a new designed Trombe wall in combination with solar chimney and water spraying system. *Energy Build.* **2015**, *102*, 45–57. [CrossRef]

28. Hu, Z.; Zhang, S.; Hou, J.; He, W.; Liu, X.; Yu, C.; Zhu, J. An experimental and numerical analysis of a novel water blind-Trombe wall system. *Energy Convers. Manag.* **2020**, *205*, 112380. [CrossRef]

29. Szyszka, J. Experimental evaluation of the heat balance of an interactive glass wall in a heating season. *Energies* **2020**, *13*, 632. [CrossRef]

30. Szyszka, J. Simulation of modified Trombe wall. In Proceedings of the VII Conference SOLINA, Polańczyk, Poland, 19–23 June 2018.

31. Zalewski, L.; Joulin, A.; Lassue, S.; Dutil, Y.; Rousse, D. Experimental study of small-scale solar wall integrating phase change material. *Sol. Energy* **2012**, *86*, 208–219. [CrossRef]

32. Fiorito, F. Trombe walls for lightweight buildings in temperate and hot climates. Exploring the use of phase-change materials for performances improvement. *Energy Procedia* **2012**, *30*, 1110–1119. [CrossRef]

33. Zhu, N.; Li, S.; Hu, P.; Lei, F.; Deng, R. Numerical investigations on performance of phase change material Trombe wall in building. *Energy* **2019**, *187*, 116057. [CrossRef]

Article

Multi-Objective Analysis of a Fixed Solar Shading System in Different Climatic Areas

Jessica Settino *, Cristina Carpino, Stefania Perrella and Natale Arcuri

Department of Mechanical, Energy and Management Engineering, University of Calabria, Via P. Bucci, 87036 Arcavacata di Rende (CS), Italy; cristina.carpino@unical.it (C.C.); perrellastefania@alice.it (S.P.); natale.arcuri@unical.it (N.A.)
* Correspondence: jessica.settino@unical.it

Received: 30 April 2020; Accepted: 16 June 2020; Published: 23 June 2020

Abstract: This study tackles the analysis of fixed external solar shading systems. The geometry of a building and of the shading system has been parametrically defined and a genetic optimization analysis has been carried out to identify an architectural solution that would allow the increase of energy savings, through a suitable window-to-wall ratio and an accurate design of the shading device. A multi-objective analysis has been performed with the aim of minimizing the energy consumption for space heating, cooling and artificial lighting, while ensuring the visual comfort of the occupants. The main goal of the study is to explore the influence of climatic context on the optimal design of shading devices. The analysis has been performed for three different latitudes across Europe. In all analyzed cases, a reduction of the annual energy consumption could be achieved, up to 42% if the optimal shading configuration is used. Moreover, the possibility of integrating the shading system with photovoltaic (PV) panels has been considered and the electricity production has been estimated.

Keywords: shading systems; multi-objective optimization; energy savings; visual comfort

---

1. Introduction

During recent decades, the energy demand grew exponentially. The energy consumption in the building sector accounts for 40% of the whole energy demand in Europe, and this value is expected to increase due to the higher comfort levels required, if mitigation strategies are not implemented. In this context, the concept of 'nearly zero energy buildings' (nZEBs) has been introduced and clearly defined [1].

NZEBs should combine the use of renewable energy systems with the best available energy efficiency strategies to minimize the building energy consumption [2]. Moreover, envelope energy requirements are minimized by employing different solutions, such as green roofs [3], Trombe walls [4] or direct gain systems [5]. Glazed surfaces and shading systems, in fact, play a crucial role in minimizing the building energy requirements. Highly glazed surfaces are common in modern building design. Solar radiation, which penetrates the indoor environment through transparent surfaces, represents an energy gain during the cold months and positively contributes to the building's energy performance growth by limiting both heating loads and the electricity consumptions for artificial lighting. Nevertheless, during the cooling season, solar radiation causes overheating by contributing to increase cooling loads and energy consumptions [6]. In order to find a good compromise between the heating and cooling seasons, shading systems play a fundamental role in the management of solar radiation and the selection of suitable screens should take place at an early stage of the design process, since it significantly affects the energy balance of the building. An analysis carried out by Friess et al. [7] in hot climatic conditions confirms the importance of an 'energy-optimized structure', showing the possibility of 20% energy savings through a proper building orientation and thermal insulation while up to 55% energy savings can be obtained thanks to an appropriate glazing type and orientation. An overview of different options to decrease insolation and increase energy savings has been provided by Valladares-Rendón et al. [8]. The authors analyzed the use of different shading devices, window-to-wall-ratios and building orientations obtaining potential energy savings ranging from 4.6% to 76.6%. An optimal window-to-wall ratio (WWR) allows to achieve 54.2% of energy savings while complex design of shading devices (SDs) can provide 66% of energy savings.

Zhang et al. [9] investigated the building envelope optimization for low-energy buildings in low latitudes of China. The energy-saving effects of different building characteristics such as the building shape, building area, aspect ratio and WWR have been evaluated by numerical simulation, and the cooling load reduction indexes have been calculated by considering different thermal performances of the envelope and local climatic conditions. Based on the results of the analysis, horizontal shadings of 500–600 mm allow to reduce cooling energy demand by 95% if used on the southern exposure and by 15% if applied on the north facing windows.

Palmero-Marrero and Oliveira [10] investigated the effect on the building energy demand of louver shading devices applied to the South, West and East facing windows, for different latitudes. An overview of different shading systems has been provided by Bellia et al. [11] whereas Mandalaki et al. [12] evaluated the energy performance of thirteen different types of fixed shading devices with integrated photovoltaic (PV) for a South oriented surface. Regarding the control aspects, Manzan [13] used genetic optimization to analyze an external shading system for a South facing window. The author considered a flat panel, parallel to the window, and performed an optimization analysis by varying four parameters: height, width, angle and distance from the wall. The simulations show that significant energy savings could be achieved in hot climatic areas up to 30%.

Jayathissa et al. [14] investigated the use of a dynamic integrated photovoltaic shading system to simultaneously optimize the building energy demand and the PV generation. Bellia et al. [15] analyzed the effect of an external shading system on the energy consumption, considering both space heating, cooling and lighting.

Tsangrassoulis et al. [16] proposed the use of a genetic algorithm to design a slat-type shading system. Rapone and Saro [17] used a Particle Swarm Optimisation (PSO) algorithm addressed to carbon emissions minimization. Khoroshiltseva et al. [18] carried out a multi-objective optimization analysis for a building located in Madrid to design an energy-efficient shading system. The authors considered the trade-off between the increase of the energy demand due to heating and lighting in winter and the reduction of the overheating in summer as a consequence of the shading devices.

Several studies have been performed to determine the energy savings that could be achieved thanks to an appropriate design of the shading systems. Nevertheless, only few research papers consider the visual comfort of the occupants.

Mandalaki et al. [19] analyzed different integrated PV shading systems for Mediterranean countries, considering both energy efficiency and visual comfort.

The study developed by Xue et al. [20] introduces a workflow for the optimization of WWR with sunshades by considering both daylighting performance and energy consumption. Weather conditions of three Chinese coastal cities are selected for testing a case building. A reference WWR is first defined based on the luminous requirements and the optimal solution is later identified according to thermal requirements, focusing particularly on the reduction of cooling loads. The results of the analysis indicate that 'comprehensive' sunshades offer the best energy performance, benefiting from the combined effect of horizontal and vertical overhangs.

Shan [21] proposed a methodology based on genetic algorithms to determine the optimal shading solution to minimize the energy required for heating, cooling and lighting. Even though daylight has been considered by the author, it is used to estimate the artificial light required to reach an illuminance threshold value of 500 lux and is not considered as an objective function. Vera et al. [22] used GenOpt to optimize a fixed exterior fenestration system considering both energy savings and visual comfort. The authors considered a fully glazed surface facing South in the locations of Montreal, Boulder and Miami and facing North in Santiago. The opaque walls, including ceiling and roof were considered adiabatic and to perform the optimization study, perforations, louver spacing and tilt angle have been varied. Nicoletti et al. [23] proposed a control algorithm for venetian blinds to minimize the annual energy consumption, maximizing solar gains in winter and minimizing the solar radiation in summer, while ensuring a suitable level of natural lighting to guarantee the visual comfort of the occupants.

This study focuses on external fixed shading systems. External shading devices (SDs) have higher performance than internal systems [24] while fixed SDs represent a more economical solution than dynamic systems and no manual adjustments are required.

Unlike existing studies that considered an office-room bordering other air-conditioned environment and with glazed surfaces exclusively facing South, the present study faced a more complex case characterized by a larger space, with glazed surfaces on all the exposures and with all dispersing envelope components. Moreover, a parametric analysis was performed in order to investigate the variability of the shading configuration according to the climatic context. Therefore, the main goal of this study is to evaluate how the optimization of fixed solar shading systems is influenced by the climatic characteristics of the site.

Beside the energy savings, the visual comfort of the occupants is considered as objective function in the multi-objective optimization analysis. The optimization is performed not only for South facing windows but the optimal design of glazed surfaces and shading device for all façades is determined.

The shape and size of the shading system are optimized through an evolutionary algorithm developed with the Grasshopper plug-in of Rhinoceros [25]. The aim of the optimization analysis carried out in this study is to minimize the overall energy required for air-conditioning and artificial lighting, while simultaneously maximizing the daylight. Dynamic simulations of the solar radiation and dynamic lighting simulations have been performed using the Lady-bug plug-in [26] and the Honey-Bee plug-in [27] of Rhinoceros. Moreover, the optimization software has been coupled with Energy Plus [28] to evaluate the building performance and determine the thermal energy demands with real weather data.

A test-case building located in three different European cities has been considered to analyze the effect of different latitude and climatic conditions on the optimal configuration of the shading system. The selected locations are: Crotone (South Italy, latitude 39°5′), Milan (North Italy, latitude 45°28′) and Copenhagen (Denmark, latitude 55°40′).

According to the Köppen–Geiger climate classification [29], the first site is characterized by hot and dry climate (Csa), the second is representative of hot and humid conditions (Cfa) while the third selected city is included in the temperate-humid zone (Cfb).

In Section 2, the methodology used to carry out the multi-objective analysis for the different scenarios is described, in Section 3 the model assumptions are discussed. The main results are highlighted in Section 4 and discussed in Section 5. In Section 6, the main outcomes are summarized.

2. Methodology

The genetic optimization analysis has been performed using the Grasshopper, a Rhinoceros plug-in. By imitating natural selection and evolution, multi-objective optimization uses parameters and optimization functions to quickly explore thousands of design variations and identify the best solution. It mimics natural selection, crossover and mutation of genes. At each iteration, the best candidates from the current population are selected, by applying a selection criterion (fitness function), to populate the next generation and new candidate solutions are generated randomly by crossover and mutation rules. Iterated over many generations, the average quality of the solutions in the pool of candidates gradually increases and the population evolves toward an optimized solution that meets the design objectives within the constrained parameters.

To perform the analysis, the optimization software needs to be coupled with software tools that allow simulations of the building's performance and that determine the thermal energy required for both heating and cooling. In this study, the dynamic simulation software Energy Plus has been selected.

This calculation engine consists of different models working together to determine building energy requirements for heating and cooling with a variety of systems and energy sources. Heat diffusion through the building components is simulated considering one-dimensional heat transfer. Construction materials properties and indoor air temperature are assumed to be constant. The heat diffusion equation is solved using the conduction transfer functions (CTF), a transient calculation method based on instantaneous energy balance usually performed on an hourly basis for each thermal zone defined in the simulation model. Further details are provided in Section 2.1.

Moreover, to consider the visual comfort of the occupants, dynamic simulations of the solar radiation and lighting have been performed using the Lady-bug plug-in and the Honey-Bee plug-in.

The Lady-bug plug-in allows importing in Grasshopper the weather files of Energy Plus, providing the possibility to analyze the sun path, the distribution of wind speed and direction (wind rose), the solar radiation and shadows. The Honey-Bee plug-in allows the Grasshopper to interface with Open Studio [30]. This is a platform that provides the possibility to simulate the building energy demand using Energy Plus and perform daylight analysis using DaySim [31]. The multi-objective analysis is performed using Octopus [32], a Grasshopper plug-in that extends the functionalities of the Galapagos for genetic algorithm optimization. A schematic diagram of the tools used to carry out the multi-objective optimization is provided in Figure 1.

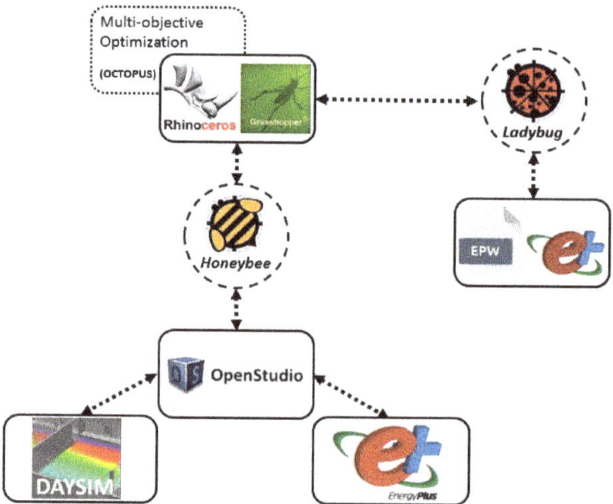

Figure 1. Schematic representation of the system architecture and of the software tools used.

2.1. Building Air Conditioning Load

The building air conditioning load (Q_{AC}) is determined by a thermal balance on an internal air node considering the heat transfer through the envelope (Q_{env}) and the glazed surfaces (Q_w), the infiltration (Q_v) and the internal energy sources (Q_{int}), as clearly described by Zhang et al [9].

$$Q_{AC} = Q_{env} + Q_w + Q_v + Q_{int} \tag{1}$$

Q_{env}, Q_w and Q_v strongly depend on the external climatic conditions.

$$Q_{env} = \sum_{j=1}^{n} K_j A_j (T_S - T_a) \tag{2}$$

A_j is the area of the envelope surface j, K_j is the convective heat transfer coefficient on the j wall internal surface, T_S is its temperature and T_a the indoor temperature. The surface temperature is determined by a thermal balance carried out on the structure by involving the infrared radiant exchange on the external and internal sides and the thermal flux delivered through the wall. The latter is evaluated by adopting the function transfer method exclusively for opaque surfaces with thermal mass.

The internal surface temperatures are affected also by the internal radiant field, in particular by the heat transfer through the glazed surfaces that can be determined through the following equation:

$$Q_w = A_w I_t \tau + A_w N_i (\alpha I_t) + K A_w (T_e - T_a) \tag{3}$$

where I_t is the incident solar radiation, A_w is the glazed surface area, τ is the transmission coefficient and α the absorption coefficient of the glazed surface. N_i is the fraction of the incident solar radiation that is absorbed and subsequently released in the indoor environment. It is worth noticing that the software considers short-wave and infrared radiative exchange separately: for the first, the definition of the internal view factors inside the conditioned cavity allows to evaluate the fraction of solar radiation that is reflected outwards through the same windowed surfaces. This fraction does not become a thermal load for the internal environment and, consequently, it is not considered for the evaluation of the building energy demand.

Usually, the solar heat gain coefficient (F) is defined, considering both the radiation directly transmitted inward and the fraction absorbed and then released.

$$F = \tau + \alpha N_i \tag{4}$$

Hence:

$$Q_w = A_w \cdot [I_t F + K(T_e - T_a)] \tag{5}$$

The second term on the right-hand side of the equation considers the energy losses due to the transmittance of the glazed surface and the air temperature difference between indoor and outdoor environment at temperature T_e.

Q_v can be determined based on the air flow rate m_v and the temperature difference between the internal and external air.

$$Q_v = m_v c_{pa} (T_e - T_a) \tag{6}$$

The internal energy source term, Q_{int}, depends on the devices used, lighting system and on activities carried out in the building space.

Energy Plus allows to consider all the described terms and accurately estimate the building thermal energy required. It has been selected for its acknowledged potential to reliably simulate buildings' thermal behavior and accurately estimate energy performance, along with its capability of effectively integrating shading systems [33].

Its development is funded by the U.S. Department of Energy's (DOE) Building Technologies Office (BTO) [34] and it is considered a well validated energy simulation engine [20].

2.2. Optimization Algorithm

The optimization analysis performed has two objective functions:

1. Minimize the electrical energy requirements for heating, cooling and artificial lighting
2. Maximize natural lighting to fulfill the requirements for visual comfort. To this purpose, the percentage of hours, during which illuminance values of 300 lux are ensured on a working surface located at 1 m from the floor, has been determined.

Every multi-objective optimization problem can be described as a minimization problem, Equation (7).

$$Min\left(F\left(\vec{x}\right)\right) \tag{7}$$

$$F(\vec{x}) = \{f_1(\vec{x}),\ f_2(\vec{x})\ldots\ f_{ob}(\vec{x})\}$$ (8)

The subscript *ob* represents the number of objective functions to be minimized.
Subject to the following constraints:

$$g_i(\vec{x}) \geq 0$$ (9)

with $i = 1, 2, \ldots, m$, where m represents the number of inequality constraints

$$h_i(\vec{x}) \geq 0$$ (10)

with $i = 1, 2, \ldots, n$, where n represents the number of equality constraints

$$x_{min} \leq x_i \leq x_{max}$$ (11)

Equation (11) represents the boundaries of the *i*th variable.

One of the main difficulties in a multi-objective optimization is that often there is a trade-off between the different objectives. Hence, it is not possible to identify a single optimal solution but a pareto front, that is a set of non-dominated solutions which provides a better result for one objective but worse for the others.

The parameters used in the optimization analysis and the set of boundary conditions are described in Section 3.

3. Model

A simplified geometry has been selected for the single zone building. The shape is rectangular, 10×6 m with a height of 4 m, as schematically represented in Figure 2a. The building was located in the areas shown in Figure 2b. In order to compare the results obtained for the different sites, the same building orientation has been considered: North–South for the largest surfaces and East–West for the smallest surfaces.

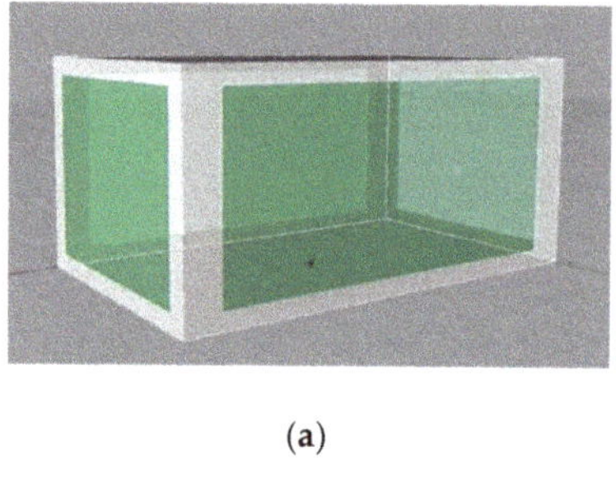

(a)

(b)

Figure 2. (a) Schematic representation of the building; (b) Selected locations

The main characteristics of the building envelope are summarized in Table 1.

Table 1. Building envelope characteristics.

Wall	
Thickness (cm)	40
Thermal transmittance (W/m^2·K)	0.4
Density (kg/m^3)	2500
Heat Capacity (J/kg·K)	800
Floor and Roof	
Thickness (cm)	40
Density (kg/m^3)	2000
Heat Capacity (J/kg·K)	800
Glazed Surface	
Refractive Index	1.52
Thermal transmittance (W/m^2·K)	1.4
Normal Solar Heat Gain Coefficient (SHGC)	0.75
Visible Transmittance (VT)	0.8

In this study, it is assumed that the building is used as an office and mainly computer-based works are carried out. The CEN EN 12464-1 [35] 'Light and lighting-Lighting of work places-Part 1: Indoor work places' recommends a range of 100–300 lux for computer-based tasks. For these reasons a value of 300 lux has been considered in the present study.

The office is considered occupied from Monday to Friday and from 9 am to 18 pm. Since the heating and cooling periods depend on the climatic conditions, in order to compare the results detected across the considered locations, the same duration for the provision of heating and cooling was considered. Therefore, the analysis has been performed considering the cooling season from the 1st of June to the 30th of September and the heating period from the 15th of November to 15th of March.

Weather files from the EnergyPlus data set were used for the simulation. Source weather data formats include IWEC and IGDG data [36,37] representing a typical year obtained by assembling measured or modeled data, to match the long-term data from a certain location using a particular-statistical measure.

Climatic data for the city of Crotone report an external air dry bulb temperature varying between a minimum of T_{min} = −2.2 °C and a maximum of T_{max} = 36.4 °C; for Milan T_{min} = −9.4 °C and T_{max} = 33.6 °C while for Copenhagen T_{min} = −9.6 °C and T_{max} = 26.8 °C. The global horizontal radiation is equal to 1354.4 kWh/m^2 year for Crotone, 1070.6 kWh/m^2 year for Milan and 980.3 kWh/m^2 year for Copenhagen.

The design temperatures for the cooling and heating seasons have been set equal to 26 and 20 °C respectively. For the artificial lighting system, two LED plafonieras with a nominal electric power of 5 W were assumed installed on the ceiling.

A parametric approach has been used, based on following assumptions:

- The fixed solar shading devices are parallel to each façade;
- Rectangular slats with a thickness of 1 cm are used.

The other geometrical properties of the slats are parametrically defined in the Grasshopper to determine the optimal design.

A schematic representation of the main geometrical properties is reported in Figure 3.

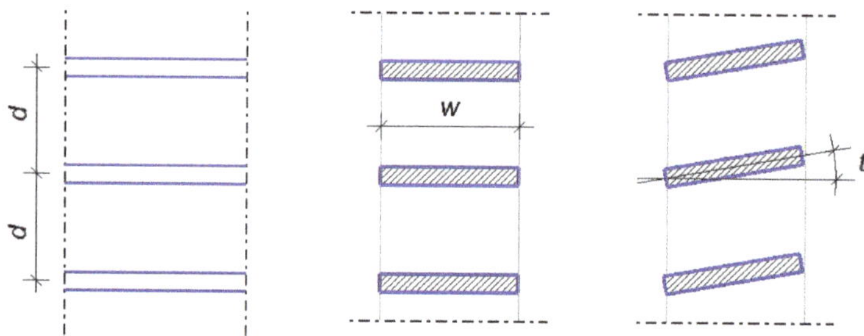

Figure 3. Geometrical parameters of the blades: distance between two consecutive slats—d, depth—w and inclination of the slats—t.

To optimize the office building design and the shading device features, the following variables have been considered:

- The window-to-wall ratio for each building façade can vary between 10% to 80%;
- The slats can be arranged either in horizontal or vertical direction;
- Number of slats;
- Distance between two consecutive slats (pitch 'd') can be varied from a minimum of 400 mm to a maximum of 1200 mm;
- Inclination of the slats with respect to their longitudinal axis (t) varies from −45° to +45°, with a step of 5.

The analysis is performed considering different depth (w) of the slats: 200, 400 and 600 mm.

4. Results

Preliminary simulations have been performed considering the minimum and maximum window-to-wall ratio (WWR) of 10% and 80% for all façades. These scenarios will be considered as reference cases. The energy demands have been determined, as well as the percentage of hours in which the minimum natural lighting of 300 lux on the working surface could be guaranteed.

The results obtained for Crotone are summarized in Table 2.

Table 2. Energy requirements for air-conditioning and percentage of hours in which the minimum lighting requirements are fulfilled, considering the minimum and maximum window-to-wall aspect ratio. The analysis refers to Crotone.

WWR (South)	WWR (East)	WWR (West)	WWR (North)	%h > 300 lux	Cooling Demand (kWh)	Heating Demand (kWh)	Consumption Artificial Lighting (kWh$_e$)
80%	80%	80%	80%	86	8277.3	22.2	9
10%	10%	10%	10%	81	3433.6	1	9.5

It is clear that, due to the Mediterranean climatic conditions, Crotone is a cooling dominated location with negligible heating requirements. However, with the augment of the WWR, the heating demand slightly increases due to the predominance of thermal losses on solar gains.

Afterwards, the optimization analysis has been performed allowing the WWR to vary between 10% to 80% to obtain a window-to-wall ratio for each façade that guarantees the visual comfort of the occupants, maximizing the daylight, while simultaneously decreasing the annual energy demand (required for summer cooling, winter heating and artificial lighting).

Due to the trade-off between these two objectives, a unique solution is difficult to find. Table 3 shows the best results obtained for different WWR.

Table 3. Individuals of the population obtained using multi-objective optimization for Crotone.

WWR (South)	WWR (East)	WWR (West)	WWR (North)	%h > 300 Lux	Cooling Demand (kWh)	Heating Demand (kWh)	Consumption Artificial Lighting (kWh$_e$)
50%	10%	30%	70%	86	6333.3	5.2	9
20%	10%	10%	60%	85	4734.2	3.5	9.1
10%	10%	10%	40%	84	3991.3	2.5	9.2
10%	10%	10%	20%	83	3632.9	1.1	9.3

It is evident that increasing the WWR, the number of hours in which the minimum lighting requirements are fulfilled also increases, at the expenses of higher cooling requirements. The results reported in Table 3 show that solutions with the WWR above 50% on the South surface, 10% for the East surface, 30% for the West direction and 70% for the North are not recommended. Conversely, the role of the South WWR is decisive to drastically reduce cooling demands; lower WWR percentages allow to reduce the energy required for space cooling, keeping almost unvaried the daylight performance.

Considering the case with a maximum WWR of 80% for all façades, reported in Table 2, not only the percentage of hours with more than 300 lux does not increase (86%), but also the cooling demand is considerably higher, 8277.3 instead of 6333.3 kWh. A good trade-off can be represented by a WWR of 10% for the South, East and West surface and 20% for the North direction. This solution allows the achievement of a significant reduction of the annual energy requirements of 56%, with only a 3% decrease in the number of hours with more than 300 lux.

As described in Section 3, the fixed solar shading system has been described parametrically to allow the variation of the geometric parameters. Tables 4–6 summarize the results of the optimization analysis for the considered slat depths.

Table 4. Optimization analysis considering a depth of the slats of 200 mm for Crotone.

Orientation	Number of Slats	Slat Direction	Tilt Angle
South	12	Horizontal	45°
East	12	Horizontal	45°
West	12	Horizontal	45°
North	12	Vertical	−10°
Cooling Demand		2192 kWh	
Heating Demand		10.3 kWh	
%h > 300 lux		75	
Consumption for Artificial Lighting		9.9 kWh$_e$	

Table 5. Optimization analysis considering a depth of the slats of 400 mm for Crotone.

Orientation	Number of Slats	Slat Direction	Tilt Angle
South	7	Horizontal	40°
East	7	Horizontal	45°
West	7	Horizontal	45°
North	13	Vertical	35°
Cooling Demand		2186 kWh	
Heating Demand		12.3 kWh	
%h > 300 lux		74	
Consumption Artificial Lighting		10.1 kWh$_e$	

Table 6. Optimization analysis considering a depth of the slats of 600 mm for Crotone.

Orientation	Number of Slats	Slat Direction	Tilt Angle
South	5	Horizontal	45°
East	5	Horizontal	40°
West	5	Horizontal	40°
North	10	Vertical	−45°
Cooling Demand		2104.2 kWh	
Heating Demand		15.5 kWh	
%h > 300 lux		68	
Consumption Artificial Lighting		10.3 kWh$_e$	

The results show that the configuration with a slats depth of 600 mm is the one that allows to achieve the highest energy savings. The total thermal requirement is 2119.7 kWh, i.e., 41.6% lower compared to the value of 3634 kWh obtained for the same configuration and glazed areas without shading devices. Moreover, a minimum natural lighting of 300 lux on the work surface is ensured in 68% of the hours with a correspondent slight increase of the electrical consumption. This result is due to the shading effect provided by the larger slats that allows to reduce the cooling demand without affecting negatively the heating demand. The overall energy saving which can be obtained with an optimized building structure is 38.3% compared to the configuration with the minimum WWR and without shading devices. Table 6 shows an increase in electricity consumption by the artificial lighting system of 10%, but it is small when compared to the achievable energy savings for the summer conditioning (with an appropriate conversion factor). It is worth mentioning the importance to install vertical slats on the North exposure to exploit daylight rationally by favoring the attainment of the required lux on the working surface, but the inclination varies significantly with the slat's width. Furthermore, the importance of producing shadows on the glazing is demonstrated by the tilt angle ranging between 40° and 45° for the surfaces exposed to the solar radiation.

Once the optimal configuration has been determined, it was considered the possibility of integrating the shading system with photovoltaic cells in accordance with the schematic representation showed in Figure 4. The main characteristics of the photovoltaic panels are summarized in Table 7.

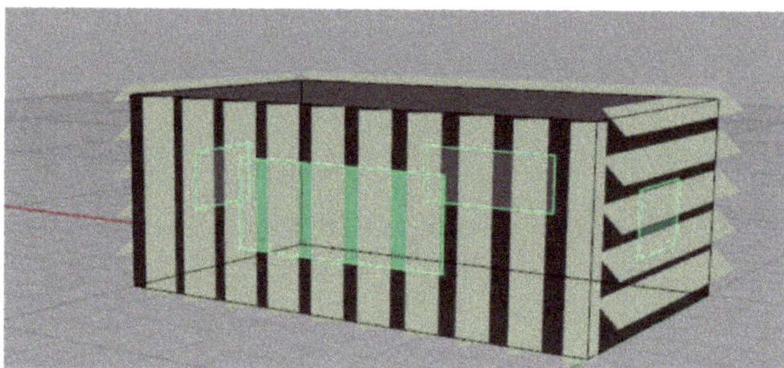

Figure 4. Schematic representation of the shading system.

Table 7. Photovoltaic panel characteristics.

Parameter	Value
Type of PV panel	Poly-Si
Maximum Power at STC	165 W
Current at Max Power (A)	4.77
Voltage at Max Power (V)	34.6

The available surface to install PV cells is 90 m^2, so an annual electricity production by the renewable source of 6526 kWh is achievable, sufficient to provide the electricity required for heating, cooling and lighting (about 700 kWhe assuming an air to air electric heat pump with a mean performance index of 3).

Similar analyses have been performed for Milan and Copenhagen. The results obtained for Milan are summarized in Tables 8–12. Table 8 highlights the building performance parameters for both the minimum and maximum value of the WWR.

Table 8. Energy requirements for air-conditioning and percentage of hours in which the minimum lighting requirements are fulfilled, considering the minimum and maximum window-to-wall aspect ratio. The analysis refers to Milan.

WWR (South)	WWR (East)	WWR (West)	WWR (North)	%h > 300 Lux	Cooling Demand (kWh)	Heating Demand (kWh)	Consumption Artificial Lighting (kWh$_e$)
80%	80%	80%	80%	83	5857.7	1088.3	8.1
10%	10%	10%	10%	65	2045.2	340	13.3

Table 9. Individuals of the population obtained using genetic algorithms for Milan.

WWR (South)	WWR (East)	WWR (West)	WWR (North)	%h > 300 Lux	Cooling Demand (kWh)	Heating Demand (kWh)	Consumption Artificial Lighting (kWh$_e$)
50%	10%	30%	70%	81	4575.7	984.4	9
30%	10%	10%	60%	79	3401.2	911.5	9.5
20%	10%	10%	50%	78	3312.5	750.8	9.9
20%	10%	10%	40%	76	2882.8	648.6	10.4

Table 10. Optimization analysis considering a depth of the slats of 200 mm for Milan.

Orientation	Number of Slats	Slat Direction	Tilt Angle
South	13	Horizontal	10°
East	12	Horizontal	45°
West	13	Horizontal	40°
North	32	Vertical	−45°
Cooling Demand		1239.9 kWh	
Heating Demand		908.1 kWh	
%h > 300 lux		65	
Consumption Artificial Lighting		13 kWh$_e$	

Table 11. Optimization analysis considering a depth of the slats of 400 mm for Milan.

Orientation	Number of Slats	Slat Direction	Tilt Angle
South	7	Horizontal	20°
East	7	Horizontal	45°
West	7	Horizontal	45°
North	15	Vertical	−40°
Cooling Demand		1218.2 kWh	
Heating Demand		944.5 kWh	
%h > 300 lux		62	
Consumption Artificial Lighting		13.4 kWh$_e$	

Table 12. Optimization analysis considering a depth of the slats of 600 mm for Milan.

Orientation	Number of Slats	Slat Direction	Tilt Angle
South	5	Horizontal	45°
East	5	Horizontal	40°
West	5	Horizontal	45°
North	10	Vertical	−45°

Cooling Demand	1267.9 kWh
Heating Demand	843.6 kWh
%h > 300 lux	69
Consumption Artificial Lighting	11.3 kWh$_e$

In this case, an evident increase in the heating demand, with a correspondent decrease of the cooling requirements, can be observed due to the continental climatic conditions. Again, for the highest WWR, the thermal losses through the transparent surfaces prevail on the increase in solar gains. The results of the multi-objective optimization analysis are reported in Table 9 showing the best individuals of the population obtained using genetic algorithms.

To minimize the thermal energy demand, a WWR of 20% for the South, 40% for the North and 10% for the East and West surfaces are required. It is worth noting that South and North exposures affect thermal demands considerably, while the electrical consumptions for artificial lighting are only slightly affected by the window-to-wall ratio.

The results of the optimization analysis performed for the shading system are summarized in Tables 10–12 considering different slats depth.

The analysis performed for Milan shows that the configuration with a slat depth of 600 mm is the one that guarantees the highest energy savings with a tilt angle of 45° for the South exposure. Conversely, by limiting the slats depth, the tilt angle decreases. This ensures a more profitable use of the solar radiation during winter but also determines an increase of the cooling demand. Additionally, in this case, a tilt angle of 40–45° for East and West exposures is recommended as well as vertical slats with negative tilt angle for the North orientation. The annual energy requirement is 2111.5 kWh, thus 40.2% lower compared to the 3531.5 kWh required for the same configuration without shading systems. Natural lighting is sufficient to maintain adequate illumination levels in 69% of the hours.

It is worth noticing that, despite the slight increase in electricity for artificial lighting, the optimized configuration with shading devices provides a more evident reduction in the annual energy requirements, highlighting the prevalent role of air-conditioning for the reduction of the overall energy consumption.

Assuming that the shading surface is covered by PV panels, the annual electricity production can be obtained. It is equal to 4851.7 kWh and it is enough to cover the energy needs for heating, cooling and the artificial lighting consumption considering the use of an electric heat pump with a mean performance index of 3 (about 700 kWh$_e$).

Simulations and optimization analysis have been also performed for Copenhagen. In this case the energy required for heating is higher than the cooling demand, as shown in Table 13. Nevertheless, increasing the WWR, not only the heating demand is higher due to the higher thermal losses but also the cooling demand increases. Energy demands for heating, cooling and electricity for artificial lighting are provided in Table 13, considering a WWR of 10% and 80%.

Table 13. Energy requirements for air-conditioning and percentage of hours in which the minimum lighting requirements are fulfilled, considering the minimum and maximum window-to-wall aspect ratio. The analysis refers to Copenhagen.

WWR (South)	WWR (East)	WWR (West)	WWR (North)	%h > 300 Lux	Cooling Demand (kWh)	Heating Demand (kWh)	Consumption Artificial Lighting (kWh$_e$)
80%	80%	80%	80%	75	1249.6	1394.7	20.5
10%	10%	10%	10%	59	232.1	711.8	29

The results of the optimization analysis are reported in Table 14. The best individuals of the population obtained using genetic algorithms are highlighted.

Table 14. Individuals of the population obtained using genetic algorithms for Copenhagen.

WWR (South)	WWR (East)	WWR (West)	WWR (North)	%h > 300 lux	Cooling Demand (kWh)	Heating Demand (kWh)	Consumption Artificial Lighting (kWh$_e$)
50%	10%	20%	70%	70	963.6	941.9	22.1
20%	10%	10%	60%	69	666	762.7	23.4
20%	20%	10%	40%	67	559.7	687.1	25.9
20%	10%	10%	10%	65	364.2	605.1	26.6

To minimize the thermal energy demand, a WWR of 20% for the South and 10% for the East, North and West surfaces is required. This is due to the need of limiting the thermal losses as much as possible due to the limited availability of solar radiation that could be exploited to reduce the heating demand.

The results of the optimization analysis for the shading device carried out for Copenhagen are summarized in Tables 15–17.

Table 15. Optimization analysis considering a depth of the slats of 200 mm for Copenhagen.

Orientation	Number of Slats	Slat Direction	Tilt Angle
South	13	Horizontal	−10°
East	13	Horizontal	40°
West	13	Horizontal	40°
North	27	Vertical	35°

Cooling Demand	125.5 kWh
Heating Demand	689.2 kWh
%h > 300 lux	61
Consumption Artificial Lighting	28.6 kWh$_e$

Table 16. Optimization analysis considering a depth of the slats of 400 mm for Copenhagen.

Orientation	Number of Slats	Slat Direction	Tilt Angle
South	7	Horizontal	−10°
East	7	Horizontal	45°
West	7	Horizontal	45°
North	16	Vertical	−20°

Cooling Demand	121.5 kWh
Heating Demand	683.0 kWh
%h > 300 lux	59
Consumption Artificial Lighting	29.2 kWh$_e$

Table 17. Optimization analysis considering a depth of the slats of 600 mm for Copenhagen.

Orientation	Number of Slats	Slat Direction	Tilt Angle
South	5	Horizontal	−10°
East	5	Horizontal	40°
West	5	Horizontal	45°
North	10	Vertical	45°
Cooling Demand		106.9 kWh	
Heating Demand		697.3 kWh	
%h > 300 lux		60	
Consumption Artificial Lighting		30.1 kWh$_e$	

The analysis performed shows that even in this case the use of a slat depth of 600 mm represents the optimal configuration. It is worth noticing that the optimal angle is −10° with horizontal direction for the slats exposed toward South, in order to exploit the scarce solar radiation as much as possible. Compared to the previous cases, the analyzed location is characterized by a higher latitude and a lower solar altitude. Therefore, the described configuration allows a better exploitation of the beam solar radiation. Tilt angles of 40–45° are preferred for East and West, and vertical slats for the North exposure. In the optimal situation, the annual energy demand for air conditioning is 804.2 kWh with an energy saving of 17% compared to the 969.2 kWh required for the same configuration without shading elements. Natural daylight is worse than the prior scenarios due to the lower WWR values, however the level is sufficient to maintain an adequate illumination level in 60% of the hours.

The optimized configuration with shading devices provides a reduction in the overall energy consumption by 14.8% compared to the solution with the minimum glazed areas and no shading systems. As for all the other examined cases, the electrical consumption for artificial lighting increases by 13% but it is negligible compared to reduction in terms of thermal requirements for space cooling and heating.

Once again, assuming that the shading surface is covered by PV panels, the annual electricity production was determined. It is equal to 3741.8 kWh, i.e., enough to cover the energy needs for heating, cooling and the artificial lighting consumption assuming the use of electric heat pumps with a mean performance index of 3 (about 300 kWh$_e$).

5. Discussion

Simulations have been performed to determine the optimal window-to-wall ratio and the optimal design of external fixed shading systems. The highest energy savings have been obtained for the locations of Crotone and Milan, characterized by a higher solar radiation and, consequently, by higher cooling requirements. Solar gains in winter, instead, are counterbalanced by a major increase of the thermal losses detected with the WWR growth. For this reason, in Copenhagen lower WWR are recommended. The main contribution of external shading devices to energy saving is due to the reduction of energy requirements for cooling. Obviously, an increase of the energy consumption for space heating is registered but the overall energy needs for both the heating and cooling seasons are significantly decreased by a percentage of 41.6% for Crotone and 40.2% for Milan. Different results have been obtained for Copenhagen, due to the different climatic conditions. While in Crotone and Milan the energy required for cooling in summer is substantial, for Copenhagen space heating represents the most critical aspect and, consequently, the optimization was carried out mainly to limit the energy consumptions during winter. Despite such a difference, a solution has been identified, that guarantees 10% energy saving.

Even though a point by point comparison with similar studies is not possible since different building layouts, construction materials, glazed surface properties, methodology, locations and climatic conditions are considered, it is anyway evident a similar trend. As highlighted by Manzan [13], the energy savings increase in climatic areas characterized by higher levels of solar radiation and, consequently, higher energy cooling demand. As underlined by Zhang et al. [9], the use of horizontal shadings of 500–600 mm on the southern exposure allows to significantly reduce cooling energy demand.

As far as concerning the visual comfort, satisfactory results have been obtained. As pointed out by Vera at al. [22], different results are obtained whether the optimization analysis is performed considering both energy requirements and visual comfort or only the energy consumption criteria. In the multi-objective analysis this aspect has been considered and the solutions that would decrease as little as possible the percentage of daily hours covered by natural lighting have been selected. Even if in some cases a reduction of 15% of sunlight hours was reached, due to the presence of shading devices, the increase in electrical consumption due to artificial lighting is limited while considerable energy savings for air conditioning could be achieved.

It is worth noticing that the simulations, aimed at optimizing the glazed surfaces, have highlighted that a minimum glazed surface is preferable for the South, East and West façades, for both Crotone and Milan where the summer solar radiation is considerable. On the other hand, higher percentages of glazed surfaces are obtained for the North direction to capture only diffuse radiation. The opposite situation occurs in Copenhagen, where the WWR is higher for the South oriented surface by exploiting slats with negative tilt angle due to the lower solar elevation angle, and lower for the North to contrast thermal losses. A synoptic representation concerning the energy demand for air-conditioning and artificial lighting is provided in Figures 5 and 6 respectively, for the considered localities, with and without the installation of the optimized shading devices. It is clear that a comparison cannot be carried out among the localities, because different envelope configurations have been obtained. Nevertheless, in Crotone the 42% of energy savings was attained by limiting exclusively the cooling demands, this is the reason why the lowest WWR is employed on the surfaces mainly exposed to the solar radiation. Moreover, the widest slats are preferred, but with a higher distance and a tilt angle of 45° for a better shading, with no significant negative impact on the daylight. Regarding the location of Milan, being higher the heating requirements, a greater WWR on the South exposure and a minimum value of WWR on East and West facades, allow a better use of winter solar gains. This effect prevails on the worsening of the cooling demand, achieving a winter energy saving of 56% that overcomes the worsening of the cooling demand (-30%). The slats' configurations are similar to those obtained for Crotone but the energy consumptions are slightly worsened, due to the 40% WWR resulting for North facade. Finally, in Copenhagen, characterized by the most severe winter climatic conditions, it is essential to limit the WWR on the North exposure in order to reduce the thermal losses, sacrificing the daylight and obtaining double electrical consumption compared to the prior localities. A WWR of 20% with negative slats tilt angle allows to exploit solar radiation rationally, and WWR of 10% for East and West orientations allows to obtain a summer energy saving of 70% that prevails on the worsening of the winter demand (-15%).

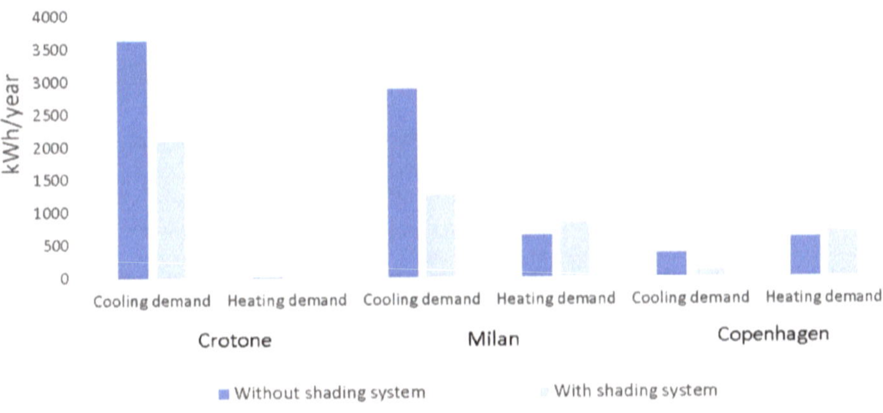

Figure 5. Energy demands for air-conditioning detected among the considered localities, assuming the optimized envelope configurations with and without external shading devices.

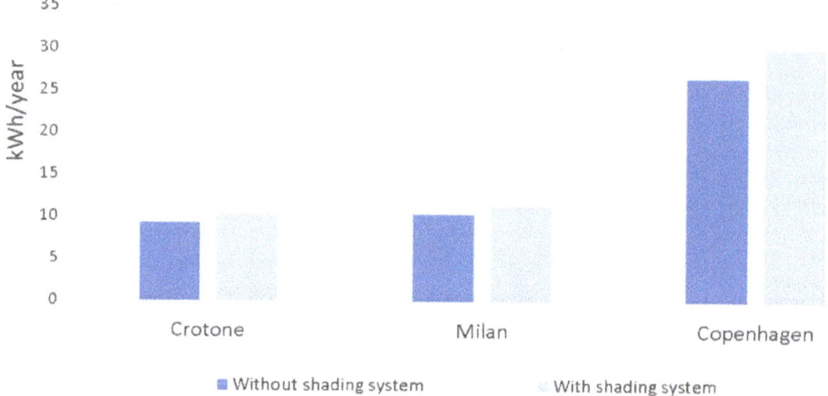

Figure 6. Energy demands for artificial lighting detected among the considered localities, assuming the optimized envelope configurations with and without external shading devices.

6. Conclusions

The analysis performed in this study shows that the shading device needs to be optimized for the climatic conditions in order to find the best compromise between the heating and cooling demands, which are much higher than the energy required by the artificial lighting system. In particular, the following main findings emerged from the analysis:

- The use of a 60 cm slat represents the best performing solution in all locations. This slat width allows appreciable shading effects by adjustment of the slats' tilt angle, which limits the degree of required elements;
- For surfaces oriented towards the North direction, vertical slats represent the optimal solution with different tilt angles;
- For surfaces oriented towards the South, East and West direction, horizontal slats are recommended;
- South oriented surfaces are very sensitive to the variations of the building location, for Crotone the optimal tilt angle is 45° while for Copenhagen it is −10°;
- The glazed areas facing South increase with latitude, while on the contrary those facing North decrease in order to find the best compromise between thermal losses and solar gains;
- The energy savings for air-conditioning are substantial and overcome the increase in electricity consumption for artificial lighting;
- PV energy production can supply the energy demand for air conditioning and artificial lighting in all analyzed scenarios.

The study highlighted the dependence of the shielding systems on the latitude of the site, which directly affects the inclination of the slats. The relation emerged particularly for the South exposure. At high latitudes, in fact, the sun is lower on the horizon and a greater tilt angle of the slats is needed to protect from solar radiation. On the contrary, at lower latitudes, the sun is higher in the sky and the slats must be tilted slightly to obtain the shading effect. In regards to the East and West exposures, instead, no significant differences arose among the analyzed sites because for these exposures the sun is always in a low position for any latitude.

Regarding the trade-off between sun-shades and illuminance, the optimal solution is the one that allows the minimization of energy requirements of the air-conditioning and to maximize the illuminance level on the working surface. In particular, it is worth noting that in winter, screens and daylight can operate synchronously. In fact, during the heating season, it is possible to increase the transmitted solar radiation to better exploit solar gains and improve daylight, at the same time.

In summer, however, the two objective functions show conflicting effects, because shielding should be used as much as possible to reduce the energy for air-conditioning and this negatively affects daylight. Conversely, to favor daylight, the sunshades should be kept as open as possible, but this increases the transmitted solar radiation, causing the overheating of the indoor ambient and worsening the air-conditioning demand.

Optimization is, therefore, essentially required due to the opposing effect between the two factors. The test case used in the analysis with glass surfaces on all exposures is aimed at stressing the impact of the cooling requirement, which has become a serious issue in the energy consumption of buildings due to exponential increases in recent years [38].

This study reported a preliminary investigation. Future research will be aimed at obtaining correlations considering the optimal design of fixed solar shading systems according to latitude and parameters that qualify the solar radiation, as well as climatic conditions of the site.

Author Contributions: Conceptualization, J.S. and N.A.; methodology, J.S. and C.C.; formal analysis N.A., J.S. and C.C.; investigation, S.P., C.C. and J.S.; resources, N.A.; data curation, J.S., C.C. and S.P.; writing—original draft preparation, J.S. and C.C.; writing—review and editing, J.S., C.C., N.A. and S.P. All authors have read and agreed to the published version of the manuscript.

Funding: This research was funded by Ministero dell'Università e della Ricerca—PON Ricerca e Innovazione 2014–2020 Progetti di ricerca industriale e sviluppo sperimentale. "Avviso per la presentazione di progetti di Ricerca Industriale e Sviluppo Sperimentale nelle 12 aree di specializzazione individuate dal PNR 2015–2020" COGITO—A COGnItive dynamic sysTem to allOw buildings to learn and adapt CUP H56C18000100005.

Conflicts of Interest: The authors declare no conflict of interest.

References

1. *Directive 2010/31/EU of the European Parliament and of the Council of 19 May 2010 on the Energy Performance of Buildings*; The Publications Office of the European Union: Luxembourg, 2010; pp. 13–35. Available online: https://eur-lex.europa.eu/LexUriServ/LexUriServ.do?uri=OJ:L:2010:153:0013:0035:EN:PDF (accessed on 21 June 2020).
2. Bruno, R.; Bevilacqua, P.; Cuconati, T.; Arcuri, N. Energy evaluations of an innovative multi-storey wooden near Zero Energy Building designed for Mediterranean areas. *Appl. Energy* **2019**, *238*, 929–941. [CrossRef]
3. Bevilacqua, P.; Bruno, R.; Arcuri, N. Green roofs in a Mediterranean climate: Energy performances based on in-situ experimental data. *Renew. Energy* **2020**, *152*, 1414–1430. [CrossRef]
4. Bevilacqua, P.; Benevento, F.; Bruno, R.; Arcuri, N. Are Trombe walls suitable passive systems for the reduction of the yearly building energy requirements? *Energy* **2019**, *185*, 554–566. [CrossRef]
5. Bruno, R. Optimization of glazing systems in Non-Residential buildings: The role of the optical properties of air-conditioned environments. *Build. Environ.* **2017**, *126*, 147–160. [CrossRef]
6. Bruno, R.; Bevilacqua, P.; Arcuri, N. Assessing cooling energy demands with the EN ISO 52016-1 quasi-steady approach in the Mediterranean area. *J. Build. Eng.* **2019**, *24*, 100740. [CrossRef]
7. Friess, W.A.; Rakhshan, K. A review of passive envelope measures for improved building energy efficiency in the UAE. *Renew. Sustain. Energy Rev.* **2017**, *72*, 485–496. [CrossRef]
8. Valladares-Rendón, L.G.; Schmid, G.; Lo, S.-L. Review on energy savings by solar control techniques and optimal building orientation for the strategic placement of façade shading systems. *Energy Build.* **2017**, *140*, 458–479. [CrossRef]
9. Zhang, T.; Wang, D.; Liu, H.; Liu, Y.; Wu, H. Numerical investigation on building envelope optimization for low-energy buildings in low latitudes of China. *Build. Simul.* **2020**, *13*, 257–269. [CrossRef]
10. Palmero-Marrero, A.I.; Oliveira, A.C. Effect of louver shading devices on building energy requirements. *Appl. Energy* **2010**, *87*, 2040–2049. [CrossRef]
11. Bellia, L.; Marino, C.; Minichiello, F.; Pedace, A. An overview on solar shading systems for buildings. *Energy Procedia* **2014**, *62*, 309–317. [CrossRef]
12. Mandalaki, M.; Zervas, K.; Tsoutsos, T.; Vazakas, A. Assessment of fixed shading devices with integrated PV for efficient energy use. *Sol. Energy* **2012**, *86*, 2561–2575. [CrossRef]
13. Manzan, M. Genetic Optimization of external fixed shading devices. *Energy Build.* **2014**, *72*, 431–440. [CrossRef]

14. Jayathissa, P.; Luzzatto, M.; Schmidli, J.; Hofer, J.; Nagy, Z.; Schlueter, A. Optimising building net energy demand with dynamic BIPV shading. *Appl. Energy* **2017**, *202*, 726–735. [CrossRef]

15. Bellia, L.; De Falco, F.; Minichiello, F. Effects of solar shading devices on energy requirements of standalone office buildings for Italian climates. *Appl. Therm. Eng.* **2013**, *54*, 190–201. [CrossRef]

16. Tsangrassoulis, A.; Bourdakis, V.; Geros, V.; Santamouris, M. A genetic algorithm solution to the design of slat-type shading system. *Renew. Energy* **2006**, *31*, 2321–2328. [CrossRef]

17. Rapone, G.; Saro, O. Optimisation of curtain wall façades for office buildings by means of PSO algorithm. *Energy Build.* **2012**, *45*, 189–196. [CrossRef]

18. Khoroshiltseva, M.; Slanzi, D.; Poli, I. A Pareto-based multi-objective optimization algorithm to design energy-efficient shading devices. *Appl. Energy* **2016**, *184*, 1400–1410. [CrossRef]

19. Mandalaki, M.; Tsoutsos, T.; Papamanolis, N. Integrated PV in shading systems for Mediterranean countries: Balance between energy production and visual comfort. *Energy Build.* **2014**, *77*, 445–456. [CrossRef]

20. Xue, P.; Li, Q.; Xie, J.; Zhao, M.; Liu, J. Optimization of window-to-wall ratio with sunshades in China low latitude region considering daylighting and energy saving requirements. *Appl. Energy* **2019**, *233*, 62–70. [CrossRef]

21. Shan, R. Optimization for heating, cooling and lighting load in building façade design. *Energy Procedia* **2014**, *57*, 1716–1725. [CrossRef]

22. Vera, S.; Uribe, D.; Bustamante, W.; Molina, G. Optimization of a fixed exterior complex fenestration system considering visual comfort and energy performance criteria. *Build. Environ.* **2017**, *113*, 163–174. [CrossRef]

23. Nicoletti, F.; Carpino, C.; Cucumo, M.A.; Arcuri, N. The Control of Venetian Blinds: A Solution for Reduction of Energy Consumption Preserving Visual Comfort. *Energies* **2020**, *13*, 1731. [CrossRef]

24. Wienold, J.; Frontini, F.; Herkel, S.; Mende, S. Climate Based Simulation of Different Shading Device Systems for Comfort and Energy Demand. In Proceedings of the 12th Conference of International Building Performance Simulation Association, Sydney, Australia, 14–16 November 2011.

25. Available online: https://www.rhino3d.com/6/new/grasshopper (accessed on 18 June 2020).

26. Available online: https://www.ladybug.tools (accessed on 18 June 2020).

27. Available online: https://www.ladybug.tools/honeybee.html (accessed on 18 June 2020).

28. Available online: https://energyplus.net (accessed on 18 June 2020).

29. Available online: http://koeppen-geiger.vu-wien.ac.at/present.htm (accessed on 18 June 2020).

30. Available online: https://www.openstudio.net (accessed on 18 June 2020).

31. Available online: https://daysim.software.informer.com/4.0/ (accessed on 18 June 2020).

32. Available online: https://www.food4rhino.com/app/octopus (accessed on 18 June 2020).

33. Al Touma, A.; Ouahrani, D. Quantifying savings in spaces energy demands and CO_2 emissions by shading and lighting controls in the Arabian Gulf. *J. Build. Eng.* **2018**, *18*, 429–437. [CrossRef]

34. Crawley, D.B.; Lawrie, L.K.; Winkelmann, F.C.; Buhl, W.F.; Huang, Y.J.; Pedersen, C.O.; Strand, R.K.; Liesen, R.J.; Fisher, D.E.; Witte, M.J.; et al. EnergyPlus: Creating a new-generation building energy simulation program. *Energy Build.* **2001**, *33*, 319–331. [CrossRef]

35. *CEN EN 12464-1, Light and Lighting-Lighting of Work Places—Part 1: Indoor Work Places*; European Committee for Standardization: Brussels, Belgium, 2011; Available online: https://lumenlightpro.com/wp-content/themes/lumenlightpro/assets/EN_12464-1.pdf (accessed on 21 June 2020).

36. ASHRAE. *International Weather for Energy Calculations (IWEC Weather Files) Users Manual and CD-ROM*; ASHRAE: Atlanta, GA, USA, 2001.

37. Available online: https://energyplus.net/sites/all/modules/custom/weather/weather_files/italia_dati_climatici_g_de_giorgio.pdf (accessed on 18 June 2020).

38. Available online: https://www.iea.org/reports/tracking-buildings (accessed on 18 June 2020).

Article

Economic Comparison Between a Stand-Alone and a Grid Connected PV System vs. Grid Distance

Concettina Marino, Antonino Nucara, Maria Francesca Panzera, Matilde Pietrafesa * and Alfredo Pudano

Department of Civil, Energetic, Environmental and Material Engineering, Mediterranea University of Reggio Calabria, Reggio Calabria 89122, Italy; concettina.marino@unirc.it (C.M.); antonino.nucara@unirc.it (A.N.); francesca.panzera@unirc.it (M.F.P.); alfredopudano@gmail.com (A.P.)
* Correspondence: matilde.pietrafesa@unirc.it; Tel.: +39-0965-169-3293

Received: 24 June 2020; Accepted: 23 July 2020; Published: 27 July 2020

Abstract: The limitation of fossil fuel uses and GHG (greenhouse gases) emissions reduction are two of the main objectives of the European energy policy and global agreements that aim to contain climate changes. To this end, the building sector, responsible for important energy consumption rates, requires a significant improvement of its energetic performance, an obtainable increase of its energy efficiency and the use of renewable sources. Within this framework, in this study, we analysed the economic feasibility of a stand-alone photovoltaic (PV) plant, dimensioned in two configurations with decreasing autonomy. Their *Net Present Value* at the end of their life span was compared with that of the same plant in both grid-connected and storage-on-grid configurations, as well as being compared with a grid connection without PV. The analysis confirms that currently, for short distances from the grid, the most suitable PV configuration is the grid-connected one, but also that the additional use of a battery with a limited capacity (storage on grid configuration) would provide interesting savings to the user, guaranteeing a fairly energetic autonomy. Stand-alone PV systems are only convenient for the analysed site from distances of the order of 5 km, and it is worth noting that such a configuration is neither energetically nor economically sustainable due to the necessary over-dimensioning of both its generators and batteries, which generates a surplus of energy production that cannot be used elsewhere and implies a dramatic cost increase and no corresponding benefits. The results have been tested for different latitudes, confirming what we found. A future drop of both batteries' and PV generators' prices would let the economic side of PV stand-alone systems be reconsidered, but not their energetic one, so that their use, allowing energy exchanges, results in being more appropriate for district networks. For all PV systems, avoided emissions of both local and GHG gases (CO_2) have been estimated.

Keywords: renewable energy sources; PV systems configurations; energy storage; net present value; emission reduction

---

1. Introduction

Today, the limitation of fossil fuel use, a major cause of the present day climate change, along with greenhouse gas emission reduction [1–3], represent the main energy challenges that must be faced, being fundamental aspects of international agreements (COP 2015) and EU (European Union) Directives [4,5], which are particularly addressed in order to reduce primary energy consumption and increase the share of renewable energy sources (RES) [6–9].

A significant rate of primary energy (with the associated greenhouse gas emissions) is consumed in buildings: with reference to Italy, potential savings are about 40% in energy end-use and 36% in greenhouse gases (GHG). They could be significantly reduced by adopting sustainable design, increasing energy efficiency [10–14] and making wider RES use [15–19]. Regional RES implementation examples are reported in [20,21].

Nowadays, the construction sector is in fact required to guarantee high levels of living comfort [22–26] with low primary energy consumption and greenhouse gases emissions, in a cost-optimal vision [27,28]: in the EU, this has allowed for the legislative implementation of the ambitious nZEB (nearly Zero Energy Building) [29,30], with a particularly high energy performance. Recent progress, future challenges and a roadmap on this matter are reported in [31–33].

Thanks to such increasingly challenging energy targets, the energy requalification of the building stock currently offers important opportunities for innovation [34–36], both concerning the envelope [37,38] and the plants. In Italy, in particular, the *Italian National Energy Strategy* [39,40] adopts both challenging thermal characteristics for the envelope (such as minimum transmittances) and the use of RES.

The integration of renewable energy systems with storage systems, both electrical [41–43] and thermal [44–46], would completely satisfy the energy demand, so that storage techniques, including innovative ones [47–51], currently represent a peculiar interest for both researchers and technicians. Anyway, the present cost of energy storage and other critical issues within the systems (over-dimensioning and energy surplus) currently make them neither energetically nor economically sustainable, unless the cost of grid connection is higher for elevated distances [52–55].

Consequently, the new energy paradigm to be established in the future would mainly be based on energy districts, nZEBs, grid-connected RES [56–59] endowed with energy storage, smart grids [60–62] and electric mobility [63,64], rather than on completely autonomous renewable systems.

To this end, in this paper we analysed different configurations of PV plants in a residential building (grid-connected, storage-on-grid and stand-alone) in terms of energy rates, costs and emission reductions, as a function of the distance from the grid.

The main aim of the study is the evaluation of the current cost of building energetic self-sufficiency and thus of user independency in the national electric network. As mentioned above, the use of stand-alone PV systems is not in fact sustainable, as both their generators and batteries must be dimensioned so as to guarantee an energy load in the worst condition, which is to say on the day of the year with the least daylight and energy production (i.e., winter solstice), which requires PV generators and batteries with an overflowing power and capacity: this implies a dramatic cost increase and no corresponding benefits for most months of the year, as it is not possible to use the respective surplus of energy production elsewhere. This means that PV stand-alone configurations are, in general, not advisable, apart from in isolated areas where the cost of a grid connection would result in being greater that of an autonomous system.

The most suitable system is, on the contrary, a grid-connected configuration [65], also using a battery of limited capacity, with no PV generator over-dimensioning, which therefore markedly reduces the system cost, allowing an energy surplus sale and its purchase within only months, with minimum irradiation. This concern about PV systems is not markedly underlined in the literature.

Within this field, the case study compares the energy rates, costs and avoided emissions of a stand-alone PV system located in the Italian city of Reggio Calabria, dimensioned in two different configurations with decreasing autonomy (and the consequent power and cost), guaranteeing a load for one or two days respectively, with those of both a grid-connected configuration of the same plant, also endowed with a small battery (storage on grid), and a simple connection to the grid (Figure 1) as a function of the distance from the grid.

To compare the five configurations, their *Net Present Value* was evaluated at the end of the plant life, identifying the relative cost/benefit items (the latter assessed on the basis of the Italian Government incentivizing fares), and the drawn vs. distance from the grid.

The analysis has been conducted by setting the same case study in three Italian areas with different latitudes (and therefore different irradiations), showing the same results.

Moreover, for all PV configurations, the avoided emissions of both GHG (CO_2) and local pollutants (NO_x, SO_2, PM_{10}) have been estimated.

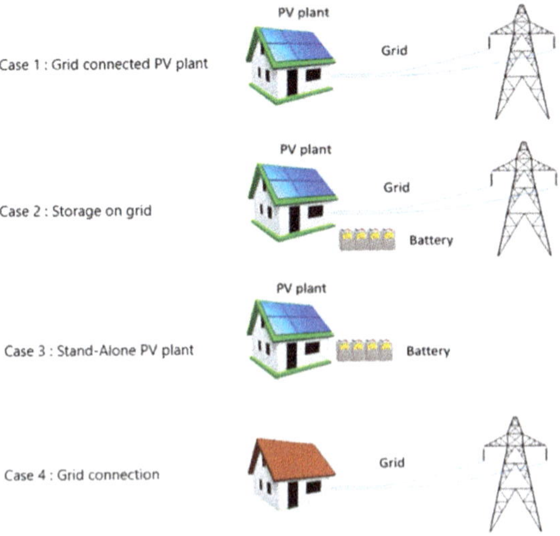

Figure 1. Assessed patterns of power supply.

2. PV System Sizing

2.1. Daily Load Assessment

The electric energy daily demand (D), simulated using the software PVSol (PV*Sol 2020 R7, Valentine Software GmbH, Stralauer Platz 34, 10243 Berlin, Germany) (by Valentine Software German company), was assessed in order to determine the yearly load, in order to size up both the nominal power of the PV array and, when present, the battery capacity. Further information required in order to dimension the stand-alone generator was the daily maximum load in the year.

2.2. Solar Irradiance on the Panels

The purpose of this step is the maximization of solar irradiance striking the panels' surfaces, corresponding to an optimal orientation and tilt angle at the site's latitude. Starting from the values of solar irradiation on the horizontal surface of the site (UNI 10349 [66]), the corresponding values on panel's surface, I, have been derived.

2.3. Nominal Power Assessment–Grid-Connected and Storage-on-Grid PV plant

The peak power of the grid-connected PV systems was estimated on the basis of the user consumption and solar irradiation in the site. It has been assessed by referring to the peak solar hours t_{eq}, the equivalent period of time with a constant irradiance $I_s = 1 \text{ kW/m}^2$ providing the same actual radiation I (kWh/m^2) striking the module's surface [67]:

$$t_{eq} = \frac{I}{I_S} \tag{1}$$

The nominal power delivered by the PV array in the absence of power losses P_{id} is calculated as:

$$P_{id} = \frac{D}{t_{eq}} \tag{2}$$

where D is the yearly energy demand.

The nominal power under the real condition P, considering power losses in the PV system (depending on cell temperatures, shading effects, solar radiation reflections, dirt on module surfaces, etc.) and in its electronic devices (inverter, batteries, charge regulator, link cables, etc.) is then calculated as:

$$P = \frac{P_{id}}{\eta} \tag{3}$$

where η is the system's overall efficiency.

2.4. Nominal Power Assessment–Stand-Alone PV Plant

Stand-alone systems must provide power to loads, even when solar irradiance is insufficient or during night time: during such periods, batteries must support the generation system, so their proper capacity must be sized up.

The power of the PV array must consequently be determined so that during sunshine periods it can allow both load feeding and battery charging. The energy that is to be delivered daily and stored in batteries is a function of the number N of autonomy days of the system and of the load to satisfy.

Two different configurations, with increasing autonomy (referred to as *case a* and *case b*), have been considered. The maximum user's daily demand, D_{max}, is to be guaranteed for one day ($N = 1$) in *case a* and for two days ($N = 2$) in *case b*.

The nominal power of the PV array in the absence of power losses P_{id} that is able to generate the energy required to meet the load and completely charge the battery is given by:

$$P_{id} = \frac{D_{max} + ND_{max}}{t^*_{eq}} = \frac{(1+N)D_{max}}{t^*_{eq}} \tag{4}$$

where t^*_{eq} are the equivalent hours of the minimum irradiance day (21st December). The nominal power under the real condition P is subsequently obtained from Equation (3).

2.5. Battery Sizing

The battery capacity C_a is:

$$C_a = \frac{ND_d}{V(1 - L_c)} \tag{5}$$

where:

D_d is the design load,
V is the rated voltage of the array (V),
L_c is the minimum allowed depth of the battery discharge (%).

2.6. Charge Controller Dimensioning

In stand-alone and storage-on-grid systems, an essential component for their operation and lifetime is the charge controller, which adjusts the energy flow and protects batteries from an insufficient or excessive charge, increasing their life. Its selection depends on the maximum producible current from the PV field:

$$I_{max} = nI_{cc,modulemax} \tag{6}$$

with n being the number of strings and $I_{cc,module}$ the short circuit current.

2.7. Shadows Pattern

The shadow patterns of the installation site are fundamental for the sizing procedure; they can both reduce solar irradiance on panels and produce dissymmetry within their operation, modifying the solar irradiance profiles and strictly affecting the system arrangement. To this end, knowledge of the yearly sun paths at the site's latitude is required.

The string distance d, minimizing reciprocal shadows, has been determined assuming that it avoids shadows in the worst case, that is on 21 December (winter solstice) at 12 a.m. [68]:

$$d = L \cos \delta \left(1 + \frac{\tan \delta}{\tan \alpha}\right) \tag{7}$$

where:

L   module installation height (m),
δ module tilt (°),
α sun altitude at 12 a.m. on 21 December (°).

3. Energy Production

The energy production E has been determined with an hourly step through PVSol software using the following equation:

$$E = \eta SI \tag{8}$$

where

η panel efficiency,
S panels surface (m^2),
I solar irradiance on the panel (kWh/m^2),
η is a function of the temperature (it decreases as the temperature increases):

$$\eta = \eta_r [1 - \beta(t_c - t_r)] \tag{9}$$

where

η_r panel efficiency at reference temperature,
β panel temperature coefficient (%/°C),
t_c cell temperature (°C),
t_r reference temperature (25 °C).

The cell temperature is estimated using the expression:

$$t_c = t_a + \frac{NOCT - 20}{800} I_s \tag{10}$$

where
t_a air temperature (°C),
$NOCT$ Nominal Operating Cell Temperature (°C),
I_s solar radiation power on the panel (W/m^2).

4. Case Study

The described methodology has been applied to a residential user, placed in Reggio Calabria (38°06′ N, 15°39′ E) in the South of Italy. The generator lies facing south, inclined at 28° (optimal inclination for the site). No shadows are present over the PV system in order to consider the most general case.

The load profile is reported in Figure 2; the total yearly demand is 3035 kWh.

For the presence of an energy-consuming air conditioning system, the maximum load is observed in summer, whereas the lower demand in winter is due to the use of a gas boiler for heating purposes.

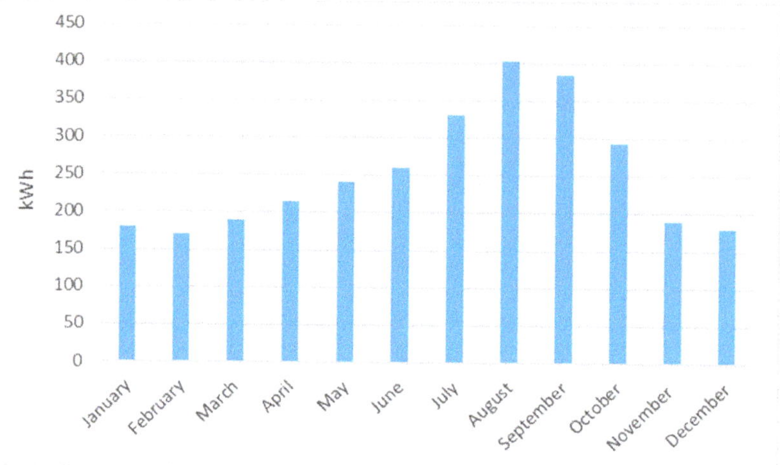

Figure 2. Electric energy load.

4.1. PV System Sizing

The UNI 10349 standard provides the values of the monthly average solar irradiation on a horizontal surface for the city of Reggio Calabria; from these, the corresponding values on the generator surface facing south, inclined at 28°, have been derived (Figure 3).

The selected PV modules are in monocrystalline silicon, with an efficiency of 0.205, and have a module power $P_m = 280$ W$_p$.

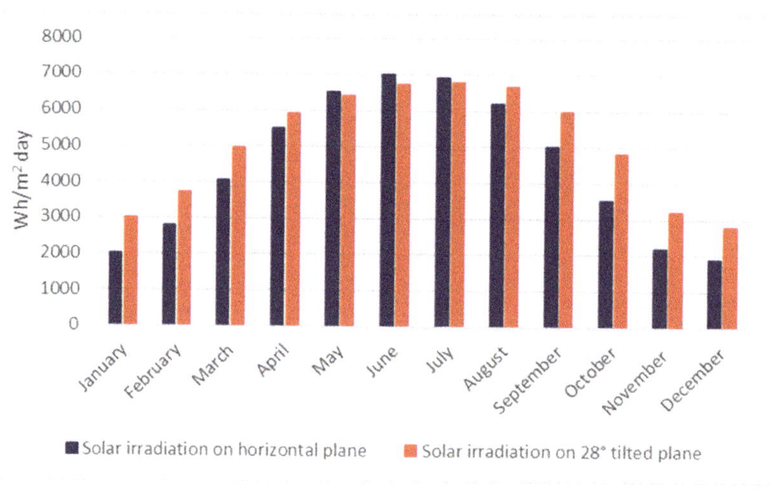

Figure 3. Monthly solar irradiation in Reggio Calabria.

4.1.1. Case 1: Grid-Connected PV System

Using Equation (3), the value of the peak power of the plant has been obtained (2.1 kW$_p$). The minimum number of modules is:

$$n = \frac{P}{P_m} = \frac{2.1 \text{ kW}_p}{0.28 \text{ kW}_p} \cong 8 \tag{11}$$

For the conversion DC/AC, an inverter has been inserted, selecting a pure sine wave one; it has a 2.8 kW power, assessed on the basis of the maximum absorption.

In Figure 4, a schematic representation of the PV systems is reported. The panels are arranged in two strings of four modules each; the distance between the modules is 3.38 m, calculated by Equation (7).

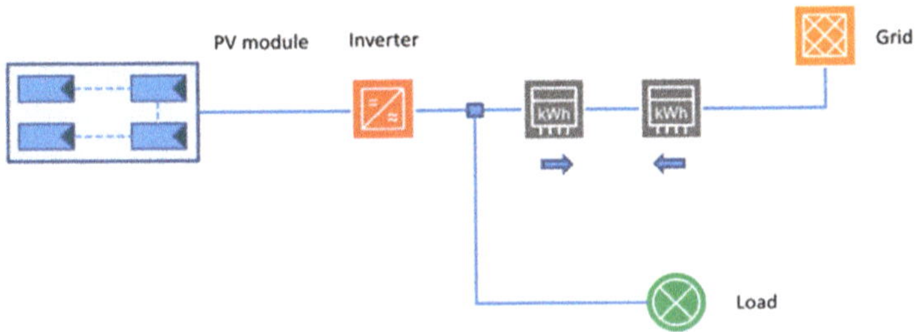

Figure 4. Schematic representation of the grid-connected PV system.

4.1.2. Case 2: Storage-on-Grid PV System

This represents an intermediate case, realised with a grid-connected plant equipped with a battery pack dimensioned only so as to satisfy the average night load (4 kWh/day), without the necessity of returning to the national grid. Consequently, the generator power, the number of panels and their arrangement in strings, together with the inverter characteristics, are the same as the grid-connected configuration. In Figure 5, a schematic representation of the configuration is reported.

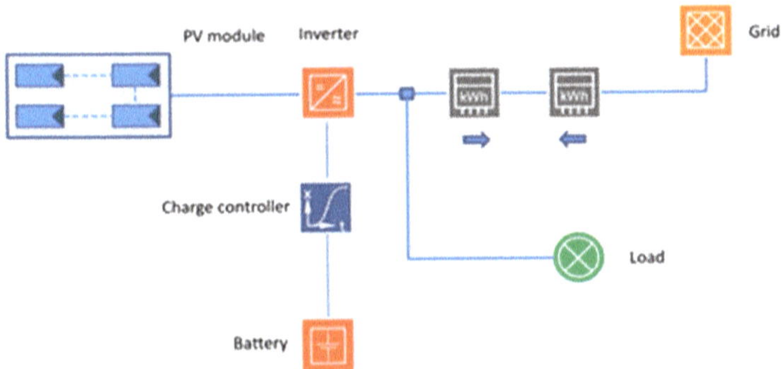

Figure 5. Schematic representation of the storage on the grid PV system.

The batteries have a gel electrolyte, 12 V voltage and a capacity of 4.8 kWh, considering a minimum discharge level of 20%. From (5), the capacity in Ah results in:

$$C_a = \frac{ND_{max}}{V(1 - L_c)} = 406 \text{ Ah} \tag{12}$$

An acid control valve allows for a long duration (average lifetime over 10 years), with no maintenance.

4.1.3. Case 3: Stand-Alone System

In the case of stand-alone systems with no grid support, a failure of energy occurring particularly in winter due to insufficient sunlight must be avoided by increasing the plant power in order to store part of the energy production. To this end, the determinations of both the daily maximum consumption in the year D_{max} (13.1 kWh) and the minimum number of peak solar hours (2.7 h) are necessary.

If only a day's autonomy is to be guaranteed ($N = 1$, *case a*), Equation (4) provides:

$$P_{id,a} = \frac{(1+N)D_{max}}{t_{eq}} = \frac{(1+1)13.1 \text{ kWh}}{2.7 \text{ h}} = 9.7 \text{ kW}_p \tag{13}$$

Assuming an overall plant efficiency equal to 0.8, its peak power P results in:

$$P_a = \frac{P_{id,a}}{\eta} = \frac{9.7 \text{ kW}_p}{0.8} = 12.1 \text{ kW}_p \tag{14}$$

and the minimum number of modules to be installed is:

$$n = \frac{P}{P_m} = \frac{12.13 \text{ kW}_p}{0.28 \text{ kW}_p} \cong 44 \tag{15}$$

Two arrays of 22 modules each connected in parallel have been chosen. The distance between the modules is unvaried with respect to the grid-connected configuration. The load being unvaried, the same inverter as for the grid-connected case has been adopted.

The batteries for energy storage have a 7.6 kWh capacity. In order to guarantee the demand, two batteries with a capacity of 1365 Ah are required.

As for the charge controller, given that $I_{cc,modulemax} = 8.33$ A, we have $I_{max} = 4 \times 8.33$ A $= 33$ A from (6). Four charge controllers, connected in parallel, have been adopted, allowing a 40 A input current each.

When considering two days of autonomy ($N = 2$, *case b*), we have a peak power of 18.2 kW$_p$, and the number of panels to install is 66, arranged in six arrays of 11 that are connected in parallel and are spaced as in previous configurations.

The storage capacity is higher, requiring four batteries. As for the charge controllers, for 66 modules, a maximum producible current of 50 A is obtained: six controllers have been used, allowing a 50 A input current each.

In Figure 6, a scheme of the configured stand-alone system is represented.

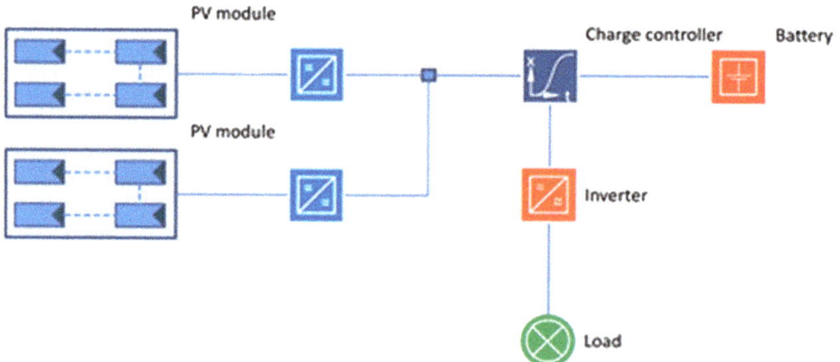

Figure 6. Schematic representation of the stand-alone PV system.

4.1.4. Case 4: Connection to the Grid

In order to evaluate the economic convenience of the configured systems, a comparison of their cost with that resulting from a simple connection to the electric grid has been conducted.

5. Energetic Analysis

The energetic behaviour of the plant typologies has been simulated with an hourly step using PV-Sol software.

In Figure 7, referring to the grid-connected configuration, the energy production and energy shares (directly used energy, grid intake and energy provided by the grid) are reported. Further information that can be drawn from the figure are the load and energy production:

load = direct use + from grid

produced energy = direct use + to grid

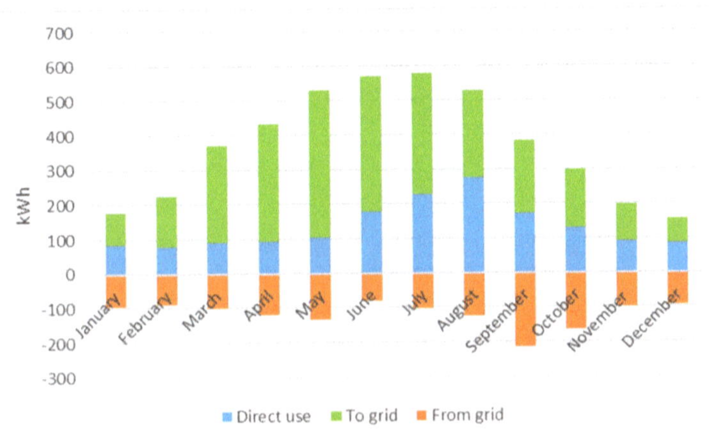

Figure 7. Monthly energy flows for the grid-connected PV configuration.

Similarly, in Figure 8, referring to the storage-on-grid configuration, the energy production and energy shares (directly used energy, energy stock in batteries, grid intake of surplus and energy provided by the grid) are reported.

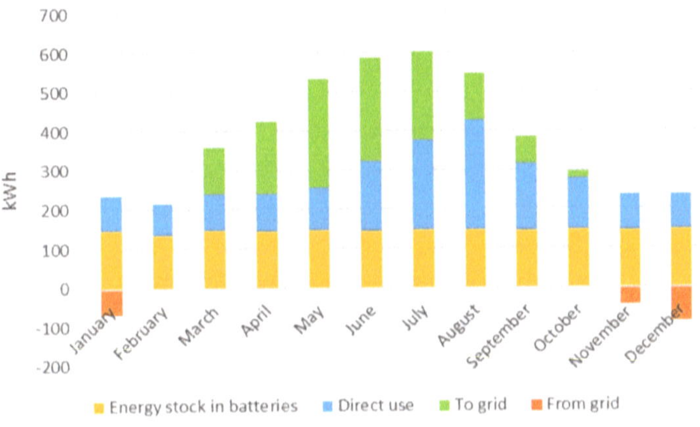

Figure 8. Monthly energy flows for the storage-on-grid PV configuration.

Further information that can be drawn from the figure are the load and energy production:

load = direct use + battery withdrawal + from grid

produced energy = direct use + battery income + to grid

From an energetic point of view, the presence of the battery leads to a reduction of the energy withdrawn from the grid and at the same time a lower energy exported to the grid.

Finally, in Figures 9 and 10, the energy production shares (load, battery recharge and over-production) are reported with reference to the two stand-alone configurations. From the figures, it is possible to draw the following information:

produced energy = load + battery charge + over production

As one can note, in stand-alone configurations, the production amply satisfies both the load and battery recharge needs; obviously, such sizing conditions markedly increase the system cost and over-production, which is equal to 21,496 kWh in *case a* and appreciably higher, at 33,762 kWh, in *case b*.

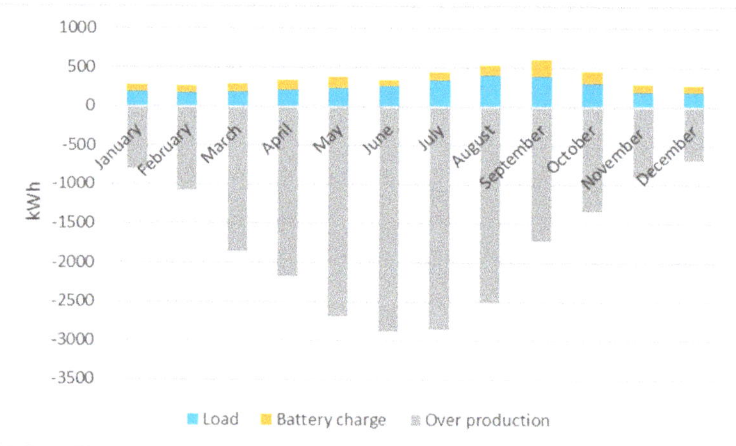

Figure 9. Energy flows for the stand-alone PV configuration (case *a*).

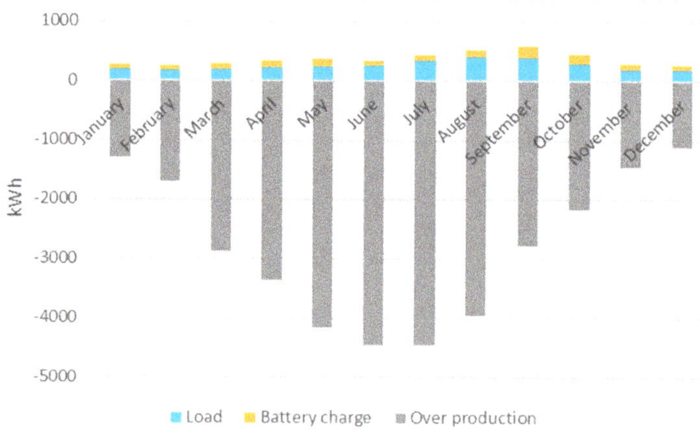

Figure 10. Energy flows for the stand-alone PV configuration (case *b*).

6. Economic Analysis

In order to assess the financial viability of the alternative configurations, an economic analysis was carried out through the assessment of the indicator *Net Present Value* (*NPV*), defined as:

$$NPV = \sum_{k=0}^{n} \frac{C_k - B_k}{(1+r)^k} + I_k \tag{16}$$

where:

k year,
C annual operating costs,
B annual benefits,
I investment costs,
r annual return rate.

which has been calculated vs. grid distance and is the particular parameter used in order to determine the economic convenience of the different choices.

For each case, the main cost/benefit items [69] have been identified, referring to both investment and annual operating costs (Table 1) and to income tax deduction and on-site exchange benefits provided by the Italian Government [70] (Table 2).

Generally, the investment cost is entirely supported at the beginning of the plant life, but PV plants also include inverters and other components (the stand-alone also includes batteries), whose economic life is usually smaller.

Given that an inverter's average life is about five years, it needs to be replaced three times during the plant's life (25 years); batteries, being a long-lasting type, are supposed to be replaced once.

Table 1. Investment and operating costs.

Design Configuration	Investment Costs	Annual Operating Costs
PV plant — Grid	Photovoltaic panels Inverters and accessories Connection to the grid	Energy purchases Maintenance facility
PV plant — Grid — Battery	Photovoltaic panels Inverters and accessories Batteries Charge Controller Connection to the grid	Energy purchases Maintenance facility
PV plant — Battery	Photovoltaic panels Inverters and accessories Batteries Charge Controllers	Maintenance facility
Grid	Connection to the grid	Energy purchases

Table 2. Benefits.

Design Configuration	Investment Benefit	Energy Benefit
PV plant — Grid	Income tax deduction	On-site energy exchange mechanism
PV plant — Grid / Battery	Income tax deduction	On-site energy exchange mechanism
PV plant / Battery	Income tax deduction	

In contrast, in the grid connection, the most relevant burden consists of the connection cost, given as the sum of a fixed cost and a variable one, as a function of the power that is available and of the distance. Both of these costs are given by ARERA (Authority for the Regulation of Energy, Grids and Environment) [71], and their sum is shown in Figure 11.

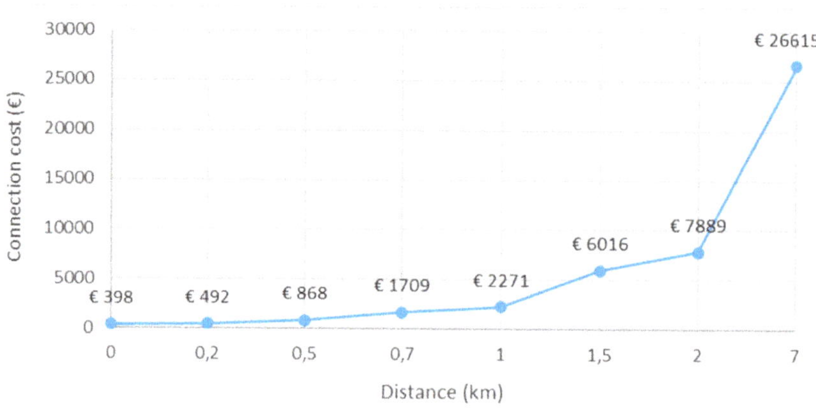

Figure 11. Grid connection cost vs. distance.

The benefits refer to both the investment cost, with an income tax deduction equal to 50% of the initial cost, and to the energy purchase, for which an on-site exchange mechanism allows for an in-taken energy surplus to be drawn from the grid in periods of PV production lack. The cost items are reported in Tables 3 and 4 [72,73], whereas the benefits are shown in Table 5.

As concerns the interest rate provided by the *Bank of Italy*, in accordance with the methods specified by the *Special Data Dissemination Standard (SDDS)* and promoted by the *International Monetary Fund* [74], a value of 0.69% has been assumed [75].

Table 3. Investment costs.

Configuration	Investment Item	Unit	Quantity	Unit Cost (€/unit)	Total Cost (€)
PV plant / Grid	Photovoltaic panels	each	8	195	1560
	Inverters	each	1	1000	1000
	Cables, installation, support structures, etc.				2000
PV plant / Grid / Battery	Photovoltaic panels	each	8	195	1560
	Inverters	each	1	1000	1000
	Cables, installation, support structures, etc.				2000
	Battery and charge controller	each	1	3500	3500
PV plant / Battery (*case a*)	Photovoltaic panels	each	44	195	8580
	Inverters	each	4	1000	4000
	Batteries	each	2	5000	10,000
	Charge controllers	each	4	200	800
	Cables, installation support structures, etc.				3000
PV plant / Battery (*case b*)	Photovoltaic panels	each	66	195	12,870
	Inverters	each	6	1000	6000
	Batteries	each	4	5000	20,000
	Charge controllers	each	6	300	1800
	Cables, installation support structures, etc.				4000
Grid	Connection to grid: a) power rate b) distance rate	km			Figure 11

The diagram in Figure 12 reports *NPV* vs. grid distance.

It can be seen that, relative to the five configurations, *NPV* functions show intersection points, for given distances from the grid, at which their total cost is equal.

In particular, the analysis points out the existence of two critical distances (4 km and 5.7 km): below 5.7 km, the cheapest PV configurations result in being both of the grid-connected ones, which show almost the same trend, whereas over such a distance the stand-alone with a lower autonomy (*case a*) becomes the most convenient; in contrast, the stand-alone *case b* is never convenient due to the high cost of its batteries.

A further result is that the two grid-connected plants are cheaper than the grid connection for any distance from the grid, which, starting from 4 km, becomes even more expensive than the stand-alone *case a*.

Table 4. Annual operating costs.

Configuration	Annual Cost Item	Unit	Quantity	Unit Cost (€/unit)	Total Cost (€)
PV plant / Grid	Maintenance		1	100	100
	Energy purchase	kWh	1406	0.22	309.3
PV plant / Grid / Battery	Maintenance		1	100	100
	Energy purchase	kWh	197	0.22	43.4
PV plant / Battery	Maintenance	€	1	100	100
Grid	Energy purchase	kWh	3035	0.22	667.7

Table 5. Annual benefits.

Design Configuration	Benefit Item	Unit Benefit (€/year)	Total Benefit (€/year)
PV plant — Grid	Income tax deduction On-site energy exchange mechanism	228 (years 1–10) 226.8	454.8
PV plant — Grid — Battery	Income tax deduction On-site energy exchange mechanism	403 (years 1–10) 50.8	453.8
PV plant — Battery	Income tax deduction (*case a*) Income tax deduction (*case b*)	1319 2233.5	1319 2233.5

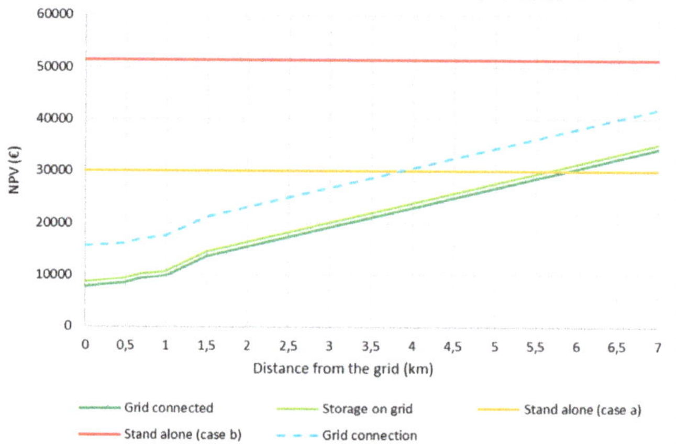

Figure 12. *NPV* of the configurations.

7. Environmental Analysis

An environmental analysis was conducted considering the pollutant emissions that, on the basis of the adopted RES plants, can be avoided, with respect to an equivalent energy production provided by traditional fossil fuels.

The energy amount that is referred to is only the used part (load) of the whole PV production in the case of stand-alone plants and the shares that are directly used plus the grid in-taken shares in the case of the grid-connected ones.

Tables 6 and 7 show, respectively, the emission factors of the main pollutants generated during the combustion of fossil fuels both with global and local effects, and the avoided emissions by analysed plant.

Table 6. Emission factors, reproduced from [76], ISPRA 2018.

Pollutant	Emission Factor (g/kWh)
CO_2	303.5
NO_x	0.2376
SO_2	0.0716
PM_{10}	0.0057

Particularly concerning CO_2 emissions, it can be observed that stand alone configuration allows to avoid 0.92 t/year, whereas the two grid connected ones 1.36 t/year.

Table 7. Avoided emissions.

Design Configuration	PV Energy (kWh/year)	Avoided Emission (kg/year)			
		CO_2	NO_x	SO_2	PM
PV plant — Grid	4468 (produced energy)	1356	1.1	0.3	0.03
PV plant — Grid — Battery	4468 (produced energy)	1356	1.1	0.3	0.03
PV plant — Battery	3035 (load)	921	0.7	0.2	0.02

8. Discussion of Main Results and Achievements

The analysis has shown that the use of stand-alone PV systems is only convenient for the analysed site (it has also been verified for different latitudes) from distances of the order of 5.7 km, confirming that such a configuration is neither energetically nor economically sustainable, requiring both its generator and batteries to be over-dimensioned (with an overflowing power and capacity, respectively) in order to satisfy the worst conditions, which occur on the winter solstice on the day of the year with the least daylight and energy production.

This implies a dramatic cost increase and no corresponding benefits during most months of the year: the surplus of energy production present during these months cannot, in fact, be used elsewhere, but must be dispersed; moreover, batteries are only filled to their maximum level during a few months of the year.

The result, according to which PV stand-alone configurations are in general not advisable, apart from in isolated areas where the cost of a grid connection would result in being greater that of an autonomous system, can be extended to other geographical areas, but the peculiarity of the critical distances which make PV stand-alone configurations convenient should be further analysed.

On the contrary, the most suitable system, at the moment, would use a grid-connected configuration, allowing for an energy surplus sale and its purchase within only months, with minimum irradiation; it is even endowed with a battery with a limited capacity, markedly reducing the system cost. Moreover, for any distance from the grid, it always results in being cheaper than the grid connection without PV.

A future drop of both battery and PV generator prices, indispensable for a RES market extension, would allow for the economic aspect of PV stand-alone systems to be reconsidered, but not their energetic aspect, so that their use could be preferably evaluated in district networks, allowing for energy exchanges.

9. Conclusions

The integration of RES plants with energy storage systems allows for the satisfactory guarantee of energy demands in buildings, with different costs depending on the presence of a grid connection or batteries so as to satisfy production lacks.

In order to evaluate the current cost of energetic self-sufficiency in buildings, in this paper a case study analyses the economic feasibility, as a function of the distance from the grid, of a stand-alone photovoltaic plant located in the Italian site of Reggio Calabria, and which is designed through two configurations with different autonomies (guaranteeing, respectively, a load for one or two days); this is compared to the same plant in a grid-connected configuration, which is also endowed with a small battery (storage on grid), and to a grid connection in the absence of PV. In fact, stand-alone configurations, due to high investment costs, are currently disadvantaged; nevertheless, they can become convenient as the distance from the grid increases.

The energetic behaviour of the plant typologies has been simulated with an hourly step, allowing for an energetic, environmental and economic assessment, the latter being evaluated through the *Net Present Value*, which previously identified the respective main cost/benefit items.

The analysis has shown that the use of stand-alone PV systems is only convenient for the analysed site (and was confirmed for two different Italian latitudes) from distances higher than 5 km (5.7 km) and with a one-day autonomy, confirming that such a configuration is neither energetically nor economically sustainable (given that both its generator and batteries are over-dimensioned, respectively, with an overflowing power and capacity) for satisfying loads during the worst conditions, which occur on the winter solstice on the day of the year with the least daylight and energy production. This implies a dramatic cost increase and no corresponding benefits during most months of the year: the surplus of energy production present during these months cannot, in fact, be used elsewhere, but must be dispersed; moreover, batteries are only filled to their maximum level during a few months of the year.

A future drop of both battery and PV generator prices, indispensable for a RES market extension, would allow the economic aspect of PV stand-alone systems to be reconsidered, but not their energetic aspect, so that their use could be preferably evaluated in district networks, allowing for energy exchanges.

On the contrary, the most suitable system, at the moment, would use a grid-connected configuration only allowing for an energy surplus sale and its purchase during months with a minimum irradiation and also endowed with a low-capacity battery to reduce the grid dependence (storage on grid), thus markedly reducing the system cost. Such a configuration is also more convenient than a grid connection without PV for any distance, as from a 4-km distance the latter also becomes more expensive than the previous stand-alone configuration.

A further result of the analysis is the environmental benefit provided by all of the plants, which in particular allow one to avoid more than 15 t_{CO2}/year.

Author Contributions: Methodology A.N.; software, M.F.P.; data curation, C.M.; writing: A.P.; review and editing, M.P.; funding acquisition, M.P. All authors have read and agreed to the published version of the manuscript.

Funding: This research was funded by the Italian Ministry of Education, University and Research, grant number n. 201594LT3F, within the research project "Research for SEAP: a platform for municipalities taking part in the Covenant of Mayors", PRIN (Programmi di Ricerca Scientifica di Rilevante Interesse Nazionale) program, https://bandi.miur.it/bandi.php/public/fellowship/id_fellow/143962.

Conflicts of Interest: Authors declare no conflict of interest.

Nomenclature

t_{eq}	equivalent period of time with constant irradiance
I	irradiance
I_s	solar radiation power on the panel
P_{id}	nominal power delivered by the PV array in the absence of power losses
D	yearly energy demand
P	nominal power under real condition
η	system overall efficiency
D_{max}	maximum daily load
N	autonomy days of the system
t^*_{eq}	equivalent hours of the minimum irradiance day

D'_{max}	maximum evening load
C	battery capacity
V	rated voltage of the array
L_c	minimum allowed depth of battery discharge
I_{max}	maximum producible current from the PV field
n	number of modules
$I_{cc,module\ max}$	short circuit current
d	string distance minimizing reciprocal shadows
L	module installation height
δ	module tilt
α	sun altitude at 12 a.m. on 21 December
E	energy production
S	panels surface
η_r	panel efficiency at reference temperature
β	panel temperature coefficient
t_c	cell temperature
t_r	reference temperature
t_a	air temperature
$NOCT$	Nominal Operating Cell Temperature
P_m	module power
NPV	Net Present Value
k	year
C	annual operating costs
B	annual benefits
I	investment costs
r	an

References

1. Apergis, N.; Payne, J.E. CO$_2$ emissions, energy usage, and output in Central America. *Energy Policy* **2009**, *35*, 4772–4778. [CrossRef]
2. Birge, D.; Bergera, A.M. Transitioning to low-carbon suburbs in hot-arid regions: A case-study of Emirati villas in Abu Dhabi. *Build. Env.* **2018**, *147*, 77–96. [CrossRef]
3. Arsalis, A.; Alexandrou, A.N.; Georghiou, G.E. Thermoeconomic modelling of a completely autonomous, zero-emission photovoltaic system with hydrogen storage for residential applications. *Renew. Energy* **2018**, *126*, 354–369. [CrossRef]
4. Official Journal of European Union. Directive 2009/28/EC of the European Parliament and of the Council of 23 April 2009, On the promotion of the use of energy from renewable sources and amending and subsequently repealing Directives 2001/77/EC and 2003/30/EC. *Off. J. Eur. Union* **2009**, *11*, 39–85.
5. Official Journal of European Union. Directive 2010/31/EC of the European Parliament and of the Council of 19 May 2010, Energy performance of buildings. *Off. J. Eur. Union* **2010**, *3*, 124–146.
6. Narayanan, A.; Mets, K.; Strobbe, M.; Develder, C. Feasibility of 100% renewable energy-based electricity production for cities with storage and flexibility. *Renew. Energy* **2019**, *134*, 698–709. [CrossRef]
7. Foley, A.; Olabi, A.G. Renewable energy technology developments, trends and policy implications that can underpin the drive for global climate change. *Renew. Sustain. Energy Rev.* **2017**, *68*, 1112–1114. [CrossRef]
8. Gonçalves da Silva, C. Renewable energies: Choosing the best options. *Energy* **2010**, *35*, 3179–3193. [CrossRef]
9. Hvelplund, F. Renewable energy and the need for local energy markets. *Energy* **2006**, *31*, 2293–2302. [CrossRef]
10. Official Journal of European Union. Directive 2006/32/EC of the European Parliament and of the Council of 5 April 2006, Efficiency in the final use of energy and energy systems. *Off. J. Eur. Union* **2006**, *2*, 222–243.
11. Official Journal of European Union. Directive 2012/27/EC of the European Parliament and of the Council of 25 October 2012, Energy efficiency. *Off. J. Eur. Union* **2012**, *2*, 202–257.
12. D'Agostino, D.; Cuniberti, B.; Bertoldi, P. Energy consumption and efficiency technology measures in European non-residential buildings. *Energy Build.* **2017**, *153*, 72–86. [CrossRef]

13. Marino, C.; Nucara, A.; Pietrafesa, M. Does window-to-wall ratio have a significant effect on the energy consumption of buildings? A parametric analysis in Italian climate conditions. *J. Build. Eng.* **2017**, *13*, 169–183. [CrossRef]

14. Murano, G.; Corgnati, S.P.; Cotana, F.; D'Oca, S.; Pisello, A.L.; Rossi, F. A Cost-Effective Human-Based Energy-Retrofitting Approach, Cost-Effective Energy Efficient Building Retrofitting. *Mater. Technol. Optim. Case Stud.* **2017**, 219–255. [CrossRef]

15. Parida, B.; Iniyan, S.; Goic, R. A review of solar photovoltaic technologies. *Renew. Sustain. Energy Rev.* **2011**, *15*, 1625–1636. [CrossRef]

16. Razykov, T.M.; Ferekides, C.S.; Morel, D.; Stefanakos, E.; Ullal, H.S.; Upadhyaya, H.M. Solar photovoltaic electricity: Current status and future prospects. *Solar Energy* **2011**, *85*, 1580–1608. [CrossRef]

17. Cotana, F.; Rossi, F.; Nicolini, A. Evaluation and optimization of an innovative Low-Cost photovoltaic solar concentrator. *Int. J. Photoenergy* **2011**. [CrossRef]

18. Malara, A.; Marino, C.; Nucara, A.; Pietrafesa, M.; Scopelliti, F.; Streva, G. Energetic and economic analysis of shading effects on PV panels energy production. *Int. J. Heat Technol.* **2016**, *34*, 465–472. [CrossRef]

19. Vad Mathiesen, B.; Duić, N.; Stadler, I.; Rizzo, G.; Guzović, Z. The interaction between intermittent renewable energy and the electricity, heating and transport sectors. *Energy* **2012**, *48*, 406–414. [CrossRef]

20. Kyriakopoulos, G.; Arabatzis, G.; Tsialis, P.; Ioannou, K. Electricity consumption and RES plants in Greece: Typologies of regional units. *Renew. Energy* **2018**, *127*, 134–144. [CrossRef]

21. Lund, H. The implementation of renewable energy systems. Lessons learned from the Danish case. *Energy* **2010**, *35*, 4003–4009. [CrossRef]

22. Marino, C.; Nucara, A.; Pietrafesa, M. Proposal of comfort classification indexes suitable for both single environments and whole buildings. *Build. Environ.* **2012**, *57*, 58–67. [CrossRef]

23. Kotopouleas, A.; Nikolopoulou, M. Evaluation of comfort conditions in airport terminal buildings. *Build. Environ.* **2017**, *130*, 162–178. [CrossRef]

24. Luoa, M.; Cao, B.; Zouy, X.; Li, M.; Zhang, J.; Ouyang, Q.; Zhu, Y. Can personal control influence human thermal comfort? A field study in residential buildings in China in winter. *Energy Build.* **2013**, *72*, 411–418. [CrossRef]

25. Marino, C.; Nucara, A.; Pietrafesa, M. Thermal comfort in indoor environment: Effect of the solar radiation on the radiant temperature asymmetry. *Solar Energy* **2017**, *144*, 295–309. [CrossRef]

26. Marino, C.; Nucara, A.; Peri, G.; Pietrafesa, M.; Pudano, A.; Rizzo, G. A MAS-based subjective model for indoor adaptive thermal comfort. *Sci. Technol. Built Environ.* **2015**, *21*, 114–125. [CrossRef]

27. Guazzi, G.; Bellazzi, A.; Meroni, I.; Magrini, A. Refurbishment design through cost-optimal methodology: The case study of a social housing in the northern Italy. *Int. J. Heat Technol.* **2017**, *35*, S336–S344. [CrossRef]

28. Piccolo, A.; Marino, C.; Nucara, A.; Pietrafesa, M. Energy performance of an electrochromic switchable glazing: Experimental and computational assessments. *Energy Build.* **2018**, *165*, 390–398. [CrossRef]

29. Marszal, A.J.; Heiselberg, P.; Bourrelle, J.S.; Musall, E.; Voss, K.; Sartori, I.; Napolitano, A. Zero Energy Building—A review of definitions and calculation methodologies. *Energy Build.* **2011**, *43*, 971–979. [CrossRef]

30. Sartori, I.; Napolitano, A.; Voss, K. Net zero energy buildings: A consistent definition framework. *Energy Build.* **2012**, *48*, 220–232. [CrossRef]

31. Kapsalaki, M.; Leal, V. Recent progress on net zero energy buildings. *Adv. Build. Energy Res.* **2011**, *5*, 129–162. [CrossRef]

32. Attia, S.; Eleftheriou, P.; Xeni, F.; Morlot, R.; Ménézo, C.; Kostopoulos, V.; Betsi, M.; Kalaitzoglou, I.; Pagliano, L.; Cellura, M.; et al. Overview and future challenges of nearly Zero Energy Buildings (nZEB) design in Southern Europe. *Energy Build.* **2017**, *155*, 439–458. [CrossRef]

33. Kolokotsa, D.; Rovas, D.; Kosmatopoulos, E.; Kalaitzakis, K. A roadmap towards intelligent net zero-and positive-energy buildings. *Solar Energy* **2011**, *85*, 3067–3084. [CrossRef]

34. AbuGrain, M.Y.; Alibaba, H.Z. Optimizing existing multistory building designs towards net-zero energy. *Sustainability* **2017**, *9*, 399. [CrossRef]

35. Lund, H.; Marszal, A.; Heiselberg, P. Zero energy buildings and mismatch compensation factors. *Energy Build.* **2011**, *43*, 1646–1654. [CrossRef]

36. Ballarini, I.; Dirutigliano, D.; Primo, E.; Corrado, V. The significant imbalance of nZEB energy need for heating and cooling in Italian climatic zones. *Energy Procedia* **2017**, *126*, 258–265. [CrossRef]

37. Romano, R.; Gallo, P. The SELFIE Project Smart and efficient envelope' system for nearly zero energy buildings in the Mediterranean Area. *GSTF J. Eng. Technol.* **2018**, *4*, 562–569. [CrossRef]
38. Trombetta, C.; Milardi, M. Building future Lab: A great infrastructure for testing. *Energy Procedia* **2015**, *78*, 657–662. [CrossRef]
39. Ministero dello Sviluppo Economico e Ministero dell'Ambiente, della Tutela del territorio e del Mare – Strategia Energetica Nazionale (SEN). 2017. Available online: https://www.mise.gov.it/images/stories/documenti/Testo-integrale-SEN-2017.pdf (accessed on 10 July 2020).
40. Ministero dello Sviluppo Economico, Ministero dell'Ambiente, della Tutela del territorio e del Mare e Ministero delle Infrastrutture e dei Trasporti–Piano Nazionale Integrato per l'Energia e per il Clima. 2019. Available online: https://www.mise.gov.it/images/stories/documenti/PNIEC_finale_17012020.pdf (accessed on 10 July 2020).
41. Beaudin, M.; Zareipour, H.; Schellenberglabe, A.; Rosehart, W. Energy storage for mitigating the variability of renewable electricity sources: An updated review. *Energy Sustain. Dev.* **2010**, *14*, 302–314. [CrossRef]
42. Pearre, N.S.; Swan, L.G. Renewable electricity and energy storage to permit retirement of coal-fired generators in Nova Scotia. *Sustain. Energy Technol. Assess.* **2013**, *1*, 44–53. [CrossRef]
43. Lorestani, A.; Ardehali, M.M. Optimization of autonomous combined heat and power system including PVT, WT, storages, and electric heat utilizing novel evolutionary particle swarm optimization algorithm. *Renew. Energy* **2018**, *119*, 490–503. [CrossRef]
44. Cammarata, G.; Monaco, L.; Cammarata, L.; Petrone, G. A numerical procedure for PCM thermal storage design in solar plants. In Proceedings of the 7th AIGE Conference, Cosenza, Italy, 10–11 June 2013.
45. Sharma, A.; Tyagi, V.V.; Chen, C.R.; Buddhi, D. Review on thermal energy storage with phase change materials and applications. *Renew. Sustain. Energy Rev.* **2009**, *13*, 318–345. [CrossRef]
46. Lopez-Sabiron, A.M.; Royo, P.; Ferreira, V.J.; Aranda-Uson, A.; Ferreira, G. Carbon footprint of a thermal energy storage system using phase change materials for industrial energy recovery to reduce the fossil fuel consumption. *Appl. Energy* **2014**, *135*, 616–624. [CrossRef]
47. Marino, C.; Nucara, A.; Pietrafesa, M.; Pudano, A. An energy self-sufficient public building using integrated renewable sources and hydrogen storage. *Energy* **2013**, *57*, 95–105. [CrossRef]
48. Avril, S.; Arnaud, G.; Florentin, A.; Vinard, M. Multi-objective optimization of batteries and hydrogen storage technologies for remote photovoltaic systems. *Energy* **2010**, *35*, 5300–5308. [CrossRef]
49. Marino, C.; Nucara, A.; Pietrafesa, M. Electrolytic hydrogen production from renewable source, storage and reconversion in fuel cells: The system of the Mediterranea University of Reggio Calabria. *Energy Procedia* **2015**, *78*, 818–823. [CrossRef]
50. Krajacic, G.; Loncar, D.; Duic, N.; Zeljko, M.; Arantegui, R.L.; Loisel, R.; Raguzin, I. Analysis of financial mechanisms in support to new pumped hydropower storage projects in Croatia. *Appl. Energy* **2013**, *101*, 161–171. [CrossRef]
51. Foley, A.; Lobera, D. Impacts of compressed air energy storage plant on an electricity market with a large renewable energy portfolio. *Energy* **2013**, *57*, 85–94. [CrossRef]
52. Al-akayshee, A.S.; Kuznetsov, O.N.; Sultan, H.M. Viability Analysis of Large Photovoltaic Power Plants as a Solution of Power Shortage in Iraq. In Proceedings of the 2020 IEEE Conference of Russian Young Researchers in Electrical and Electronic Engineering (EIConRus), St. Petersburg and Moscow, Russia, 27–30 January 2020. [CrossRef]
53. Yang, Y.; Lian, C.; Ma, C.; Zhang, Y. Research on Energy Storage Optimization for Large-Scale PV Power Stations under Given Long-Distance Delivery Mode. *Energies* **2020**, *13*, 27. [CrossRef]
54. Bernal, E.A.; Lopez, M.B.; Ardila, A. Solar Microgrids to Enhance Electricity Access in Remote Rural Areas. 2019. Available online: https://www.researchgate.net/publication/335430858_Solar_Microgrids_to_Enhance_Electricity_Access_in_Remote_Rural_Areas (accessed on 10 July 2020).
55. Nwaigwe, K.N.; Mutabilwa, P.; Dintwa, E. An overview of solar power (PV systems) integration into electricity grids. *Mater. Sci. Energy Technol.* **2019**, 629–633. [CrossRef]
56. Kaviani, A.K.; Baghaee, H.R. Gholam Hossein Riahy, Optimal Sizing of a Stand-alone Wind/Photovoltaic Generation Unit using Particle Swarm Optimization. *Soc. Modeling Simul. Int.* **2009**, *85*. [CrossRef]
57. Barbaro, G.; Foti, G.; Labarbera, F.; Pietrafesa, M. Energetic and economic comparison between systems for the production of electricity from renewable energy sources (hydroelectric, wind generator, photovoltaic). *Accad. Peloritana Pericolanti Cl. Phys. Math. Nat. Sci.* **2019**, *97*. [CrossRef]

58. Bruno, R.; Bevilacqua, P.; Longo, L.; Arcuri, N. Small Size Single-axis PV Trackers: Control Strategies and System Layout for Energy Optimization. *Energy Procedia* **2015**, *82*, 737–743. [CrossRef]
59. Bevilacqua, P.; Bruno, R.; Arcuri, N. Comparing the performances of different cooling strategies to increase photovoltaic electric performance in different meteorological conditions. *Energy* **2020**, *195*. [CrossRef]
60. Baghaee, H.R.; Mirsalim, M.; Gharehpetian, G.B.; Taleb, H.A. Reliability/cost-based multi-objective Pareto optimal design of stand-alone wind/PV/FC generation microgrid system. *Energy* **2016**, *115*, 1022–1041. [CrossRef]
61. Baghaee, H.R.; Mirsalim, M.; Gharehpetian, G.B. Multi-objective optimal power management and sizing of a reliable wind/PV microgrid with hydrogen energy storage using MOPSO. *J. Intell. Fuzzy Syst.* **2017**, *32*, 1753–1773. [CrossRef]
62. Nikolovski, S.; Baghaee, H.R.; Mlakić, D. ANFIS-Based Peak Power Shaving/Curtailment in Microgrids Including PV Units and BESSs. *Energies* **2018**, *11*, 2953. [CrossRef]
63. Gattuso, D.; Greco, A.; Marino, C.; Nucara, A.; Pietrafesa, M.; Scopelliti, F. Sustainable Mobility: Environmental and Economic Analysis of a Cable Railway, Powered by Photovoltaic System. *Int. J. Heat Technol.* **2016**, *34*, 7–14. [CrossRef]
64. Sinigaglia, T.; Lewiski, F.; Martins, M.E.S.; Siluk, J.C.M. Production, storage, fuel stations of hydrogen and its utilization in automotive applications-a review. *Int. J. Hydrog. Energy* **2017**, *42*, 24597–24611. [CrossRef]
65. Marino, C.; Nucara, A.; Panzera, M.F.; Pietrafesa, M. Towards the nearly zero and the plus energy building: Primary energy balances and economic evaluations. *Therm. Sci. Eng. Prog.* **2019**, *13*. [CrossRef]
66. UNI 10349. *Riscaldamento e Raffrescamento Degli Edifici: Dati Climatici, Ente Italiano per L'unificazione*; UNI 10349: Milano, Italy, 1994.
67. Vart, T.M.; Castaner, L. *Practical Handbook of Photovoltaic: Fundamentals and Applications*; Elsevier: Amsterdam, The Netherlands, 2006.
68. Kreider, J.F.; Kreith, F. *Solar Energy Handbook*; McGraw-Hill: New York, NY, USA, 1981.
69. Allegrezza, M.; Giuliani, G.; Mileto, G.; Pietrafesa, M.; Summonte, C. Impact of increased efficiency on cost of WP. In Proceedings of the 27th EU PVSEC European Photovoltaic Solar Energy Conference, Messe Frankfurt, Germany, 24–28 September 2012. [CrossRef]
70. Deliberazione dell'Autorità Per L'energia Elettrica E Il Gas ARG/elt 74/08, Testo Integrato per lo Scambio sul Posto. Available online: https://www.arera.it/allegati/docs/08/074-08argall2.pdf (accessed on 26 July 2012).
71. Allegato, C.; ARERA, Authority for the Regulation of Energy, Grid and Enrivonment. Testo Integrato Delle Condizioni Economiche per l'erogazione del Servizio di Connessione (TIC) (2020–2023). Available online: https://www.arera.it/allegati/docs/19/568-19allc.pdf (accessed on 15 January 2020).
72. Italian Regulatory Authority for Electricity and Gas, Electric Energy Prices. 2013. Available online: http://www.autorita.energia.it/it/dati/eep35.htm (accessed on 15 January 2020).
73. Italian National Institute of Statistics, Consumer prices. 2013. Available online: http://www.istat.it/it/archivio/14413. (accessed on 15 January 2020).
74. Bank of Italy, Interest Rates. 2013. Available online: http://www.bancaditalia.it/statistiche/SDDS/stat_fin/tassi_int/ (accessed on 15 January 2020).
75. Ministero dello Sviluppo Economico—Decreto ministeriale 20 dicembre 2019—Tasso da Applicare per le Operazioni di Attualizzazione e Rivalutazione ai Fini della Concessione ed Erogazione delle Agevolazioni in Favore delle imprese. Available online: https://www.mise.gov.it/index.php/it/incentivi/impresa/strumenti-e-programmi/tasso-di-attualizzazione-e-rivalutazione. (accessed on 15 January 2020).
76. Istituto Superiore per la Protezione e la Ricerca Ambientale (ISPRA). *Fattori di Emissione Atmosferica di Gas a Effetto Serra e Altri Gas Nel Settore Elettrico*; Rapporto; ISPRA: Roma, Italy, 2018.

Article

Updated Typical Weather Years for the Energy Simulation of Buildings in Mediterranean Climate. A Case Study for Sicily

Vincenzo Costanzo, Gianpiero Evola *, Marco Infantone and Luigi Marletta

Department of Electric, Electronic and Computer Engineering, University of Catania, Via Santa Sofia 64, 95123 Catania, Italy; vincenzo.costanzo@unict.it (V.C.); m.infantone@yahoo.it (M.I.); luigi.marletta@unict.it (L.M.)

* Correspondence: gevola@unict.it

Received: 14 July 2020; Accepted: 3 August 2020; Published: 9 August 2020

Abstract: Building energy simulations are normally run through Typical Weather Years (TWYs) that reflect the average trend of local long-term weather data. This paper presents a research aimed at generating updated typical weather files for the city of Catania (Italy), based on 18 years of records (2002–2019) from a local weather station. The paper reports on the statistical analysis of the main recorded variables, and discusses the difference with the data included in a weather file currently available for the same location based on measurements taken before the 1970s but still used in dynamic energy simulation tools. The discussion also includes a further weather file, made available by the Italian Thermotechnical Committee (CTI) in 2015 and built upon the data registered by the same weather station but covering a much shorter period. Three new TWYs are then developed starting from the recent data, according to well-established procedures reported by ASHRAE and ISO standards. The paper discusses the influence of the updated TWYs on the results of building energy simulations for a typical residential building, showing that the cooling and heating demand can differ by 50% or even 65% from the simulations based on the outdated weather file.

Keywords: weather data; typical weather year; building energy simulations; residential building; energy demand

---

1. Introduction

The energy consumption in buildings, which is in large part due to heating, ventilation, and air-conditioning, covers a high share of the overall energy balance worldwide. According to the International Energy Agency (IEA), construction and operation of buildings accounted for 36% of the global final energy use and 39% of the energy-related CO_2 emissions in 2017 [1]. In particular, the energy use in urban areas amounts to over two thirds of the world's energy consumption and is responsible for more than 70% of the global CO_2 emissions [2].

Then, in light of the rising concern towards environmental issues, and in accordance with the growing number of national and international regulations aiming at a severe reduction in the depletion of non-renewable primary energy sources, the need of detailed studies concerning the energy behavior of buildings has recently become strongly felt.

To this aim, software tools for dynamic energy simulation of buildings are nowadays commonly available and widely used by researchers, engineers, and others involved in the design and optimization of the energy performance of buildings. Indeed, these tools allow a detailed evaluation of both the time-dependent thermal loads and the indoor thermal comfort conditions, while taking into account the inertial effects due to the thermal capacity of the building enclosure [3].

However, reliable energy simulations depend strongly on the availability of accurate weather data. Most building energy simulation tools make use of weather files based on a Typical Weather Year (TWY), that is to say a full year of typical local hourly weather data generated by statistically averaging long-term weather measurements, issued by weather stations commonly placed in peripheral or rural zones. The use of TWYs has gained wide consensus in the building simulation community because they depict average climate trends, which in turn are related to the thermal behavior of a building in average conditions, while disregarding extreme weather events [4].

Now, the users of a building energy simulation tool should be aware that several drawbacks may arise when using a TWY as an input to energy simulation. Indeed, a TWY may not reflect the actual weather conditions experienced by a building in a specific site due to year-to-year climate fluctuations: Hence, a TWY does not account for the yearly variations in the energy needs, which would only be highlighted through a multi-year energy simulation process [5,6], nor can it be reliably used to calculate the peak loads occurring in design conditions [3]. Moreover, many available TWYs are out-of-date, meaning that they are based on weather measurements dating back several decades. This is likely to affect the calculated energy needs of a building, in a range between 5% and 10% according to Wang et al. [7], as well as the predicted indoor thermal comfort [8]; similar conclusions were drawn by Pernigotto et al., who found out that even the energy rating of a building can be influenced by the use of inappropriate or outdated weather files [9]. Lou et al. suggest that the weather files should be updated every 12 years [10]. These results question the suitability of many existing—and universally used—TWYs for the accurate prediction of heating and cooling loads of buildings.

In Italy, most of the currently available weather files derive from hourly measurements taken in the period 1951–1970. These data were then elaborated in the framework of the national research project entitled "Progetto Finalizzato Energetica" in 1979, and led to the release of the Italian climatic data collection "Gianni De Giorgio" (IGDG): This is a database populated by weather information from 68 weather stations evenly distributed in the national territory, later processed in order to be used in different simulation tools [11]. An update of many weather files was released in 2015 by the Italian Thermotechnical Committee (CTI), which enlarged the original domain by considering a set of 110 localities [12]. However, for many of the available locations the period of measurement is short (i.e., less than 10 years), and this may undermine the reliability of these weather files in the field of building energy simulation. It is then undeniable that new TWYs are needed based on more recent and extensive weather data.

Under these premises, this paper presents recent hourly weather data recorded from 2002 to 2019 by a weather station owned by the Sicilian Agrometeorological Information System (SIAS) in Catania (Italy). A statistical analysis compares this dataset with the old IGDG dataset available for the same city and currently largely used by the local research community thus underlining the main differences. Then, after a literature review regarding the creation of typical years in different countries worldwide, as well as their effect on the outcomes of energy simulation tools (Section 2), in Section 3 the paper describes three well-established procedures commonly used to extract a typical year from a set of weather data, which are then used to generate three new TWYs from the SIAS dataset. The aim is not only the development of the updated TWYs, but also a critical comparison of the different procedures and their outcomes, while also discussing the need to perform some pre-processing activities for data quality check and gap filling (Section 4). Finally, the paper investigates the influence of these updated weather files on the results of building energy simulations for a typical residential building located in the city of Catania.

2. Literature Review on Typical Weather Years for Energy Simulation of Buildings

A literature review on this topic shows that many different procedures have been established to extract a typical weather year from a multi-year weather dataset. However, in the field of energy simulation of buildings and their technical systems, three procedures are mainly used:

- the procedure designed by Hall et al. [13] at Sandia Laboratories in 1978, leading to the original format called Typical Meteorological Year (TMY) and then slightly modified by the National Renewable Energy Laboratory (NREL) into the more recent formats called TMY2 in 1995 [14] and TMY3 in 2005 [15];
- the ASHRAE procedure, leading to the IWEC format (International Weather for Energy Calculations), respectively in its original version of 2002 [16] and the updated IWEC2 version of 2014 [17];
- the procedure introduced by the ISO 15927-4 Standard in 2005 [18].

Other procedures proposed in the literature, such as the Festa-Ratto method [19] and the Danish method [20], did not find wide diffusion in the scientific community and are seldom used. In Canada, the weather files are instead created according to the Canadian Weather Year for Energy Calculation (CWEC) method [21].

In any case, these procedures aim to extract—for each calendar month—a typical month from all the years of observation. The twelve selected typical months, which do not necessarily belong to the same year, are then concatenated to create a typical year. This approach must not be confused with the generation of a Typical Reference Year (TRY) introduced by ASHRAE in the 1970s [22], where an entire year—with its 8760 h series of data—is selected out of the multi-year measurements.

The three previously listed methods differ from each other in various aspects, such as the list of weather parameters included in the selection process and the weights that are attributed to them. These weights are intended to describe the relative importance attributed to each weather parameter, and should in principle be coherent with the final purpose of the selected typical weather year. As an example, the original TMY was firstly developed for the simulation of solar energy systems, and for this reason a very high weight is attributed to the solar irradiation, while other parameters (wind speed, relative humidity) have a minor role [13]. Other differences concern the source of the weather data: In some cases, the values of solar irradiation (either global horizontal or direct normal) do not derive from actual measurements, but they originate from the application of suitable mathematical models. More details about the methodological differences among the selection procedures are provided in Section 3.1.

Table 1 reports a list of papers dealing with the generation of typical weather years for studying the energy performance of buildings and their energy systems, published in the last twenty years either on peer-reviewed journals or on conference proceedings [9,23–54]. The original TMY format has been adopted in 63% of the papers, and is by far the most commonly used, followed by the ISO procedure (23%) and the IWEC procedure (14%).

A few studies adopt other official procedures, such as the Festa-Ratto method and the Danish method, or even try to develop new procedures implemented in TRNSYS [44] or C++ [51]. All studies conducted in Canada rely on the CWEC procedure [26,38]. Finally, around 47% of the reported papers discusses the influence of the newly generated weather files on the results of building energy simulations, when compared with outdated weather files.

The first message emerging from this literature review is that a weather file built upon a too short weather dataset (i.e., below 10 years) is likely to produce misleading results [9]. Actually, this problem affects many of the recent weather files prepared by CTI for Italian locations, including Catania: Then, these CTI files do require further development based on more years of measurements.

As far as the different selection procedures are concerned, it is not possible to identify a single procedure that always performs better than the others. Some authors found out that the results of the energy simulations carried out with different typical weather years are similar [25,43], while other authors could find some discrepancies (up to 10%) in the calculation of the annual energy needs [35, 37,49], which may also influence the decision-making process during the design stage [47]; however, this outcome depends on the specific building and climate.

Table 1. Papers dealing with the typical weather year (TWY) generation and their effect on building energy simulation.

Authors	Ref.	Year	Selection Procedure				Number of Sites	Energy Simulation	Country
			TMY	IWEC	ISO	Other			
Argiriou et al.	[23]	1999	x	-	-	x	1	x	Greece
Kalogirou	[30]	2003	x	-	-	-	1	-	Cyprus
Sawaqed et al.	[41]	2005	x	-	-	-	7	-	Oman
Levermore and Parkinson	[33]	2006	-	x	-	-	14	-	UK
Chan et al.	[24]	2006	x	x	-	-	1	-	Hong Kong
Rahman and Dewsbury	[40]	2007	-	-	x	-	1	-	Malaysia
Skeiker	[35]	2007	x	-	-	x	1	x	Syria
Lhendup and Lhendup	[44]	2007	-	-	-	x	4	-	Australia
Yang et al.	[52]	2007	x	-	-	-	35	-	China
Janjai and Deeyai	[43]	2009	x	-	-	x	4	x	Thailand
David et al.	[51]	2010	-	-	-	x	23	-	Worldwide
Ebrahimpour and Maerefat	[42]	2010	x	-	-	-	1	-	Iran
Lee et al.	[32]	2010	-	x	-	-	7	-	South Korea
Jiang	[28]	2010	x	-	-	-	8	-	China
Oko and Ogoloma	[34]	2011	x	-	-	-	1	-	Nigeria
Bhandari et al.	[37]	2012	x	-	-	-	1	x	USA
Sorrentino et al.	[49]	2012	x	-	x	x	1	x	Italy
Kalamees et al.	[29]	2012	-	-	x	-	4	x	Finland
Ohunakin et al.	[45]	2013	x	-	-	-	5	-	Nigeria
Ohunakin et al.	[53]	2014	x	-	-	-	1	-	Nigeria
Pernigotto et al.	[9]	2014	-	-	x	-	5	x	Italy
Pernigotto et al.	[48]	2014	x	-	x	-	5	x	Italy
Bre and Fachinotti	[27]	2016	-	x	-	-	15	x	Argentina
Chan	[25]	2016	x	x	-	-	1	x	Hong Kong
Zang et al.	[46]	2016	x	-	-	x	35	-	China
Kulesza	[31]	2017	x	-	x	-	1	-	Poland
Yilmaz and Ekmekci	[36]	2017	x	-	-	-	81	-	Turkey
Pernigotto et al.	[47]	2017	-	-	x	-	2	x	Italy
Hosseini et al.	[38]	2017	-	-	-	x	1	x	Canada
Kim et al.	[50]	2017	x	-	x	-	18	x	South Korea
Arima et al.	[54]	2017	-	-	-	x	1	x	Japan
Crawley and Lawrie	[39]	2019	x	-	-	x	5	x	USA, UK
Siu and Liao	[26]	2020	-	-	-	x	1	x	Canada

Moreover, many authors agree that using a single typical weather year cannot reflect the year-by-year variability in the energy demand of a building. Indeed, this can deviate from the results of a single-year simulation, especially in the winter and for residential uninsulated lightweight buildings [9,23,27], while in the summer the deviation is less evident, unless in largely glazed buildings [9]. In commercial buildings with large flat roofs, the deviation in the seasonal energy demand for cooling may reach 12% [38]. Bevilacqua et al. highlighted that simulated heating and cooling load for a building equipped with a green roof in Southern Italy may vary by around 30% and 15%, respectively, when adopting weather files based on two consecutive actual years [55]. This suggests that further investigation is needed to develop case-dependent weighting sets in order to minimize the distance with the average multi-year simulation results [9,25,29,39,48].

Finally, several authors have underlined that the use of outdated typical years, dating back from the 1990s or even before, causes a significant overestimation in the heating demand of buildings, and a slight underestimation in the cooling demand [26,56–58]. This also suggests the need to investigate how weather data can evolve according to different climate scenarios, and to generate suitable TWYs referring to future time horizons. An interesting application of this approach is presented by Vasaturo et al. [59].

3. Methodology

3.1. Procedure for the Identification and Concatenation of the Typical Months

This section describes the methodology followed to build the typical weather years according to the different selection procedures. It is here worth highlighting that the authors made the choice of working only with measured weather data, thus not relying on the adoption of mathematical models for the estimate of the different solar irradiation components. For this reason, and because the SIAS weather station does not measure the direct normal irradiation, the only procedures considered in the paper are those not requiring this parameter through the selection process, that is to say TMY (first original version), IWEC, and ISO 15927.

The procedure for the extraction of Typical Months (TMs) from a long-term observational weather dataset can be summarized in the following sequential steps:

1. *Selection of the weather parameters for statistical analysis.* TMY and IWEC methodologies, building upon the original method by Hall et al. [13], choose the following nine parameters that are considered with daily frequency: Minimum, maximum and mean values of dry bulb air temperature (°C) and dew point temperature (°C), maximum and mean values of the wind speed (m/s), and cumulated global horizontal solar irradiation (Wh/m2). The ISO 15927-4 Standard reduces this set of parameters to only three, i.e., the daily average of dry bulb temperature (°C), relative humidity (%), and global horizontal solar irradiation (Wh/m2). In all cases, the dew point temperature can be replaced by the relative humidity because the two variables can be determined from each other by also using one more independent moist air property (dry bulb temperature) through psychrometrics.

2. *Construction of the Cumulative Distribution Functions (CDFs) for the selected weather parameters.* For every calendar month, the daily values of the selected weather parameters are sorted in ascending order and then used for creating monthly CDFs referred both to the whole period of record (long-term) and to each single year (short-term). CDFs are built according to Equation (1) in the TMY approach, or Equation (2) if the ISO 15927-4 Standard is adopted:

$$S_n(x)_{TMY} = \begin{cases} 0 & \text{for } x < x_{(1)} \\ (k-0.5)/n & \text{for } x_{(k)} \le x \le x_{(k+1)} \\ 1 & \text{for } x > x_{(n)} \end{cases} \tag{1}$$

$$S_n(x)_{ISO} = k(x)/(n+1) \tag{2}$$

In the above equations, x is the weather parameter, n is the total number of daily observations for that parameter, k is the order number for each x-value within that calendar month (short-term) or within the whole dataset (long-term). No explicit reference to the previous equations is made in the IWEC method. In this paper, the approach indicated in Equation (1) is followed in the construction of the IWEC typical year.

3. *Estimate of the closeness of single months' CDFs to the long-term CDF.* For each weather parameter and for each calendar month, the distance between long-term and short-term distributions is evaluated by means of the Finkelstein-Schafer (FS) statistics [60], which is defined as follows for the different procedures:

$$FS_{TMY\text{-}IWEC}(x) = \frac{1}{n} \cdot \sum_{k=1}^{n} \delta_k = \frac{1}{n} \cdot \sum_{k=1}^{n} \left| S_{n,year}(x_k) - S_{n,long\text{-}term}(x_k) \right| \tag{3}$$

$$FS_{ISO}(x) = \sum_{k=1}^{n} \delta_k = \sum_{k=1}^{n} \left| S_{n,year}(x_k) - S_{n,long\text{-}term}(x_k) \right| \tag{4}$$

being n the number of days of that specific month and δ_k the absolute difference between the long-term and the single month CDF values evaluated for every day of the month.

4. *Weighted sum of the FS statistics and ranking order.* Since some weather parameters may be deemed more important than others for building energy simulations, the IWEC and the TMY procedures introduce a weighted sum of the single FS statistics calculated for every parameter, as in Equation (5):

$$WS = \sum_{j=1}^{9} FS(j) \cdot W_j \tag{5}$$

The weighting factors W_j attributed to each of the nine weather parameters vary with the standards, and are summarized in Table 2. As far as the ISO 15927-4 Standard is concerned, this does not introduce any weighting process, meaning that the same importance is attributed to the three weather parameters considered in this standard.

5. *Selection of the Typical Month (TM).* Here, the IWEC approach implies—for the five months with the lowest WS—the calculation of the monthly mean and median of both the dry bulb temperature and the global horizontal irradiance, and eventually chooses the month closest to the long-term series. On the other hand, the TMY includes a persistence analysis: The candidate months with the longest run and with the highest number of runs either above the 67th long-term percentile or below the 33rd long-term percentile, as well as those with no runs, are excluded; eventually, the top ranked candidate month remaining after the persistence analysis is retained as the Typical Month. Finally, the ISO 15927-4 Standard selects the TM amongst the three best candidate months coming from step 4 as the one with the lowest deviation of the monthly mean wind speed from the long-term figure.

6. *Smoothing discontinuities at month interfaces.* Since the 12 selected Typical Months are likely to belong to different years of observation, a smoothing procedure is needed between the last hours of the preceding month and the beginning hours of the following month via curve-fitting or interpolation techniques.

Table 2. Non-dimensional weighting factors according to different procedures (T: Dry bulb temperature; RH: Relative humidity; W: Wind speed; GHI: Global horizontal irradiance).

Method	T_{min}	T_{max}	T_{avg}	RH_{min}	RH_{max}	RH_{avg}	W_{max}	W_{avg}	GHI_{cum}
IWEC	5/100	5/100	30/100	2.5/100	2.5/100	5/100	5/100	5/100	40/100
TMY	1/24	1/24	2/24	1/24	1/24	2/24	2/24	2/24	12/24

3.2. Preparation of the Weather File in the .epw Format

The final step required to use a typical weather year in a building energy simulation tool consists of arranging the weather data in a suitable format. Amongst the various available formats, this paper relies on the EnergyPlus Weather file (.epw), which can be used with several simulation tools such as EnergyPlus, DesignBuilder, TRNSYS, IES, and ESP-r. This is an ASCII file where, along with a header containing information about the site and the period of record, a series of weather variables are listed with an hourly time step for each Julian day of the year [61].

In addition to the weather variables used to select the TMs, i.e., dry bulb temperature (°C), relative humidity (%), wind speed (m/s), and global horizontal solar irradiance (W/m^2), the weather file must contain other physical quantities to correctly estimate the energy and mass exchange between the building and the surroundings. These can be classified as:

- *Compulsory parameters:* Dew point temperature (°C), atmospheric pressure (Pa), horizontal infrared radiation intensity (W/m^2), direct normal irradiance (W/m^2), diffuse horizontal irradiance (W/m^2), wind direction (degrees from north), and opaque sky cover (tenths of coverage).

- *Additional parameters:* Snow depth (cm) and liquid precipitation depth (mm).

Actually, once the atmospheric pressure is known (e.g., measured), the dew point temperature depends on the dry bulb temperature and the relative humidity, thus it can be easily calculated by means of psychrometric relations [62].

Building a weather file that can be read by energy simulation programs is then a further demanding step that adds to the generation of the TWY and may even need specific software tools [63].

In this paper, since the direct normal and diffuse horizontal components of the solar irradiance are not measured by the weather station, they are estimated starting from the global horizontal irradiance through the well-established model elaborated by Boland and Ridley [64], which is also the one suggested in the Italian technical norm UNI 10349:1 [65]. Another important parameter to consider is the horizontal infrared radiation of the sky that affects the long-wave radiant exchange of the building. Very few meteorological stations are able to record this quantity, so it has been estimated through Equation (6) [66,67]:

$$G_{IR} = \varepsilon_{SKY} \cdot \sigma \cdot T_{DB}^4 \tag{6}$$

Here, G_{IR} is the horizontal infrared radiation intensity in W/m^2, $\sigma = 5.67 \times 10^{-8}$ W/(m^2 K^4) is the Stefan-Boltzmann constant, T_{DB} is the dry bulb temperature recorded at the weather station (in K). The sky emissivity ε_{SKY} can be calculated through Equation (7) [66,67]:

$$\varepsilon_{SKY} = \left[0.787 + 0.764 \cdot \ln\left(\frac{T_{DP}}{273.15} \right) \right] \cdot \left(1 + 0.0224 \cdot CC - 0.0035 \cdot CC^2 + 0.00028 \cdot CC^3 \right) \tag{7}$$

where T_{DP} is the dew point temperature (K) and CC is the opaque sky cover, also known as *cloud cover* (tenths of sky coverage). Through this approach, the only missing variable is the sky cover.

Since the local weather station does not measure this quantity, the surface-based total cloud cover data have been acquired from the NCDC (National Climatic Data Center) Integrated Surface Database (ISD). Hourly data are freely available at the NCDC website for any user-selected time period and station [68]. Periods of missing data (2 h gaps) were linearly interpolated and adjusted to be consistent with the other meteorological factors.

3.3. Case Study Building and Dynamic Energy Simulations

The building chosen as a case study is an apartment block with four storeys located in Catania (Italy), a city with hot and humid climate and sunny days for most of the year. Figure 1 reports a view of the north facade. This residential building has a rectangular shape, and consists of two identical blocks placed side by side; each block has an overall gross floor area of 816 m^2 and a net floor area of around 725 m^2. The gross height of each storey is 3 m, while the heated gross volume of the entire building is around 2500 m^3.

There are two apartments per floor organized following the same internal distribution and number of rooms: Two or three bedrooms, a kitchen, a living room, two bathrooms, a balcony facing south, and a balcony with a veranda facing north (Figure 1). A stairwell on the north side provides access to the flats at the different floors.

The envelope is quite typical of the Italian buildings constructed in the 1970s, and this makes the building representative of the vast majority of the residential stock in Catania. In particular, the exterior walls are cavity walls with two layers of hollow clay bricks (8 cm on the inner side and 12 cm on the outer side, respectively) and a 10-cm thick air space in between. The intermediate floors and the roof are made of a 20-cm thick lightweight concrete slab, covered with 2-cm thick tiles. Internal partitions consist of 8 cm-thick hollow clay bricks covered with concrete plaster on both sides, while windows are single-glazed with an aluminum frame and no thermal break.

No insulating material is applied to the envelope in its current state. However, the dynamic energy simulations will also consider a variant of the building where the exterior walls are insulated through the addition of a 4 cm outer layer of extruded polyurethane (XPS, $\lambda = 0.030$ W m^{-1} K^{-1}), and the existing windows are replaced by double-glazed low-emissivity windows. This makes it possible to investigate the effect of the different weather files also on a more modern building with better insulation levels that comply with current regulations. The corresponding U-values of the different components, together with the SHGC of the windows, are summarized in Table 3.

Table 3. Features of the envelope for the selected building.

Building Component	U-Value		SHGC	
	Uninsulated Case [W m^{-2} K^{-1}]	Insulated Case [W m^{-2} K^{-1}]	Uninsulated Case [–]	Insulated Case [–]
External walls	0.86	0.40	-	-
Slabs	2.72	2.72	-	-
Windows and glazed doors	5.78	1.82	0.862	0.743

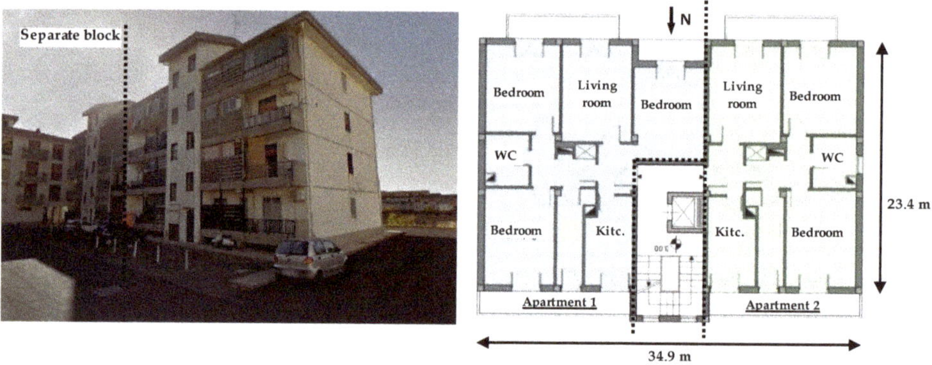

Figure 1. North view of the apartment block (**left-hand side**) and plan of a typical floor (**right-hand side**).

For energy simulation purposes, the well-known and validated EnergyPlus v.9.0.1 software is used [61]. The simulation only regards one of the two middle-storeys in the Western block that is representative of the average behaviour of the entire building: Accordingly, the adjoining surfaces with the upper and the lower apartments, as well as with the adjacent block, have been treated as adiabatic. The resulting thermal model is shown in Figure 2. In the simulations, the thermal conductivity of the various materials is kept constant; the air cavity is simulated in EnergyPlus as a thermal resistance with no inertial effect. Heat conduction is simulated through the *Conduction Transfer Function* option in EnergyPlus, with a two-minute time step; the distribution of the solar radiation over the indoor surfaces is computed through the *Full Interior and Exterior with Reflections* algorithm.

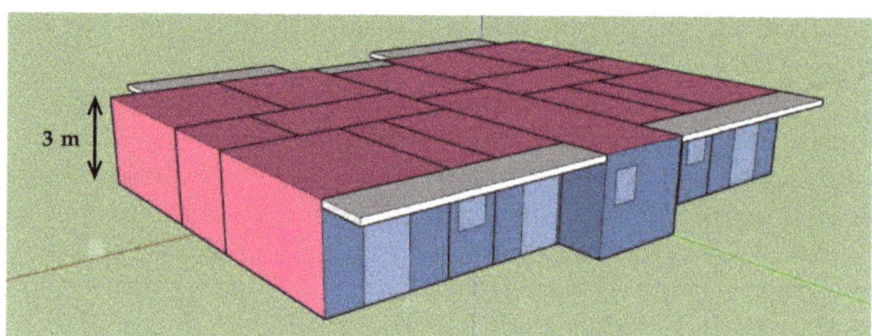

Figure 2. Axonometric view of the thermal model with two apartments per floor (pink surfaces: Adiabatic; blue surfaces: Exterior; white surfaces: Shadings).

The incoming outdoor air flow rate is supposed to be composed by a fixed value amounting to 0.3 h^{-1} (average intentional ventilation rate) plus an additional contribution related to air infiltration paths that varies with time according to wind pressure and temperature difference between indoors and outdoors. This contribution is computed through the Effective Leakage Area (ELA) method proposed in the Fundamentals volume of the ASHRAE Handbook [69]. Such an approach first defines a leakage area A_L (in cm^2) for every thermal zone as a function of its net volume V_n (in m^3) by using the following relation:

$$A_L = \frac{n_{\Delta p} \cdot V_n}{3600} \cdot \sqrt{\frac{\rho}{2 \cdot \Delta p}} \cdot 10000 \; [\text{cm}^2] \tag{8}$$

Here, $n_{\Delta p}$ is the air exchange rate (in h^{-1}) under a pressure difference Δp. If one assumes $\Delta p = 50$ Pa, the corresponding value (n_{50}) can be measured through the blower door test. In the absence of experimental measurements, n_{50} has been here set to 8 h^{-1} as suggested by the Italian technical Standard UNI 11300-1:2014 [70] for multi-family apartment blocks with high air permeability. Eventually, the rate of adventitious air flowing through the envelope is computed accounting for both the stack and wind effects by means of Equation (9):

$$\dot{V}_{inf} = \frac{A_L}{1000} \cdot \sqrt{C_s \cdot \Delta T + C_w \cdot w^2} \; [\text{m}^3/\text{s}] \tag{9}$$

Here, the *stack coefficient* C_s depends on the number of floors in the buildings, whereas the *wind coefficient* C_w is a function of the number of floors and the type of surrounding shelters. The selected values ($C_s = 1.45 \times 10^{-4}$ $(\text{L/s})^2$ $(\text{cm}^4 \cdot \text{K})^{-1}$ and $C_w = 2.71 \times 10^{-4}$ $(\text{L/s})^2$ $(\text{cm}^4 \text{ m}^2/\text{s}^2)^{-1}$) are derived from the ASHRAE Handbook [69].

As far as internal gains are concerned, different schedules are defined for the kitchen, bathroom, living room, and bedrooms, but their intensity is kept the same for all the rooms. More in detail, the average occupation rate is 0.04 person/m^2 with people involved in sedentary activities and releasing 120 W/person indoors (30% of heat is released via the radiant heat exchange). Additionally, internal lights and various electric equipment are lumped together and result in 4 W/m^2 discharged indoors with a radiant fraction of 20%.

The HVAC system is modeled as an ideal air system, that is to say a fictitious whole-air system with infinite capacity that is always able to instantaneously meet the thermal load. The heating mode runs for 24 h during the winter period (from 1 October to 31 March) at a set point temperature of 20 °C, whereas the cooling mode is in place throughout the day during the summer (from 1 June to 30 September) at a set point temperature of 26 °C. These set point values are chosen according to the provisions of EN 15,251 Standard for Category II of comfort [71].

The simplifications employed in the energy modeling, such as adopting adiabatic surfaces for the bottom and top floors and considering an ideal air system working for 24 h during the heating and cooling seasons, are justified because the aim of these simulations is to highlight the role of weather variables and the consequent differences in the energy needs for space heating and cooling according to different weather datasets only, rather than running detailed building-scale analyses.

4. The Weather Station

The building energy simulation practice currently relies on the use of weather data collected at rural weather stations for performing hourly annual simulations, thus neglecting—or in the best cases correcting this inaccuracy via morphing procedures [72]—the so called Urban Heat Island (UHI) effect, i.e., the increase in air temperature values in urban settlements if compared to rural ones [73].

For this reason, the weather data used in this study have been recorded by the rural weather station owned by SIAS (Sicilian Agrometeorological Information System), which is located at the following geographical coordinates: 37°26′31″ N, 15°04′04″ E (altitude 10 m asl). Instead, the data included in the IGDG series refer to the weather station next to Fontanarossa airport (37°27′52″ N,

15°03′30″ E, altitude 17 m asl), which is located just 3 km north of the SIAS station. As shown in Figure 3, both weather stations are placed in a rural area, overlooking the sea on the east side and mainly agricultural fields on the north and the south side. The SIAS station has a light industrial and warehousing district to its west side.

The SIAS weather dataset covers a period of 18 years (2002–2019) and contains hourly values of dry bulb temperature (in °C), relative humidity (in %), global horizontal irradiation (in MJ/m²), wind speed (in m/s), wind direction, and atmospheric pressure (in hPa). The dry bulb temperatures were measured from 2002 to 2009 with a MTX-TAM sensor (accuracy ±0.15 °C), whereas in 2010 a Vaisala HMP45 temperature and humidity probe was installed (accuracy ±0.2 °C and ±1% at 20 °C, respectively). The global horizontal irradiation is measured by a Schenk pyranometer type 8102 (accuracy < 10 W/m²). The wind speed and direction are measured by two Gill Windsonic ultrasonic anemometers (accuracy ±2% and ±2°, respectively) placed at 2 and 10 m above ground, as shown in Figure 4.

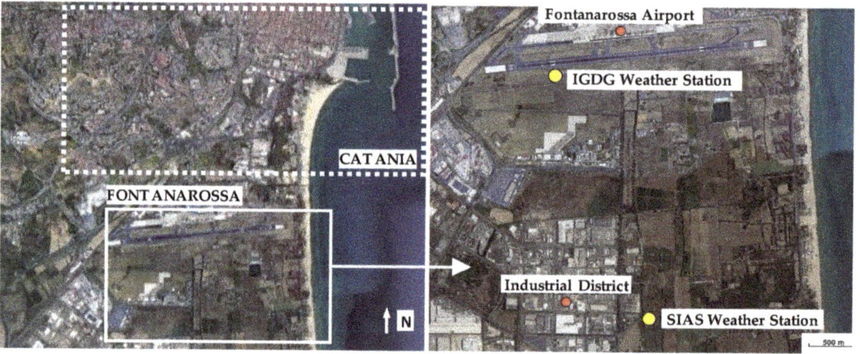

Figure 3. Position of the weather stations in relation to the city of Catania site.

Figure 4. Picture of the Sicilian Agrometeorological information system (SIAS) weather station and its probes.

The collected data were preprocessed and checked for detecting missing or invalid measurements. In particular, the criteria listed in Table 4 were used to identify invalid measurements. Table 5 reports the number of gaps in the hourly data for at least one of the following parameters: Dry bulb temperature, relative humidity, wind speed, and solar irradiation. Those months showing gaps longer than 72 consecutive h were rejected (in bold) and not considered in the further analyses for the typical weather year development, whereas different interpolation methods were used for all gaps up to 72 consecutive h.

In particular, invalid and missing data up to six consecutive h were replaced by using linear interpolation, while gaps between 6 and 72 h were filled by linear interpolation of the measured data having the same position (i.e., hour) in the neighboring days.

Table 4. Criteria leading to the identification of invalid measured data.

Dry bulb Temperature	• Values outside an acceptable range ($-30 \div 50\ °C$) • Values exceeding by 50% the 99th percentile • Constant values for more than five consecutive h • Data with a derivative larger than $\pm6\ °C/h$
Relative Humidity	• Values outside an acceptable range ($0 \div 100\%$) • Constant values for more than five consecutive h
Global Horizontal Irradiation	• Values outside an acceptable range ($0 \div 1400\ Wh/m^2$) • Positive values after sunset or before sunrise
Wind Speed	• Values outside an acceptable range ($0 \div 50\ m/s$) • Values exceeding by 50% the 99th percentile • Constant values for more than five consecutive h

Table 5. Gaps in the hourly data measured by the SIAS station, before and after the filling procedure.

Year	Jan	Feb	Mar	Apr	May	Jun	Jul	Aug	Sept	Oct	Nov	Dec
2002	744/-	672/-	744/-	720/-	744/-	720/-	**517/744**	**329/744**	-/-	528/-	720/-	448/-
2003	-/-	-/-	-/-	-/-	-/-	-/-	-/-	**287/744**	34/-	-/-	**483/720**	**744/744**
2004	**348/744**	**96/672**	-/-	-/-	-/-	**200/720**	-/-	35/-	1/-	-/-	171/-	**208/744**
2005	41/-	-/-	-/-	1/-	-/-	-/-	-/-	-/-	**107/720**	2/-	-/-	53/-
2006	**124/744**	3/-	65/-	1/-	**150/744**	-/-	-/-	-/-	16/-	**199/744**	16/-	83/-
2007	1/-	10/-	39/-	5/-	-/-	7/-	-/-	-/-	-/-	8/-	10/-	21/-
2008	1/-	**95/672**	32/-	-/-	-/-	-/-	33/-	-/-	-/-	-/-	**217/720**	12/-
2009	40/-	4/-	3/-	-/-	-/-	-/-	-/-	-/-	2/-	11/-	-/-	13/-
2010	**221/744**	45/-	**117/744**	**269/720**	34/-	1/-	17/-	10/-	4/-	1/-	8/-	2/-
2011	2/-	-/-	2/-	-/-	2/-	2/-	-/-	-/-	-/-	-/-	-/-	1/-
2012	1/-	2/-	-/-	1/-	1/-	-/-	-/-	-/-	-/-	1-/-	-/-	-/-
2013	4/-	4/-	-/-	6/-	14/-	-/-	-/-	2/-	-/-	-/-	-/-	-/-
2014	1/-	-/-	-/-	1/-	1/-	-/-	-/-	-/-	-/-	-/-	-/-	25/-
2015	2/-	-/-	-/-	-/-	-/-	-/-	-/-	2/-	2/-	-/-	-/-	-/-
2016	-/-	-/-	-/-	-/-	-/-	1/-	-/-	-/-	-/-	-/-	-/-	2/-
2017	-/-	1/-	-/-	1/-	-/-	-/-	-/-	-/-	-/-	-/-	3/-	1/-
2018	-/-	-/-	1/-	-/-	-/-	-/-	-/-	-/-	-/-	2/-	-/-	94/-
2019	-/-	-/-	-/-	-/-	1/-	-/-	-/-	-/-	1/-	-/-	-/-	1/-

Examples of the filling procedure are reported in Figure 5. The filled values were adjusted to preserve nonlinearities and appropriately scaled to match the endpoints of the interpolation interval. Most of the global horizontal irradiation data are missing in 2002: In this case, hourly data for solar irradiation were recovered from the National Aeronautics and Space Administration (NASA), MERRA-2 Project (http://gmao.gsfc.nasa.gov/reanalysis/MERRA-2).

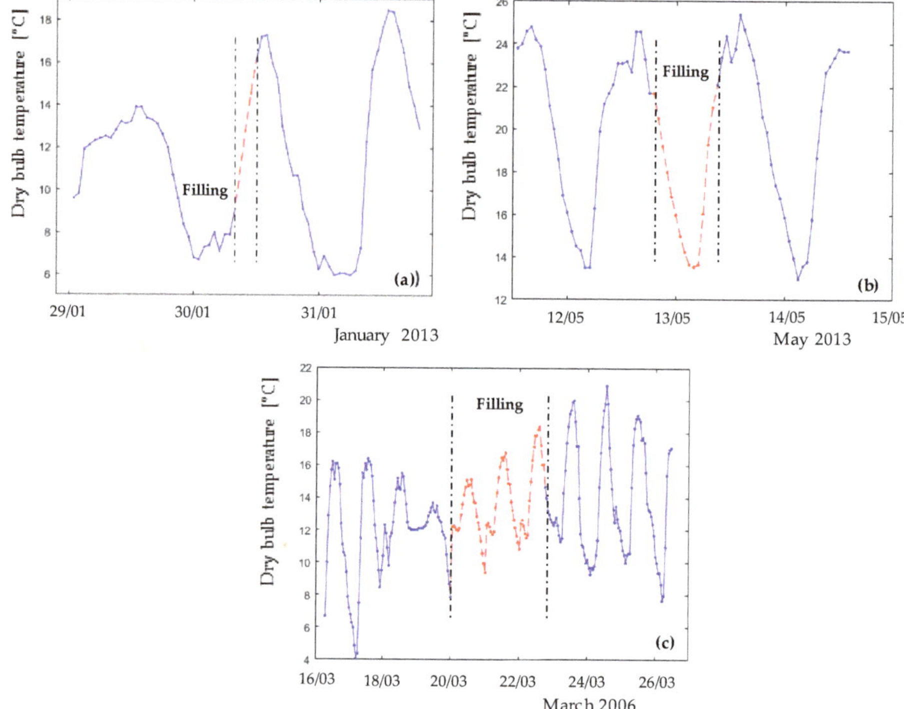

Figure 5. Examples of the filling procedure. (**a**) 0–6 h gaps. (**b**) 6–24 h gaps, (**c**) 24–72 h gaps (the dotted lines correspond to the data added through the filling procedure).

At the end of this section it is worth recalling that the data recorded by the SIAS weather station from 2002 to 2009 have been used by the Italian Thermotechnical Committee (CTI) for the release of a Test Reference Year in 2015. This reference year, which is based on a very short observation period, will be used as a term of comparison in the following.

5. Results

5.1. Statistical Analysis of the SIAS Weather Data

This section discusses the main weather data measured by the SIAS weather station between 2002 and 2019 and compares them with the data included in the weather files available for the airport "Catania Fontanarossa" in the EnergyPlus website (IGDG) and in the website of the Italian Thermotechnical Committee (CTI).

The parameters considered for this analysis are the air temperature, the daily global irradiation on the horizontal plane (GHI), the relative humidity, and the wind speed. The calculation of the daily GHI has implied the integration of the hourly values registered by the pyranometers over every day of the recording period. The analysis was performed after the filling and quality check procedure described in Section 4.

The plots reported in Figure 6 show, for each of the above parameters, the monthly average calculated for the IGDG and the CTI (marker points) along with their range of variation (min, max and mean, solid and dotted lines) for the single years of measurement from 2002 to 2019.

Starting from the dry bulb air temperature (Figure 6a), for each calendar month the mean monthly values referring to the 18 years included in the SIAS dataset can vary within a range whose amplitude is around 3 °C. The IGDG series reports systematically lower mean values, which in some cases fall below the range of variation for the SIAS dataset. This means that a slight increase in the mean air temperature has been observed in the 30 years elapsed from the recording of the IGDG data to the recording of the SIAS weather station. As an example, in July the average of the IGDG series is 24.8 °C, but it rises to 26.1 °C in the SIAS weather dataset. Coherently, a non-negligible overestimation in the mean relative humidity emerges when using the IGDG dataset, especially in the summer months, amounting to around 5% or 10% (Figure 6b).

The GHI shows even more evident differences (Figure 6c). Indeed, the GHI measured by the SIAS weather station is constantly well above the values reported in the IGDG weather file: As an example, in April the mean value is around 4700 Wh/m^2 in the IGDG dataset, but it rises to 5600 Wh/m^2 in the SIAS dataset (2002–2019), meaning an increase by 20% on average. In July, the average increases from 6300 Wh/m^2 in the IGDG dataset to 7300 Wh/m^2 in the SIAS dataset, which corresponds to 17% higher solar energy available on average.

Finally, the mean wind speed measured by the SIAS weather station is significantly below the values proposed in the IGDG typical year, especially in January and April, despite the fact that the distance between the two weather stations amounts to just a few kilometers (Figure 6d). This difference could originate from a variation in the local wind pattern in the last fifty years, but it is much more likely that the wind speed measured by the SIAS station is affected by the proximity of an industrial district with low-rise warehouses at a distance of approximately 300 m, which can be observed in Figure 3.

As far as the CTI data are concerned, these are generally quite close to the mean SIAS values: Indeed, they derive from a recent (although shorter) recording period and also from the same weather station. The only relevant difference concerns the wind speed (Figure 6d): However, it is important to observe that the wind data included in the CTI weather file refer to a height of 2 m above ground, while all other wind data have been measured at 10 m above ground. This is in the authors' opinion a serious flaw in the CTI dataset, which implies the need to perform arbitrary conversions through suitable equations in order to make comparisons—and simulations—possible.

One more comparison that is here proposed concerns the wind direction. Figure 7 shows the wind rose plotted both for the IGDG and the entire SIAS dataset: In this case, the wind rose refers to the whole period of 18 years, but similar plots can be obtained if one considers every single year. The wind roses suggest that very different wind patterns emerge from the two data sources. In particular, the prevailing wind direction in the SIAS dataset is from South-West, which is coherent with the presence of Mount Etna in the North-West quadrant, acting as a shield. On the contrary, the wind rose for the IGDG dataset suggests an even distribution for the wind direction, with a slightly lower frequency from North-West. As already discussed for the wind speed, such a significant difference is actually difficult to justify.

Finally, Figure 8 compares, on a monthly basis, the mean values of the daily minimum and maximum dry bulb temperature, and consequently the mean diurnal temperature fluctuation. These data are particularly relevant to calculate the peak daily energy demand and, especially in the summer, to assess the effectiveness of natural ventilation strategies. Now, the long-term measured data show—apart for two months—a non-negligible reduction in the amplitude of the average diurnal temperature variation: As an example, in July this decreases from 10.8 to 10.1 °C. Together with the higher daily minimum temperatures appearing in the SIAS dataset, this is likely to affect the cooling potential of nighttime natural ventilation strategies, which is then over-estimated by the use of the outdated IGDG weather file.

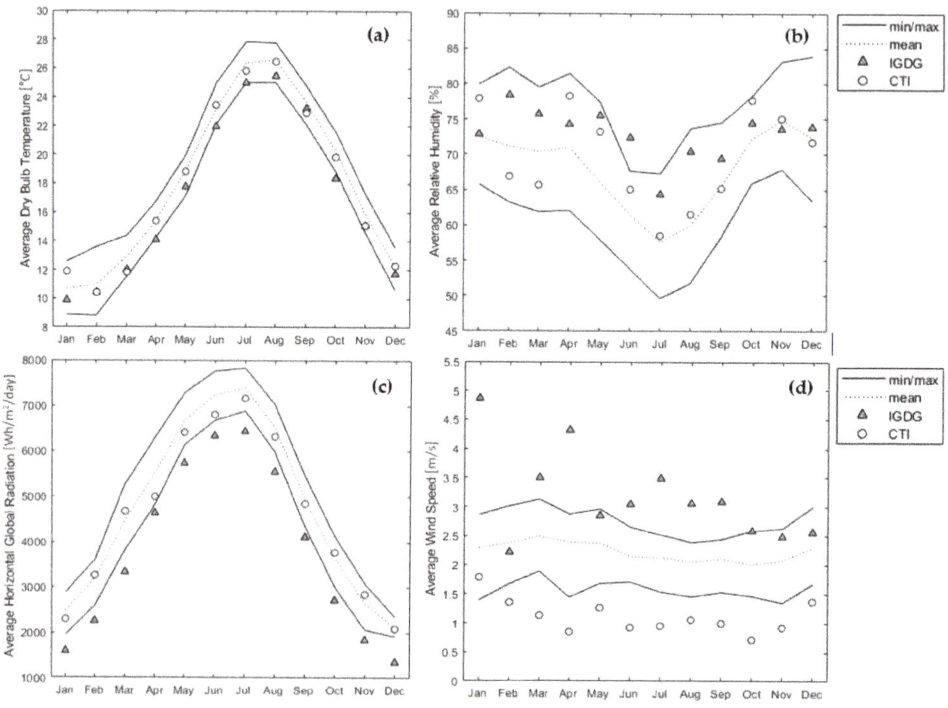

Figure 6. Average monthly values for the main weather data. The curves identify the range of variation for the single actual years. (**a**) Dry bulb temperature; (**b**) relative humidity; (**c**) daily global horizontal irradiation; (**d**) wind speed.

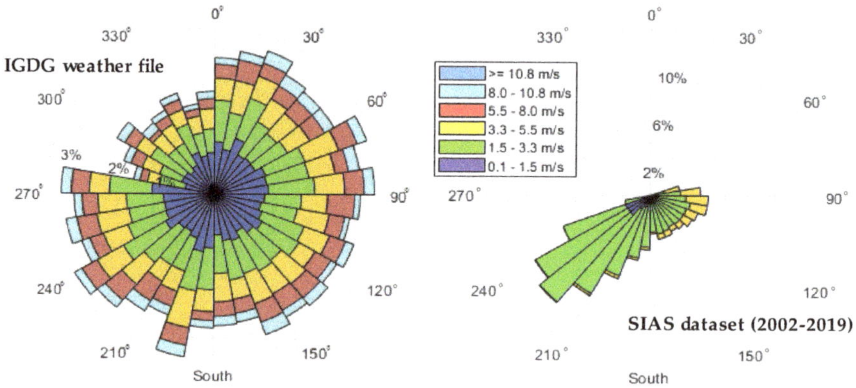

Figure 7. Wind rose for the EnergyPlus website (IGDG) weather file (**left-hand side**) and for the entire long-term SIAS dataset (**right-hand side**).

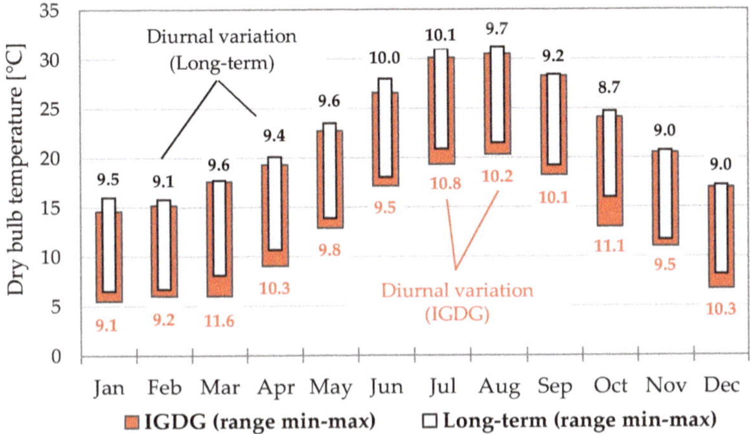

Figure 8. Monthly mean values for the daily minimum and maximum dry bulb temperatures (and corresponding diurnal temperature fluctuations).

5.2. Selection of the Typical Months and Their Comparison

This section aims at showing that the application of the different procedures outlined in Section 3.1 may lead to the selection of different typical calendar months, hence giving rise to a variety of typical years including different weather data. In fact, Table 6 suggests that for only three calendar months (February, June, and December) the three procedures lead to the identification of the same typical month (occurring in 2006, 2009, and 2008, respectively). In five cases, at least one procedure identifies a typical month that does not correspond to the one selected by the other procedures; finally, four calendar months are common for every procedure.

Now, according to the selection procedure discussed in Section 3.1, the resulting typical month takes into account the distances from the long-term cumulative distribution function of each parameter, depending on the weighing factors, and therefore it may differ from the "best" month, i.e., the one with the shortest distance from CDF, associated to each individual parameter. This can be clearly observed in Figure 9, which refers to April and compares—for the main weather parameters—the cumulative distribution functions pertaining to the typical months selected through the various procedures, as well as the best and the worst month (the latter is the month with the longest distance from the long-term cumulative distribution function).

As an example, the typical month selected by the ISO procedure (April 2011) corresponds to the best month in relation to the relative humidity (Figure 9c); however, it is quite distant from the long-term cumulative distribution function with regard to the GHI (Figure 9b). However, all the selected typical months match very well with the long-term cumulative distribution function for temperature (Figure 9a), and this reflects the different weights attributed to the weather parameters by the various procedures.

Table 6. Selection of the typical months according to the different procedures.

Method	Jan	Feb	Mar	Apr	May	Jun	Jul	Aug	Sept	Oct	Nov	Dec
IWEC	2011	2006	2012	2019	2005	2009	2005	2007	2013	2015	2019	2008
TMY	2011	2006	2012	2008	2016	2009	2008	2004	2004	2009	2004	2008
ISO	2014	2006	2011	2011	2016	2009	2004	2010	2013	2015	2010	2008

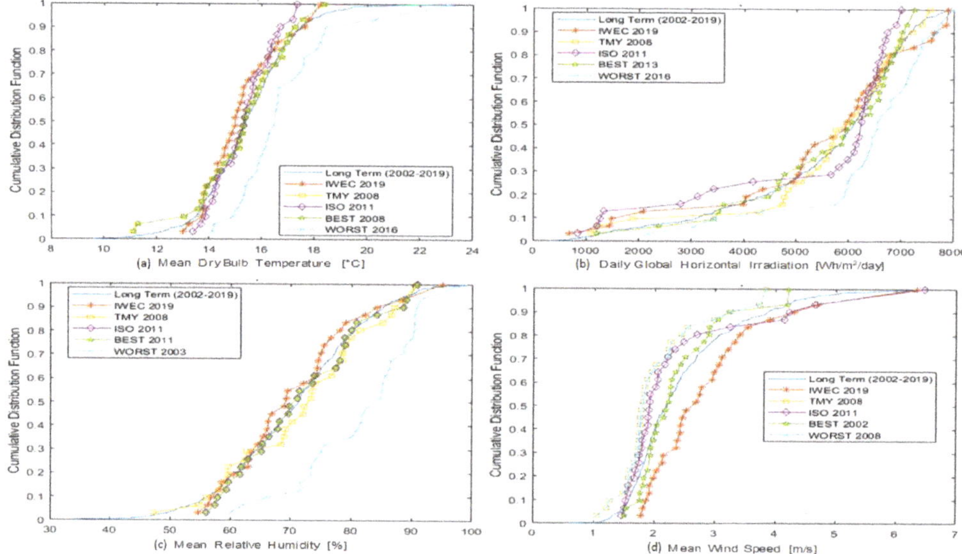

Figure 9. Cumulative distribution: Comparison among long-term measurements and selected typical months for April. (**a**) Dry bulb temperature, (**b**) daily global horizontal irradiation, (**c**) relative humidity, (**d**) wind velocity.

In order to further highlight this issue, Table 7 shows—for the four main weather variables—the values of the Mean Bias Error (MBE), the Mean Absolute Error (MAE), and the Root-Mean-Square Error (RMSE), defined as in the following equations:

$$\mathrm{MBE} = \frac{\sum_{i=1}^{12}\left(\mathrm{P_i} - \mathrm{P_{L,i}}\right)}{12} \tag{10}$$

$$\mathrm{MAE} = \frac{\sum_{i=1}^{12}\left|\mathrm{P_i} - \mathrm{P_{L,i}}\right|}{12} \tag{11}$$

$$\mathrm{RMSE} = \sqrt{\frac{\sum_{i=1}^{12}\left(\mathrm{P_i} - \mathrm{P_{L,i}}\right)^2}{12}} \tag{12}$$

Here, for each calendar month (i), p is the average of the hourly values belonging to the selected typical month, and p_L is the long-term mean value from 2002 to 2019, that is to say:

$$\mathrm{P_L} = \frac{\sum_{j=2002}^{2019}\left(\mathrm{P_j}\right)}{18} \tag{13}$$

The results reported in Table 7 basically confirm the outcomes discussed above, but they refer to the entire typical years and not only to a single month as in Figure 9. Compared with the long-term distribution, the ISO procedure allows creating a typical weather year with the lowest RMSE for the relative humidity and wind speed, but with the highest discrepancy in relation to GHI, the latter being almost 50% higher than for the IWEC and TMY typical years (i.e., 139.4 kWh/m²/day versus around 90 kWh/m²/day). The discrepancy of the IWEC and TMY typical years is very similar, even if the IWEC shows a better match with the long-term dry bulb temperature, due to the very high weight attributed to this weather parameter (overall 40% as shown in Table 2).

Table 7. Mean discrepancy between the typical years and the long-term average values for the measured weather parameters.

Variable	Error	IWEC	TMY	ISO
Dry Bulb Temperature [°C]	MBE	−0.08	−0.34	−0.05
	MAE	0.25	0.40	0.32
	RMSE	0.31	0.55	0.42
Relative Humidity [%]	MBE	−0.90	1.03	0.34
	MAE	2.49	2.21	1.43
	RMSE	3.03	2.74	1.69
Wind Speed [m/s]	MBE	0.02	−0.11	0.01
	MAE	0.22	0.19	0.14
	RMSE	0.26	0.23	0.15
Daily Global Solar Irradiation [kWh/m^2/day]	MBE	−9.4	23.6	−53.3
	MAE	80.4	69.6	112.6
	RMSE	92.7	88.0	139.4

Further processing included the calculation, for all the recording years, of the *Heating Degree Days* (HDD) and the *Cooling Degree Days* (CDD). The HDD—relative to a base outdoor temperature of 18 °C—are integrated from mid-October to mid-April, while the CDD—relative to a base outdoor temperature of 24 °C—are calculated from mid-April to mid-October. Figure 10 compares the values obtained for the IGDG and the CTI typical years against the statistical distribution of the yearly values referring to the SIAS database (whose median value is indicated by the red straight line inside the boxes). As it is possible to observe, the HDD for the IGDG weather file exceeds the maximum value of the SIAS distribution, if one excludes the only outlier that corresponds to an exceptionally cold year (2005). In particular, the HDD for the IGDG file exceeds by 14% the median of the SIAS distribution. Likewise, the CDD for the IGDG weather file is below any other value occurring in the SIAS series, and the difference with the median is around 40%. The HDD and CDD values for the IWEC typical year are very close to the long-term median, while the ISO and the TMY typical years approach respectively the first quartile of the CDD and the third quartile of the HDD, thus suggesting once again that the IWEC is the selection procedure resulting in the most reliable estimation of the average long-term dry bulb temperatures for this site.

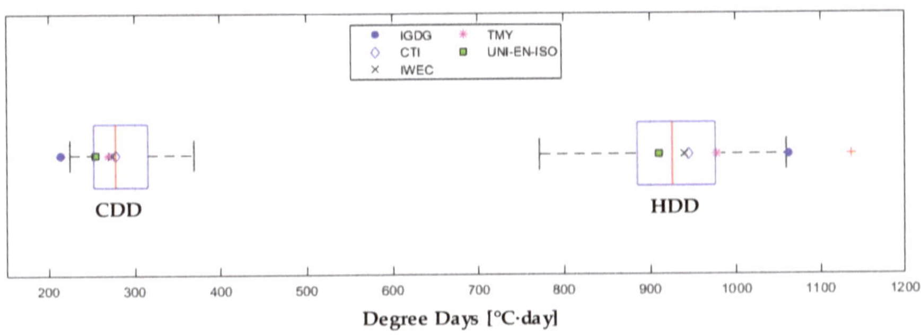

Figure 10. Heating degree days (HDD) and the cooling degree days (CDD): Comparison among typical years and long-term distribution.

5.3. Results of the Dynamic Simulations

As an example of the preparation process needed to organize the weather data in the .epw format. Figure 11 reports—for three consecutive days extracted from the TMY typical year—the trend of dry bulb air temperature, dew point temperature, and Cloud Cover (panel a), with the related sky emissivity and infrared sky irradiance (panel b). When looking at these graphs, it emerges that the sky emissivity is highly influenced by the dew point temperature and the cloud cover, with a peak value of around 0.9 for T_{DP} = 23 °C and a cloud cover of five tenths. However, sky emissivity values range between 0.82 and 0.90 during the selected days: This affects the infrared irradiance, whose fluctuations between a minimum of about 340 W/m^2 and a maximum of around 440 W/m^2 mostly depend on the dry bulb air temperature swinging between 18 and 32 °C.

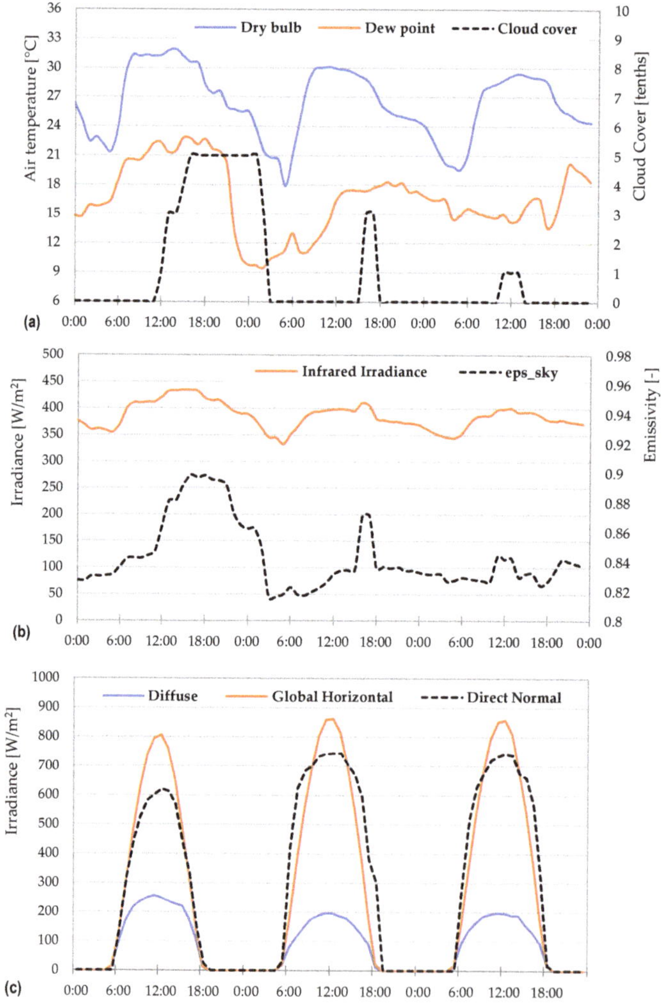

Figure 11. Time trend of some weather variables (21–23 August 2019, TMY typical year). (**a**) Input data to calculate the horizontal infrared sky irradiance; (**b**) sky emissivity and horizontal infrared sky irradiance; (**c**) components of the solar irradiance.

Moreover, Figure 11c shows how the Global Horizontal Solar Irradiance (solid orange line) is split into the diffuse (solid blue line) and the direct normal (black dashed line) components. Here, it is easy to appreciate the predominance of the direct normal component in the specific climate conditions of Catania, and the influence of the cloud cover (the minimum peak direct component is achieved in the first day when CC is equal to only five tenths). It is also important to consider that the direct normal component refers to a plane normal to the sun ray: For this reason, at sunrise and sunset it can even exceed the global irradiance measured on the horizontal plane.

Before commenting on the differences in the predicted energy needs generated by the different weather datasets, it is worth highlighting a significant methodological difference in the estimation of the solar irradiance components. In fact, while the weather files developed in this research rely on *hourly measured* values for GHI—split into their diffuse and direct normal components as per the Boland and Ridley model [64]—the IGDG dataset reports *hourly derived* solar irradiance components obtained through the application of the Erbs' and the modified Liu-Jordan models to daily (not hourly) integral GHI measurements [11]. Along with the different observation period, this aspect contributes to the big variations reported in Table 8.

Here, it is possible to observe that significant differences occur in both heating and cooling energy needs predicted through the IGDG (base case) in comparison with the other datasets: The heating energy needs are indeed halved when the simulations are run with one of the developed weather files whereas the cooling needs dramatically increase by about 67% when shifting from the IGDG to the IWEC weather file. In the simulations with the insulated envelope, the relative difference in the heating energy needs predicted by a new weather file and by the old IGDG one is in the ranges from −59% (ISO dataset) to −56% (TMY dataset), while the difference in the cooling energy needs ranges from +59% (ISO weather file) to +64% (IWEC weather file).

Table 8. Heating and cooling energy needs according to the different weather datasets.

Case	Energy Use	IGDG	TMY	IWEC	ISO
Current Building	Heating [kWh]	4128	2175	2125	2047
	(% difference)	-	−47%	−48%	−50%
	Cooling [kWh]	1885	3095	3143	3072
	(% difference)	-	+64%	+67%	+63%
Insulated Building	Heating [kWh]	2910	1293	1259	1195
	(% difference)	-	−56%	−57%	−59%
	Cooling [kWh]	1675	2698	2748	2670
	(% difference)	-	+61%	+64%	+59%

This discrepancy is high if compared with the outcomes of previous papers. As an example, Radhi found out that using a typical year based on weather data from 1961 to 1990 underestimates the electricity consumption in the cooling system by up to 14.5% for two buildings in Bahrain, compared with the adoption of recent weather data (from 1992 to 2005) [74]. Koci et al. reported a warming trend in the recent weather of Prague, Czech Republic; consequently, the simulated energy demand for a residential building decreased by 4% to 15% when using recent weather data (2013–2017) instead of historical weather data (1990–2015). The remarkable discrepancy emerging in the present paper can then be attributed not only to the increasing trend in the dry bulb temperature, but also to the high difference observed in the global horizontal irradiation and the wind speed already discussed in Section 5.1, and to the older observation period (1951–1970).

On the contrary, it is possible to state that the choice of a specific selection procedure for the typical year has a minor role, leading to a fluctuation by around 3% or 4% in the results: This is almost negligible if compared with the great inaccuracy deriving from the use of the outdated IGDG weather file.

No simulations using the CTI weather dataset have been run because, as discussed in Section 2, the observation period for this dataset is too short and covers only a sub-period of that used for developing the updated TWYs. Furthermore, several parameters that are necessary to build the weather file, such as wind direction and cloud cover, are missing in the CTI dataset: Accordingly, they should be retrieved from other sources and checked for consistency. Finally, as reported in Section 5.1, the CTI weather dataset reports the wind speed at 2 m above ground, while this is commonly measured at a height of 10 m: A conversion would then be required.

6. Conclusions

This paper has investigated the application of various procedures to the generation of Typical Weather Years representative of recent climate trends in the moderately hot and humid city of Catania (Italy). Weather data recorded from 2002 to 2019 through a stationary meteorological station owned by the Sicilian Agrometeorological Information System (SIAS) and located outside the city context have been compared to two currently available datasets: The one provided by the Italian Thermotechnical Committee (CTI), based on the data coming from the same meteorological station but limited to the period 2002–2009, and the one included in the Italian climatic data collection "Gianni De Giorgio" (IGDG) based on data measured by a different station located in the airport at a distance of just 3 km. The latter relies on outdated data (1951–1970) but is still largely in use for building simulation purposes.

The statistical analysis revealed that the recent weather data show higher values of dry bulb air temperature, lower values of relative humidity (in the range of 5% to 10%), and higher global horizontal irradiance (GHI) than in the old IGDG dataset. Furthermore, the variability in the GHI is more marked in the SIAS dataset because the hourly values included in the IGDG series are not measured but rather derived through a mathematical model starting from daily integral values. This change in the average measured weather data can be due both to a difference in the local conditions and to the global climate variation in the last decades; in any case, it remarkably affects the outcomes of building energy simulations. In fact, the predicted heating energy needs for a poorly insulated typical residential building are halved when the simulations take into account the most recent weather dataset, whereas the cooling energy needs can increase by about 65%. Similar figures recur in the case of insulated envelope, while negligible variance (below 5%) emerges when adopting the other typical years considered in this analysis.

The results of the paper are particularly relevant because they cast light on the urgent need of updating the IGDG weather dataset, which is still widely used by researchers and practitioners in the field of building energy simulation and, at least in the case of Catania, gives rise to inacceptable inaccuracy in the calculation of the energy demand for space heating and cooling. This is the first study concerning this issue and referring to the Mediterranean climate of Southern Italy.

Future work will focus on appraising the effects of using updated weather datasets on a wider set of building typologies through detailed energy simulations, while also investigating the importance of multi-year simulations.

Author Contributions: Conceptualization, V.C. and G.E.; methodology, V.C. and G.E.; software, M.I.; formal analysis, G.E.; investigation, V.C.; data curation, M.I.; writing—original draft preparation, V.C., G.E., and M.I.; writing—review and editing, L.M.; supervision, L.M.; funding acquisition, L.M. All authors have read and agreed to the published version of the manuscript.

Funding: This research received no external funding.

Acknowledgments: The authors are grateful to the Sicilian Agrometeorological Information System (SIAS) for making available the data registered by their weather station for the scopes of this research activity.

Conflicts of Interest: The authors declare no conflict of interest.

References

1. IEA. 2018 Global Status Report. Available online: https://www.iea.org/reports/2018-global-status-report (accessed on 7 April 2020).
2. Fichera, A.; Frasca, M.; Palermo, V.; Volpe, R. An optimization tool for the assessment of urban energy scenarios. *Energy* **2018**, *156*, 418–429. [CrossRef]
3. De Wilde, P. *Building Performance Analysis*, 1st ed.; Wiley Blackwell: Oxford, UK, 2018.
4. Hensen, J.L.M.; Lamberts, R. *Building Performance Simulation for Design and Operation*, 1st ed.; Taylor & Francis: Abingdon, UK, 2011.
5. De Masi, R.F.; Gigante, A.; Ruggiero, S.; Vanoli, G.P. The impact of weather data sources on building energy retrofit design: Case study in heating-dominated climate of Italian backcountry. *J. Build. Perform. Simul.* **2020**, *13*, 264–284. [CrossRef]
6. Cui, Y.; Yan, D.; Hong, T.; Xiao, C.; Luo, X.; Zhang, Q. Comparison of typical year and multiyear building simulations using a 55-year actual weather data set from China. *Appl. Energy* **2017**, *197*, 890–904. [CrossRef]
7. Wang, L.; Mathew, P.; Pang, X. Uncertainties in energy consumption introduced by building operations and weather for a medium-size office building. *Energy Build.* **2012**, *53*, 152–158. [CrossRef]
8. Erba, S.; Causone, F.; Armani, R. The effect of weather datasets on building energy simulation outputs. *Energy Procedia* **2017**, *134*, 545–554. [CrossRef]
9. Pernigotto, G.; Prada, A.; Cóstola, D.; Gasparella, A.; Hensen, J.L.M. Multi-year and reference year weather data for building energy labelling in north Italy climates. *Energy Build.* **2014**, *72*, 62–72. [CrossRef]
10. Lou, S.; Li, D.H.W.; Huang, Y.; Zhou, X.; Xia, D.; Zhao, Y. Change of climate data over 37 years in Hong Kong and the implications on the simulation-based building energy evaluations. *Energy Build.* **2020**, *222*, 110062. [CrossRef]
11. IGDG Weather Dataset. Available online: https://energyplus.net/sites/all/modules/custom/weather/weather_files/italia_dati_climatici_g_de_giorgio.pdf (accessed on 7 April 2020).
12. CTI Energia Ambiente (Comitato Termotecnico Italiano). Available online: https://try.cti2000.it (accessed on 7 April 2020).
13. Hall, I.J.; Prairie, R.; Anderson, H.; Boes, E. *Generation of a Typical Meteorological Year for 26 SOLMET Stations*; Technical Report SAND-78-1601; Sandia National Laboratories: Albuquerque, NM, USA, 1978.
14. Marion, W.; Urban, K. *User's Manual for TMY2*; National Renewable Energy Laboratory: Golden, CO, USA, 1995. Available online: http://rredc.nrel.gov/solar/pubs/tmy2 (accessed on 7 April 2020).
15. Wilcox, S.; Marion, W. *User's Manual for TMY3 Data Sets*; Technical Report NREL/TP-581-43156; National Renewable Energy Laboratory: Golden, CO, USA, 2008.
16. Thevenard, D.J.; Brunger, A.P. The development of typical weather years for international locations: Part I, algorithms. *ASHRAE Trans.* **2002**, *108*, 376.
17. Huang, Y.J.; Su, F.; Sheo, D.; Krarti, M. Development of 3012 IWEC2 weather files for international locations (RP-1477). *ASHRAE Trans.* **2014**, *120*, 340–355.
18. Hygrothermal performance of buildings—Calculation and presentation of climatic data—Part 4: Hourly data for assessing the annual energy for heating and cooling. In *ISO 15927-4:2005*; ISO: Vernier, Switzerland, 2005.
19. Festa, R.; Ratto, C.F. Proposal of a numerical procedure to select Reference Years. *Sol. Energy* **1993**, *50*, 9–17. [CrossRef]
20. Lund, H.; Eidorff, S. *Selection Methods for Production of Test Reference Years, Appendix D, Contract 284-77 ES DK*; Final Report, EUR 7306 EN; Technical University of Denmark: Lyngby, DK, 1980.
21. *Canadian Weather Energy and Engineering Data Sets (CWEEDS files) and Canadian Weather for Energy Calculations (CWEC files) Updated User's Manual*; Environment Canada and National Research Council of Canada: Ottawa, ON, Canada, 2008.
22. *ASHRAE Handbook Fundamentals*, S.I. ed.; Chapter 24; American Society of Heating, Refrigerating and Air-Conditioning Systems: Atlanta, GA, USA, 1989.
23. Argiriou, A.; Lykoudis, S.; Kontoyiannidis, S.; Balaras, C.A.; Asimakopoulos, D.; Petrakis, M.; Kassomenos, P. Comparison of methodologies for TMY generation using 20 years data for Athens, Greece. *Solar Energy* **1999**, *66*, 33–45. [CrossRef]
24. Chan, A.L.S.; Chow, T.T.; Fong, S.K.F.; Lin, J.Z. Generation of a typical meteorological year for Hong Kong. *Energy Convers. Manag.* **2006**, *47*, 87–96. [CrossRef]

25. Chan, A.L.S. Generation of typical meteorological years using genetic algorithm for different energy systems. *Renew. Energy* **2016**, *90*, 1–13. [CrossRef]
26. Siu, C.Y.; Liao, Z. Is building energy simulation based on TMY representative: A comparative simulation study on doe reference buildings in Toronto with typical year and historical year type weather files. *Energy Build.* **2020**, *211*, 109760. [CrossRef]
27. Bre, F.; Fachinotti, V.G. Generation of typical meteorological years for the Argentine Littoral Region. *Energy Build.* **2016**, *219*, 432–444. [CrossRef]
28. Jiang, Y. Generation of typical meteorological year for different climates of China. *Energy* **2010**, *35*, 1946–1953. [CrossRef]
29. Kalamees, T.; Jylhä, K.; Tietäväinen, H.; Jokisalo, J.; Ilomets, S.; Hyvönen, R.; Sakub, S. Development of weighting factors for climate variables for selecting the energy reference year according to the EN ISO 15927-4 standard. *Energy Build.* **2012**, *47*, 53–60. [CrossRef]
30. Kalogirou, S.A. Generation of typical meteorological year (TMY-2) for Nicosia, Cyprus. *Renew. Energy* **2003**, *28*, 2317–2334. [CrossRef]
31. Kulesza, K. Comparison of typical meteorological year and multi-year time series of solar conditions for Belsk, central Poland. *Renew. Energy* **2017**, *113*, 1135–1140. [CrossRef]
32. Lee, K.; Yoo, H.; Levermore, G.J. Generation of typical weather data using the ISO Test Reference Year (TRY) method for major cities of South Korea. *Build. Environ.* **2010**, *45*, 956–963. [CrossRef]
33. Levermore, G.J.; Parkinson, J.B. Analyses and algorithms for new Test Reference Years and Design Summer Years for the UK. *Build. Serv. Eng. Res. Technol.* **2006**, *27*, 311–325. [CrossRef]
34. Oko, C.O.C.; Ogoloma, O.B. Generation of a Typical Meteorological Year for Port Harcourt zone. *J. Eng. Sci. Technol.* **2001**, *6*, 204–214.
35. Skeiker, K. Comparison of methodologies for TMY generation using 10 years data for Damascus, Syria. *Energy Convers. Manag.* **2007**, *48*, 2090–2102. [CrossRef]
36. Yilmaz, S.; Ekmekci, I. The generation of Typical Meteorological Year and climatic database of Turkey for the energy analysis of buildings. *J. Environ. Sci. Eng. A* **2017**, *6*, 370–376.
37. Bhandari, M.; Shrestha, S.J. Evaluation of weather datasets for building energy simulation. *Energy Build.* **2012**, *49*, 109–118. [CrossRef]
38. Hosseini, M.; Lee, B.; Vakilinia, S. Energy performance of cool roofs under the impact of actual weather data. *Energy Build.* **2017**, *145*, 284–292. [CrossRef]
39. Crawley, D.; Lawrie, L. Should we be using just 'Typical' weather data in building performance simulation? In Proceedings of the Building Simulation International Conference, Rome, Italy, 2–4 September 2019.
40. Rahman, I.A.; Dewsbury, J. Selection of typical weather data (test reference years) for Subang, Malaysia. *Building Environ.* **2007**, *42*, 3636–3641. [CrossRef]
41. Sawaqed, N.M.; Zurigat, Y.H.; Al-Hinai, H. A step-by-step application of Sandia method in developing typical meteorological years for different locations in Oman. *Int. J. Energy Res.* **2005**, *29*, 723–737. [CrossRef]
42. Ebrahimpour, A.; Maerefat, M. A method for generation of typical meteorological year. *Energy Convers. Manag.* **2010**, *51*, 410–417. [CrossRef]
43. Janjai, S.; Deeyai, P. Comparison of methods for generating typical meteorological year using meteorological data from a tropical environment. *Appl. Energy* **2009**, *86*, 528–537. [CrossRef]
44. Lhendup, T.; Lhendup, S. Comparison of methodologies for generating a typical meteorological year (TMY). *Energy Sustain. Dev.* **2007**, *11*, 5–10. [CrossRef]
45. Ohunakin, O.S.; Adaramola, M.; Oyewola, O.M.; Fagbenle, R.O. Generation of a typical meteorological year for north-east, Nigeria. *Appl. Energy* **2013**, *112*, 152–159. [CrossRef]
46. Zang, H.; Wang, M.; Huang, J.; Wei, Z.; Sun, G. A hybrid method for generation of Typical Meteorological Years for different climates of China. *Energies* **2016**, *9*, 1094. [CrossRef]
47. Pernigotto, G.; Prada, A.; Cappelletti, F.; Gasparella, A. Impact of reference years on the outcome of multi-objective optimization for building energy refurbishment. *Energies* **2017**, *10*, 1925. [CrossRef]
48. Pernigotto, G.; Prada, A.; Gasparella, A.; Hensen, J.L.M. Analysis and improvement of the representativeness of EN ISO 15927-4 reference years for building energy simulation. *J. Build. Perform. Simul.* **2014**, *7*, 391–410. [CrossRef]
49. Sorrentino, G.; Scaccianoce, G.; Morale, M.; Franzitta, V. The importance of reliable climatic data in the energy evaluation. *Energy* **2012**, *48*, 74–79. [CrossRef]

50. Kim, S.; Zirkelbach, D.; Künzel, H.M.; Lee, J.-H.; Choi, J. Development of test reference year using ISO 15927-4 and the influence of climatic parameters on building energy performance. *Build. Environ.* **2017**, *114*, 374–386. [CrossRef]

51. David, M.; Adelard, L.; Lauret, P.; Garde, F. A method to generate Typical Meteorological Years from raw hourly climatic databases. *Build. Environ.* **2010**, *45*, 1722–1732. [CrossRef]

52. Yang, L.; Lam, J.C.; Liu, J. Analysis of typical meteorological years in different climates of China. *Energy Convers. Manag.* **2007**, *48*, 654–668. [CrossRef]

53. Ohunakin, O.S.; Adaramola, M.S.; Oyewola, O.M.; Fagbenle, R.L.; Abam, F.I. A Typical Meteorological Year Generation Based on NASA Satellite Imagery (GEOS-I) for Sokoto, Nigeria. *Int. J. Photoenergy* **2014**, *2014*, 468562. [CrossRef]

54. Arima, Y.; Ooka, R.; Kikumoto, H. Proposal of typical and design weather year for building energy simulation. *Energy Build.* **2017**, *139*, 517–524. [CrossRef]

55. Bevilacqua, P.; Bruno, R.; Arcuri, N. Green roofs in a Mediterranean climate: Energy performances based on in-situ experimental data. *Renew. Energy* **2020**, *152*, 1414–1430. [CrossRef]

56. Chiesa, G.; Grosso, M. The influence of different hourly typical meteorological years on dynamic simulation of buildings. *Energy Procedia* **2015**, *78*, 2560–2565. [CrossRef]

57. Lupato, G.; Manzan, M. Italian TRYs: New weather data impact on building energy simulations. *Energy Build.* **2019**, *185*, 287–303. [CrossRef]

58. Kočí, J.; Kočí, V.; Maděra, J.; Černý, R. Effect of applied weather data sets in simulation of building energy demands: Comparison of design years with recent weather data. *Renew. Sustain. Energy Rev.* **2019**, *100*, 22–32. [CrossRef]

59. Vasaturo, R.; van Hooff, T.; Kalkman, I.; Blocken, B.; van Wesemael, P. Impact of passive climate adaptation measures and building orientation on the energy demand of a detached lightweight semi-portable building. *Build. Simul.* **2018**, *11*, 1163–1177. [CrossRef]

60. Finkelstein, J.M.; Schafer, R.E. Improved goodness-of-fit tests. *Biometrika* **1971**, *58*, 641–645. [CrossRef]

61. US Department of Energy. EnergyPlus Version 8.9. Available online: https://energyplus.net/documentation (accessed on 7 April 2020).

62. *ASHRAE Handbook Fundamentals*, S.I. ed.; Chapter 1: Psychometrics; American Society of Heating, Refrigerating and Air-Conditioning Systems: Atlanta, GA, USA, 2013.

63. Bianchi, C.; Smith, A.D. Localized Actual Meteorological Year File Creator (LAF): A tool for using locally observed weather data in building energy simulations. *SoftwareX* **2019**, *10*, 100299. [CrossRef]

64. Boland, J.; Ridley, B. Models of Diffuse Solar Fraction. In *Modeling Solar Radiation at the Earth's Surface*; Springer: Berlin/Heidelberg, Germany, 2008.

65. UNI/TS 10349-1; *Riscaldamento e raffrescamento degli edifici—Dati Climatici—Parte 1: Medie Mensili per la Valutazione della Prestazione Termo-Energetica Dell'edificio e Metodi per Ripartire l'Irradianza Solare Nella Frazione Diretta e Diffusa e per Calcolare l'Irradianza Solare su di una Superficie Inclinata*; Ente Nazionale di Unificazione: Milan, Italy, 2016. (In Italian)

66. Walton, G.N. *Thermal Analysis Research Program Reference Manual*; NBSSIR 83-2655; National Bureau of Standards: Washington, DC, USA, 1983; p. 21.

67. Clark, G.; Allen, C. The estimation of atmospheric radiation for clear and cloudy skies. In Proceedings of the 2nd National Passive Solar Conference (AS/ISES), Philadelphia, PA, USA, 16–18 March 1978; pp. 675–678.

68. Smith, A.; Lott, N.; Vose, R. The Integrated Surface Database: Recent developments and partnerships. NOAA's National Climatic Data Center, Asheville, North Carolina. *Bull. Am. Meteor. Soc.* **2011**, *92*, 704–708. [CrossRef]

69. *ASHRAE Handbook—Fundamentals*, S.I. ed.; Chapter 26 Ventilation and Infiltration; American Society of Heating, Refrigerating and Air-Conditioning Systems: Atlanta, GA, USA, 2009.

70. UNI/TS 11300-1, *Prestazioni Energetiche Degli Edifici—Parte 1: Determinazione del Fabbisogno di Energia Termica dell'Edificio per la Climatizzazione Estiva ed Invernale*; Ente Nazionale di Unificazione: Milan, Italy, 2014. (In Italian)

71. EN 15251; *Indoor Environmental Input Parameters for Design and Assessment of Energy Performance of Buildings-Addressing Indoor Air Quality, Thermal Environment, Lighting and Acoustics*; European Committee for Standardisation: Brussels, Belgium, 2007.

72. Salvati, A.; Palme, M.; Chiesa, G.; Kolokotroni, M. Built form, urban climate and building energy modelling: Case-studies in Rome and Antofagasta. *J. Build. Perform. Simul.* **2020**, *13*, 209–225. [CrossRef]
73. Santamouris, M. Cooling the cities—A review of reflective and green roof mitigation technologies to fight heat island and improve comfort in urban environments. *Sol. Energy* **2014**, *103*, 682–703. [CrossRef]
74. Radhi, H.A. Comparison of the accuracy of building energy analysis in Bahrain using data from different weather periods. *Renew. Energy* **2009**, *34*, 869–875. [CrossRef]

MDPI

St. Alban-Anlage 66

4052 Basel

Switzerland

Tel. +41 61 683 77 34

Fax +41 61 302 89 18

www.mdpi.com

Energies Editorial Office

E-mail: energies@mdpi.com

www.mdpi.com/journal/energies